Endothelial Cells

Volume III

Editor

Una S. Ryan, Ph.D.
Professor of Medicine
Department of Medicine
University of Miami
Miami, Florida

CRC Press, Inc.
Boca Raton, Florida

Library of Congress Cataloging-in-Publication Data

Endothelial cells/editor, Una S. Ryan.
    p.   cm.
  Includes bibliographies and indexes.
  ISBN 0-8493-4988-5 (set)
  1. Endothelium — Cytology.   I. Ryan, Una, S., 1941-
  [DNLM: 1. Endothelium — anatomy & histology. 2. Endothelium —
physiology.   QS 532.5.E7 E566]
QP88.45.E53 1988
599′.087 — dc 19
DNLM/DLC

                                                                    87-32542

Direct all inquiries to CRC Press, Inc., 2000 Corporate Blvd., N.W., Boca Raton, Florida, 33431.

© 1988 by CRC Press, Inc.

International Standard Book Number 0-8493-4988-5 (set)
International Standard Book Number 0-8493-4990-7 (v.1)
International Standard Book Number 0-8493-4991-5 (v.2)
International Standard Book Number 0-8493-4992-3 (v.3)

Library of Congress Card Number 87-32542
Printed in the United States

PREFACE

In the Foreword to a book on Yankee epitaphs called *Over Their Dead Bodies** the authors remark " . . . on the rich chronicle they offered of contemporary ideas and events". This is precisely the mission of this book, although I doubt that any of the authors would wish their chapters to represent their "last words".

The endothelium itself is far from a morbid subject. It has come alive in recent years and its story is told in these three volumes. It has risen from being a ghostly substance, scarcely showing in early pathology or histology texts, of no known properties save that of lining the blood vessels, to being a collection of cells endowed with pores that could be mathematically modeled but never seen, to possessing a rich array of enzymatic and processing properties. Its stature has grown to that of a metabolically active and responsive tissue endowed with a diversity of enzymes, receptors, and transport molecules. It can be grown in culture and manipulated into postures it may never have to succumb to in vivo, its gene products have been cloned, and it has been made to reveal its relationships with other cells and molecules both near neighbors or distance targets. It is claimed as a regulator of blood pressure, a team player in hemostasis, a sparring partner with various blood cell types, and the dancing partner of the vascular smooth muscle cell. At one time seen as the innocent victim of inflammatory attack we now know that it frequently calls the tune. It is both a target and a source of hormones, growth factors, vasoactive substances, hemostatic factors, and oxygen radicals.

It binds complement components, can express receptors for immune reactions, presents antigens, and can engulf and kill microorganisms. It can be activated, excited, and primed. Activated endothelium represents a remarkable amplification surface for local immune and inflammatory reactions and is able to initiate events that lead to closing off a vessel. Activation of endothelium plays a key role in the host response yet when inappropriately expressed can underlie much of vascular pathology. In fact, it is likely that all diseases have a vascular etiology.

Proud though we are of the extraordinary accomplishments of the last 2 decades of research we know that the endothelium has not yet yielded up all its mysteries. I hope this book will stand as a monument to the success of its brilliant band of contributors and a stepping stone for the next runners in the relay race to capture the prize of understanding the vascular endothelial cell.

Una S. Ryan
Miami, October 1987

---

* Mann, T. C. and Greene, J., *Over Their Dead Bodies: Yankee Epitaphs and History*, Stephen Greene Press, Brattleboro, Vermont, 1962.

THE EDITOR

Una S. Ryan, Ph.D., is Professor of Medicine at the University of Miami School of Medicine, Chief of the Division of Vascular Cell Biology, and Director of the Hybridoma Facility.

Dr. Ryan obtained her training in England and received a B.Sc. degree from the University of Bristol in 1963 and a Ph.D. from the University of Cambridge in 1968. She was a Howard Hughes Investigator and Director of the Laboratory for Ultrastructure Studies at the Howard Hughes Medical Institute, Miami, from 1967 to 1971. She served as Instructor in Medicine from 1967 to 1972, Assistant Professor of Medicine from 1972 to 1977, Associate Professor from 1977 to 1980, Research Professor from 1980 to 1986, and Professor of Medicine from 1986.

Dr. Ryan is a member of the American Society for Cell Biology, Society for Neuroscience, Tissue Culture Association, American Heart Association, Council on Basic Research, Council on Circulation, Cardiopulmonary Council, American Physiological Society, Microcirculatory Society, European Society for Microcirculation, American Thoracic Society, New York Academy of Sciences, International Society for Heart Research, and the Royal Society of Medicine.

She was an Established Investigator of the American Heart Association from 1972 to 1977. She received the Louis Artur Lucian Award for Research in Circulatory Diseases in 1984, and the Lamport Lectureship in 1985. She received a MERIT Award from the National Institutes of Health in 1986, she is President-Elect of the Sigma Xi Miami chapter, and was elected to the Royal Society of Medicine in 1986.

She has been the recipient of numerous research grants from the National Institutes of Health and has served on and chaired many review and advisory panels, including the Pulmonary Diseases Advisory Committee, 1972 to 1976 Committee A. She is currently a member of Pathology A Study Section.

Dr. Ryan is editor of *Tissue & Cell* (an international cell biology journal) and an advisory or reviewing editor for a large number of other journals. She has edited 3 books, authored over 180 papers and over 160 abstracts. Her major research interest centers on the cell biology of the pulmonary endothelium.

CONTRIBUTORS

Magnus Bundgaard, Ph.D.
Associate Professor
Department of General Physiology and
  Biophysics
University of Copenhagen
Copenhagen, Denmark

William E. Burkel, Ph.D.
Professor
Department of Anatomy and Cell Biology
University of Michigan Medical School
Ann Arbor, Michigan

Eugene I. Chazov, M.D.
Professor and Director
USSR Cardiology Research Center
Moscow, Union of Soviet Socialist
  Republic

Christian Crone, Ph.D.
Professor and Doctor of Medicine
Department of General Physiology and
  Biophysics
University of Copenhagen
Copenhagen, Denmark

Bruce A. Freeman, Ph.D.
Associate Professor
Departments of Anesthesiology and
  Biochemistry
University of Alabama at Birmingham
Birmingham, Alabama

Linda M. Graham, M.D.
Associate Professor
Department of Surgery
Case Western Reserve University
Cleveland, Ohio

George J. Grega, Ph.D.
Associate Professor of Pharmacology
Department of Pharmacology
University of Houston
Houston, Texas

Susan M. Harding, M.D.
Associate in Medicine
Department of Medicine
University of Alabama at Birmingham
Birmingham, Alabama

Robert M. Jackson, M.D.
Assistant Professor
Department of Medicine
University of Alabama at Birmingham
Birmingham, Alabama

Anthony Johns, Ph.D.
Group Leader for Vascular Biology
Department of Pharmacology
Berlex Laboratories, Inc.
Cedar Knolls, New Jersey

Alain F. Junod, M.D.
Professor of Medicine
Department of Medicine
Hoptial Cantonal Universitaire
Geneva, Switzerland

Raouf A. Khalil, M.D.
Ph.D. Candidate
Department of Pharmacology
University of Miami School of Medicine
Miami, Florida

Yasuo Kubota, M.D.
Visiting Fellow
Dermatology Branch
National Institutes of Health
Bethesda, Maryland

David M. Larson, Ph.D.
Assistant Professor of Pathology
Mallory Institute of Pathology
Boston University School of Medicine
Boston, Massachusetts

Thomas J. Lawley, M.D.
Senior Investigator
Dermatology Branch
National Institutes of Health
Bethesda, Maryland

Sadis Matalon, Ph.D.
Professor
Departments of Anesthesiology and
  Physiology
University of Alabama at Birmingham
Birmingham, Alabama

Søren-Peter Olesen, M.D.
Research Associate
Department of General Physiology and
 Biophysics
University of Copenhagen
Copenhagen, Denmark

Carl G. A. Persson, Ph.D.
Department of Clinical Pharmacology
University Hospital, and Department of
 Research and Development
Pharmacology Laboratory
Lund, Sweden

Marlene Rabinovitch, M.D.
Associate Professor
Department of Pediatrics and Pathology
University of Toronto
Toronto, Ontario, Canada

Vadim S. Repin, M.D.
Professor and Head of Laboratory
Institute of Experimental Cardiology
U.S.S.R. Cardiology Research Center
Moscow, Union of Soviet Socialist
 Republic

Gabor M. Rubanyi, M.D., Ph.D.
Director of Pharmacology
Department of Pharmacology
Berlex Laboratories
Cedar Knolls, New Jersey

Una S. Ryan, Ph.D.
Professor of Medicine
Department of Medicine
University of Miami
Miami, Florida

Vladimir N. Smirnov, Ph.D.
Professor and Director
Institute of Experimental Cardiology
U.S.S.R. Cardiology Research Center
Moscow, Union of Soviet Socialist
 Republic

James C. Stanley, M.D.
Professor
Department of Surgery
University of Michigan Medical Center
Ann Arbor, Michigan

Erik Svensjo, Ph.D.
Associate Professor
Department of Pharmaceutical
 Pharmacology
University of Uppsala
Uppsala, Sweden

Vsevolod A. Tkachuk, Ph.D.
Head of the Laboratory
Institute of Experimental Cardiology
U.S.S.R. Cardiology Research Center
Moscow, Union of Soviet Socialist
 Republic

Cornelis Van Breemen, D.V.M., Ph.D.
Professor
Department of Pharmacology
University of Miami School of Medicine
Miami, Florida

TABLE OF CONTENTS

Volume I

TABLE OF CONTENTS

Volume II

Growth Factors and Growth Control

Hemodynamic Forces and Interactions with Blood Cells

TABLE OF CONTENTS

Volume III

Phagocytosis and Free Radical Production

Chapter 28

EFFECTS OF HYPEROXIA AND O₂ METABOLITES ON DNA

Alain F. Junod

TABLE OF CONTENTS

I. INTRODUCTION

Until recently, no systematic study on the effects of high oxygen (O_2) concentration on DNA synthesis of endothelial cells, whether of pulmonary or of systemic origin, was available in medical literature. That endothelial cell replication was impaired following exposure to toxic O_2 concentrations could only be inferred from the morphometric examination of lungs exposed to hyperoxia.[1] A decreased endothelial cell volume was noted after a few days of exposure, although, at least for primates, signs of endothelial cell necrosis were not conspicuous. Other in vivo observations were based on the morphological observation and the radioautographic study of the repair process taking place after traumatic denudation of the intima.[2] Treatment of the whole animal with lipopolysaccharides (LPS) was another condition associated with both endothelial damage and increased replication rate.[3] Recovery from exposure to hyperoxia and study of the incorporation sites of ^{3}H-thymidine (TdR) also revealed that an increase in DNA synthesis took place in pulmonary endothelial cells a few days after hyperoxic exposure had ceased.[4]

There are, however, several reasons for being actively interested in the response of endothelial cells to high oxygen tension and to O_2 metabolites released by activated phagocytes or generated by the hypoxanthine-xanthine oxidase reaction.[5,6] First, capillary and venous pulmonary endothelial cells as well as endothelial cells on the arterial site of the systemic circulation are potentially exposed to high O_2 concentrations. Second, diseases such as the adult respiratory distress syndrome are known to affect pulmonary endothelial cells at an early stage and are thought to result,[7] at least in part, from the action of activated neutrophils. This clinical syndrome and its treatment could associate therefore two potentially deleterious conditions for the endothelium with, as a consequence, the development of endothelial damage and the prevention of its repair.

The recent and easy availability of endothelial cells in culture has given several investigators the opportunity to study in more detail the effects of hyperoxic exposure and O_2 metabolites on the DNA synthesis process as well as, to a more limited extent, on its damage and repair. This review will try to summarize the main acquisitions in this field over the past few years.

II. EFFECTS OF HYPEROXIA ON DNA SYNTHESIS

During the course of our studies on the hyperoxia-related cytotoxic effect on primary cultures of porcine aortic endothelial cells at confluence,[8] it soon became apparent that values of our index quantifying cell loss (DNA content) could not be accounted for only by the cytotoxic effects of O_2 as assessed by LDH release. Decreased DNA synthesis could explain such a discrepancy, and measurements of ^{3}H-TdR incorporation into DNA confirmed this hypothesis. A 50% decrease in DNA synthesis from TdR was found already after a 24-hr exposure to 95% O_2 to reach a few percent by 5 days of exposure. This effect, unlike the O_2-induced LDH release, could not be corrected by the addition of selenomethionine, which was associated with a marked increase in glutathione peroxidase.

Other experiments in a comparative study on the effects of hyperoxia and paraquat on the same cell preparation confirmed this finding.[9] They also indicated that paraquat, a compound also thought to induce the intracellular generation of O_2 radicals after having undergone intracellular reduction,[10,11] had a similar inhibitory action on DNA synthesis from TdR.

This O_2-related effect, which is concentration dependent (Figure 1), could have been exerted at different levels: (1) transport of the precursor; (2) phosphorylation of TdR into mono-, di-, and trinucleotides (TMP, TDP, TTP); or (3) incorporation of the immediate precursor thymidine triphosphate (TTP) into DNA.

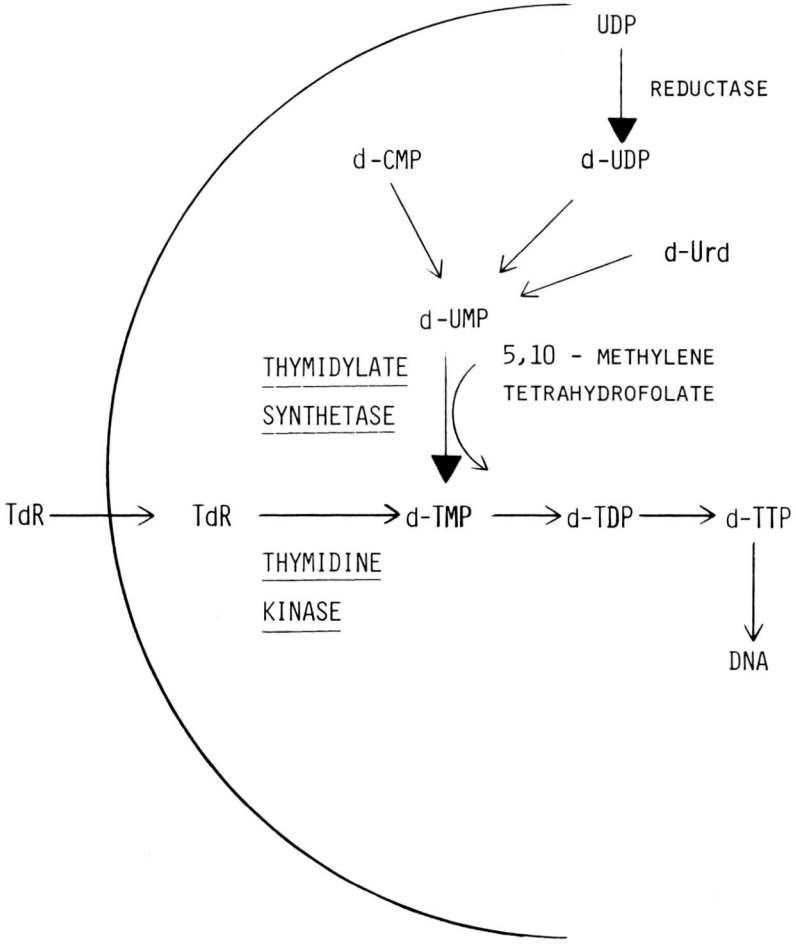

FIGURE 1. Diagram representing the various steps involved in the synthesis of TdR triphosphate, the direct precursor for DNA synthesis. The uptake and incorporation of exogenous TdR constitute the so-called salvage pathway for the formation of TdR nucleotides, whereas the synthesis of TdR monophosphate from d-uridine monophosphate (d-UMP) represents the *de novo* pathway.

A. TdR Transport

TdR uptake by mammalian cells has been studied by Plagemann et al.[12] They concluded on the existence of a facilitated diffusion; the rapid phosphorylation which follows the entrance of TdR into the cell explains why previous studies considered TdR uptake as being compatible with a saturable, high-affinity transport.

To truly measure the rate of transport of TdR by endothelial cells, the experimental protocol would have required very short incubation times and the use of metabolic inhibitors or ATP-depleted cells. Since these manipulations could have obscured the effect of oxygen, measurements of TdR uptake (1 to $100 \times 10^{-7} M$) were done over a 3-min period with a standard medium under control condition or exposure to 95% O_2 for 48 hr[13] These determinations probably included, to a certain extent, the phosphorylation of the substrate once intracellular. Hyperoxia had only a modest effect on this uptake process, although it was more marked for the lowest substrate concentrations. The different time course of the hyperoxic (95% O_2) exposure-related effect on $5 \times 10^{-7} M$ TdR uptake and incorporation into DNA (Figure 2) made it obvious that the decrease in TdR incorporation could not be explained by this reduction of uptake.

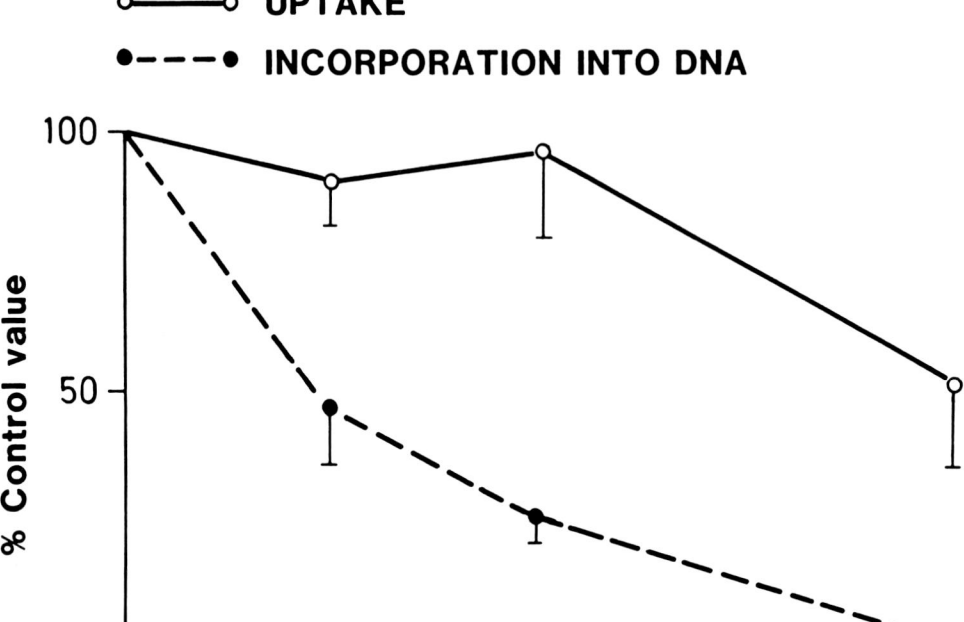

FIGURE 2.   Effects of various durations of exposure to 95% O_2 on the uptake (○—○) and incorporation (●-----●) of 5×10^{-7} *M* TdR into DNA of cultured aortic endothelial cells over a 3-min period. Results are expressed as percent of control values. Bars represent 1 SE. (From Clement, A., Hübscher, U., and Junod, A. F., *J. Appl. Physiol.*, 59, 1110, 1985. With permission.)

B. TdR Phosphorylation

Measurements of TdR phosphorylation under control conditions and hyperoxia revealed that this step was sensitive to the effect of 95% O_2.[13] Chromatographic analysis of TdR and its phosphorylated derivatives showed that more TdR and less phosphorylated TdR were present in hyperoxic cells. TdR kinase assay in cell extracts confirmed an obvious duration-dependent inhibitory effect of hyperoxic exposure (Figure 3).

Further experiments were done to test the sensitivity of endothelial cell TdR kinase to various O_2 concentrations and to compare it to that of adenosine kinase, an enzyme with a similar function on a different substrate.[14] Figures 4 and 5 show the relationship between O_2 concentration and TdR incorporation into DNA and TdR kinase activity. To find out about the possible mechanism of the inhibitory action of high O_2 concentrations, endothelial cells were treated with hypoxanthine-xanthine oxidase, which is known to generate $O_2^{\overline{}}$, H_2O_2, as well as OH· and $^1O_2$.[15] The extracts of cells previously treated with this enzyme system showed a xanthine-oxidase-dependent reduction of TdR kinase activity (Figure 6). This suggested that O_2 radicals or intermediates could be the cause of this reduction in enzyme activity. Confirmation of this hypothesis was obtained from experiments with normal cell extracts which, when directly in contact with H_2O_2, also showed reduced TdR kinase activity.

These results suggest therefore that the O_2-induced reduction in TdR kinase could result from a direct toxic effect of O_2 metabolites, particularly H_2O_2.

TdR kinase is an enzyme whose activity is known to markedly fluctuate during the cell

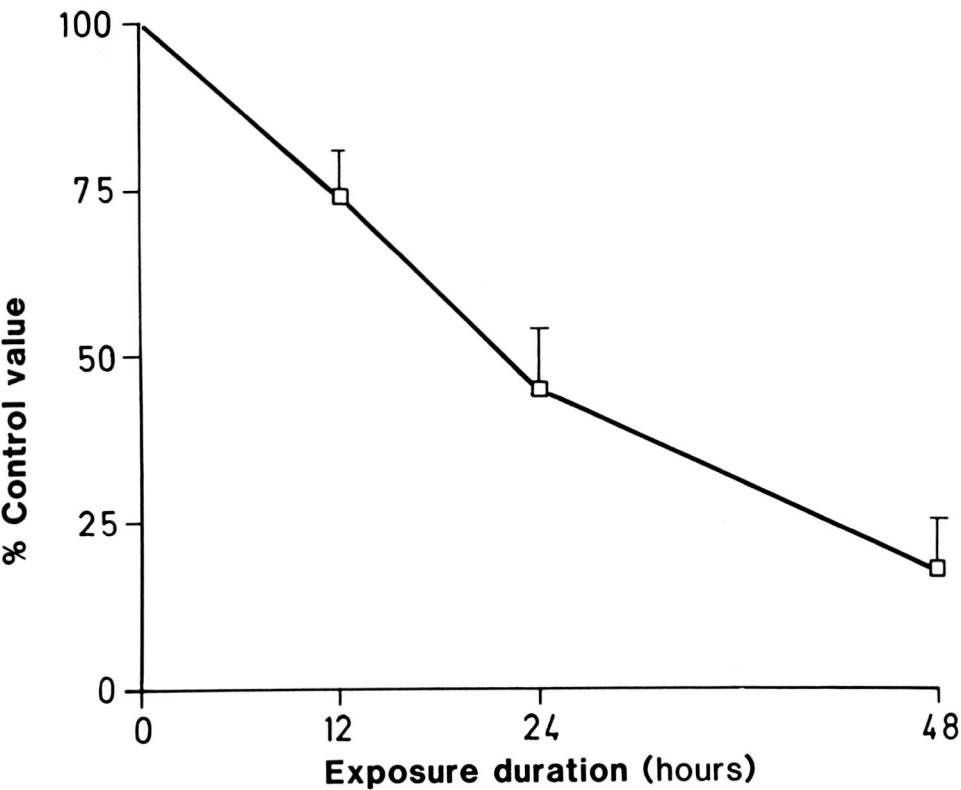

FIGURE 3. Effects of various durations of exposure to 95% O_2 on TdR kinase activity in confluent aortic endothelial cells. Results are expressed as percent of values obtained under control conditions. Bars represent 1 SE. (From Clement, A., Hübscher, U., and Junod, A. F., *J. Appl. Physiol.*, 59, 1110, 1985. With permission.)

cycle, possibly under the influence of a regulatory protein.[16,17] Its extreme and unexpected sensitivity to high O_2 concentration and to H_2O_2 makes us suspect that it may contain an active site especially vulnerable to oxidation.

Although reduction of TdR kinase under O_2 exposure could explain the decreased TdR incorporation into DNA, it did not necessarily mean that there was decreased DNA synthesis or diminished cell replication. For the synthesis of TdR nucleotides, another pathway exists — the *de novo* generation of d-TMP from d-uridine monophosphate (d-UMP) under the action of thymidylate synthetase in the presence of tetrahydrofolic acid (Figure 1). To assess the respective roles of TdR kinase and thymidylate synthetase on DNA replication of endothelial cells under 95% O_2, we used preconfluent, actively growing endothelial cells and determined cell proliferation together with the activities of the two above-mentioned enzymes.[18] These experiments indicated that thymidylate synthetase activity was unaffected by hyperoxia, at the same time as cell proliferation was stopped and TdR kinase activity reduced (Table 1). These data suggest therefore that the *de novo* synthesis of TMP does not seem to operate in endothelial cells to either accompany or replace the use of TdR for DNA synthesis. That TdR kinase activity could control, at least partly, DNA synthesis of cultured endothelial cells was indicated by the parallel changes affecting cell proliferation, TdR incorporation into DNA, and TdR kinase under various O_2 concentrations or during a recovery phase from previous hyperoxic exposure.

C. DNA Polymerases

The decrease in TMP formation, although accounting for the decrease in TdR incorporation

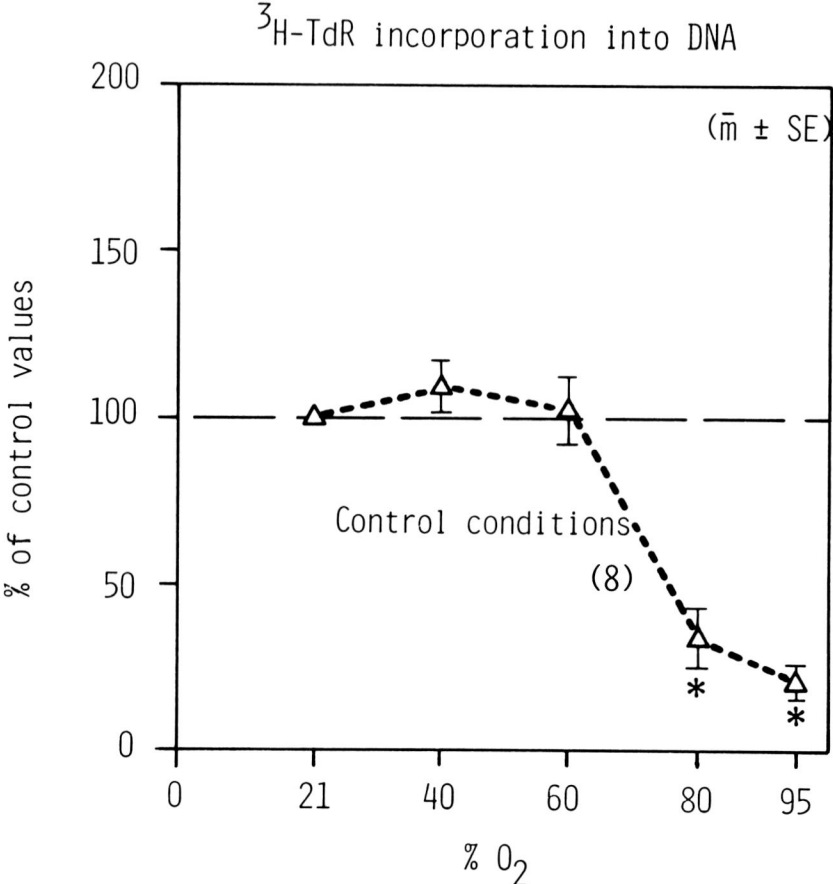

FIGURE 4.   Effects of various O_2 concentrations on the incorporation of trace amounts of ^{3}H-TdR into DNA of postconfluent endothelial cells. The duration of exposure was 48 hr. Asterisk represents values significantly different from control.

into DNA, could not rule out the existence of another O_2-related effect on the incorporation into DNA of TTP, its most direct precursor.

To answer this question, we measured the incorporation of TTP into DNA by permeabilized endothelial cells, cultured under control condition or in 95% O_2.[13] It was found to be similar in both conditions. In another set of experiments, measurements of DNA polymerases alpha and beta were made in control and hyperoxic cells. The enzymatic assay revealed that DNA polymerase alpha activity,[19] which is associated with DNA replication, was significantly decreased in hyperoxic cells; on the other hand, an increased activity of the beta form, associated with DNA repair, also resulted from hyperoxia. Thus the different sensitivity to graded O_2 concentration of endothelial cells previously exposed to hyperoxia (Figure 7) could be explained on the basis of increased repair activity.

This result suggests that hyperoxic exposure could be associated with DNA damage, a condition most often associated with irradiation, UV exposure, treatment with alkylating drugs or direct contact with H_2O_2 and especially OH·.[20]

III. EFFECTS OF EXTRACELLULARLY GENERATED O_2 INTERMEDIATES ON DNA SYNTHESIS AND REPAIR

The direct addition of H_2O_2 to culture medium or the incubation in the presence of the

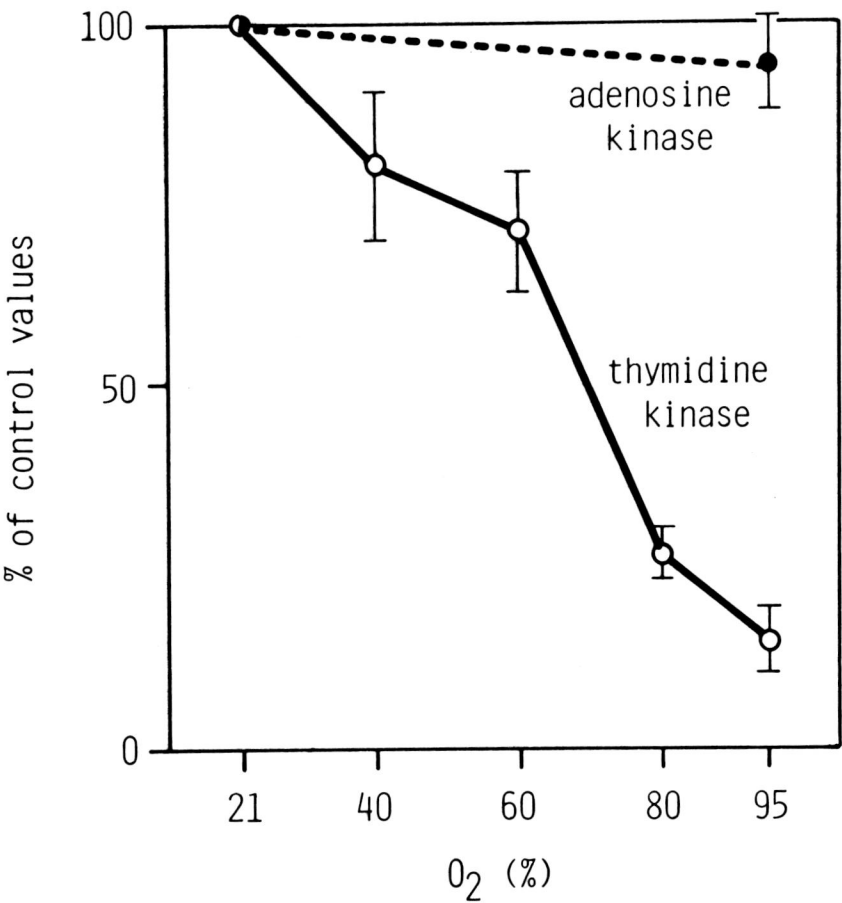

FIGURE 5. Effects of a 48-hr exposure on various O_2 concentrations on adenosine kinase and TdR kinase activities of confluent aortic endothelial cells. Results are expressed as percent of control values. Bars represent ± 1 SE. (From Junod, A. F., Clement, A., Jornot, L., and Petersen, H., *Biochim. Biophys. Acta,* 847, 20, 1985. With permission.)

enzyme system hypoxanthine-xanthine oxidase or of activated polymorphonuclear cells is accompanied by profound dose-dependent effects on DNA synthesis from TdR, which can be blocked by catalase.[21] Spragg et al.[22-24] have studied the time course of the early events following such an aggression and found that ATP levels fell to 30% of control values within 3 min. This ATP depletion in endothelial cells can be related to similar changes found in lymphoid cells and monocytes.[25] These cell preparations, exposed to the same agents, also show NAD depletion together with an increased poly(ADP)ribose polymerase activity. These profound changes, potentially leading to cell death, could be the result of DNA damage with strand breaks. DNA repair following DNA damage of various causes, can stimulate poly(ADP)-ribose polymerase, which results in the formation of poly(ADP)-ribose, a polymer which can play a still somewhat undefined role in DNA repair. But the formation of poly(ADP)ribose requires the cleavage of ADP-ribose from NAD, with the subsequent cellular depletion of NAD and ATP, leading to cell death. Berger has called this sequence the suicide response of poly(ADP)ribose polymerase.[26] Endothelial cells appear to share with other cell populations the privilege of belonging to the category that can respond in such a manner to massive DNA damage. This response should therefore be considered as a possible cause of the cytotoxicity of extracellularly generated O_2 metabolites, and lipid peroxidation may not be the only cellular damage to be involved.

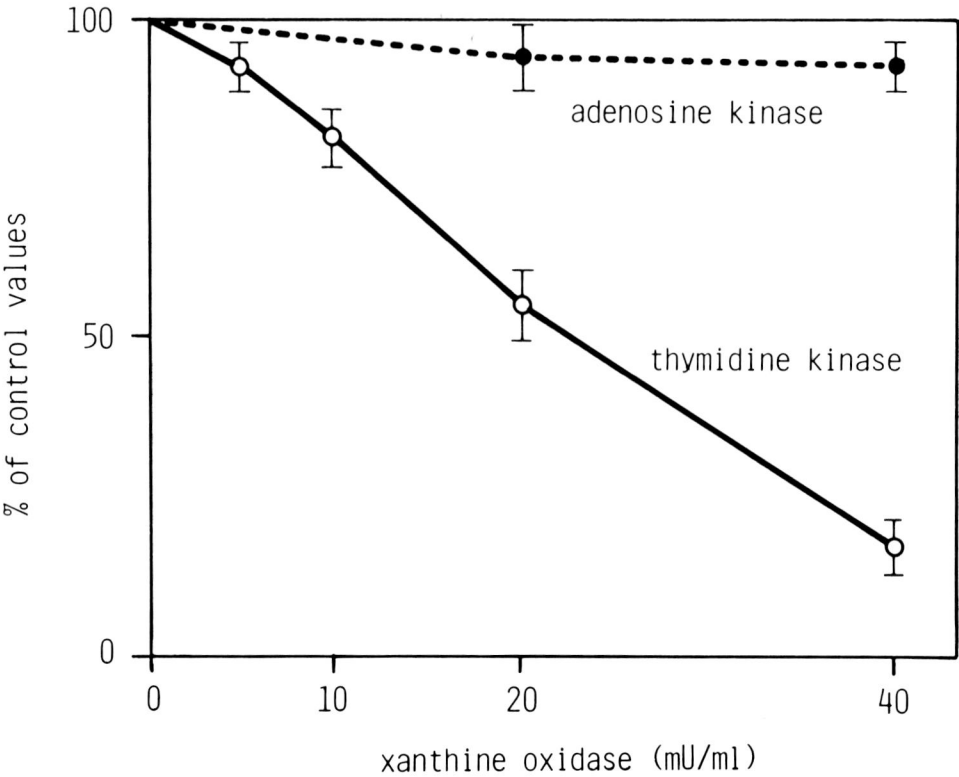

FIGURE 6.   Effects of a 30-min incubation in the presence of various amounts of xanthine oxidase (+ 2 m*M* hypoxanthine) on adenosine kinase and TdR kinase activities of confluent aortic endothelial cells. (From Junod, A. F., Clement, A., Jornot, L., and Petersen, H., *Biochim. Biophys. Acta,* 847, 20, 1985. With permission.)

Table 1
EFFECTS OF A 2-DAY EXPOSURE TO 95° O_2 ON THYMIDINE KINASE
AND THYMIDYLATE SYNTHETASE ACTIVITIES AND
PROLIFERATION OF PRECONFLUENT AORTIC ENDOTHELIAL CELLS

	Thymidine kinase (nmol TMP/mg protein)	Thymidylate synthetase (nmol TMP/mg protein)	Cell number (× 10³)
30% Confluence			
Control condition	14.5	11.1	346
+ 2 Days of culture			
Control condition	6.5	11.3	804
95% O_2 exposure	1.0	14.0	436

The release of O_2 intermediates by activated neutrophils has also been shown to provoke mutagenic lesions. Because of their close relationship with neutrophils, endothelial cells can be considered to be especially vulnerable targets. This aspect of the biology of O_2 intermediates will certainly receive a more sustained attention in the near future.

So far, only the effects of O_2 metabolites have been tested. That exposure to hyperoxia could also cause DNA damage and activate poly(ADP)ribose polymerase has not yet been studied. The magnitude of the toxic effects is likely to be modulated by both the antioxidant defenses of the cell population and the intensity of the oxidative stress.

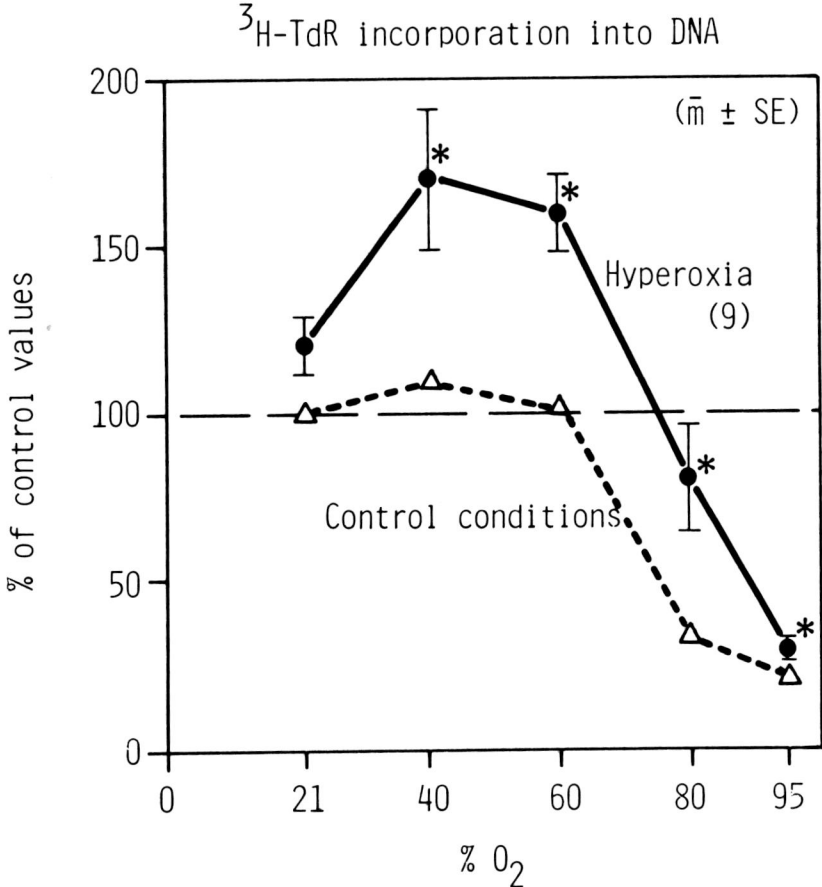

FIGURE 7. Effects of a 48-hr preexposure to 95% O_2 on the sensitivity of confluent aortic endothelial cells to various O_2 concentrations administered for another 48-hr period. The two periods of exposure to O_2 were separeted by a 24-hr period of culture under control conditions. Asterisk represents values significantly different from those obtained under control conditions (interrupted line).

IV. CONCLUSIONS

The data presented in this review illustrate how sensitive endothelial cells are to high O_2 concentrations or to extracellularly generated O_2 metabolites. Impaired DNA synthesis, DNA damage, and DNA repair can occur, with a different time course and intensity, depending on the nature and magnitude of the oxidative stress.

The occurrence of oxidative damage by activated phagocytes during inflammation, by activated xanthine oxidase during reperfusion after ischemia, and the often liberal administration of O_2 to patients may therefore have marked effects on endothelial cells in various organs, not only in terms of immediate cytotoxicity, but also for the subsequent repair of injured cells.

REFERENCES

1. **Kapanci, Y., Weibel, E. R., Kaplan, M. P., and Robinson, F. R.,** Pathogenesis and reversibility of the pulmonary lesions of oxygen toxicity in monkeys. II. Ultrastructural and morphometric studies, *Lab. Invest.,* 20, 101, 1969.
2. **Fishman, J. A., Ryan, G. B., and Karnovsky, M. J.,** Endothelial regeneration in the rat carotid artery and the significance of endothelial denudation in the pathogenesis of myointimal thickening, *Lab. Invest.,* 32, 339, 1975.
3. **Gerrity, R. G., Richardson, M., Caplan, B. A., Cade, J. R., Hirsch, J., and Schwartz, C. J.,** Endotoxin-induced vascular endothelial injury and repair. II. Focal injury, en face morphology, ^{3}H-thymidine uptake and circulating endothelial cells in the dog, *Exp. Mol. Pathol.,* 24, 59, 1976.
4. **Bowden, D. H. and Adamson, I. Y. R.,** Endothelial regeneration as a marker of differential vascular response in oxygen-induced pulmonary edema, *Lab. Invest.,* 30, 350, 1974.
5. **Babior, B. M.,** The respiratory burst of phagocytes, *J. Clin. Invest.,* 73, 599, 1984.
6. **McCord, J.,** Oxygen-derived free radicals in postischemic tissue injury, *N. Engl. J. Med.,* 312, 159, 1985.
7. **Bachofen, M. and Weibel, E. R.,** Structural alterations of lung parenchyma in the adult respiratory distress syndrome, *Clin. Chest Med.,* 3, 35, 1982.
8. **Housset, B., Ody, C., Rubin, D. B., Elemer, G., and Junod, A. F.,** Oxygen toxicity in cultured aortic endothelium: selenium-induced partial protective effect, *J. Appl. Physiol.,* 55, 343, 1983.
9. **Ody, C. and Junod, A. F.,** Direct toxic effects of paraquat and oxygen on cultured endothelial cells, *Lab. Invest.,* 52, 77, 1985.
10. **Forman, H. J., Aldrich, T. K., Posner, M. A., and Fisher, A. B.,** Differential paraquat uptake and redox kinetics of rat granular pneumocytes and alveolar macrophages, *J. Pharmacol. Exp. Ther.,* 221, 428, 1982.
11. **Keeling, P. L. and Smith, L. L.,** Relevance of NADPH depletion and mixed disulphide formation in rat lung to the mechanism of cell damage following paraquat administration, *Biochem. Pharmacol.,* 31, 3242, 1982.
12. **Plagemann, P. G. W., Marz, R., and Erbe, J.,** Transport and countertransport of thymidine in ATP depleted and thymidine kinase-deficient Novikoff rat hepatoma and mouse L cells: evidence for a high Km facilitated diffusion system with wide nucleoside specificity, *J. Cell Physiol.,* 89, 1, 1976.
13. **Clement, A., Hübscher, U., and Junod, A. F.,** Effects of hyperoxia on DNA synthesis in cultured porcine aortic endothelial cells, *J. Appl. Physiol.,* 59, 1110, 1985.
14. **Junod, A. F., Clement, A., Jornot, L., and Petersen, H.,** Differential effects of hyperoxia and hydrogen peroxide on thymidine kinase and adenosine kinase activities of cultured endothelial cells, *Biochim. Biophys. Acta,* 847, 20, 1985.
15. **Fridovich, I.,** Quantitative aspects of the production of superoxide anion radical by milk xanthine oxidase, *J. Biol. Chem.,* 245, 4053, 1970.
16. **Bello, L. J.,** Regulation of thymidine kinase synthesis in human cells, *Exp. Cell Res.,* 89, 263, 1974.
17. **Coppock, D. L. and Pardee, A. B.,** Regulation of thymidine kinase activity in the cell cycle by a labile protein, *J. Cell. Physiol.,* 124, 269, 1985.
18. **Junod, A. F., Petersen, H., and Jornot, L.,** Thymidine kinase, thymidylate synthetase and endothelial cell growth under hyperoxia, *J. Appl. Physiol.,* 62, 10, 1987.
19. **Hübscher, U.,** DNA polymerases in prokaryotes and eukaryotes: mode of action and biological implications, *Experientia,* 39, 1, 1983.
20. **Brawn, K. and Fridovich, I.,** DNA strand scission by enzymically generated oxygen radicals, *Arch. Biochem. Biophys.,* 206, 414, 1981.
21. **Ody, C. and Junod, A. F.,** Effect of a variable glutathione peroxidase activity on H_2O_2 related cytotoxicity in cultured aortic endothelial cells, *Proc. Soc. Exp. Biol. Med.,* 180, 103, 1985.
22. **Spragg, R. G., Hinshaw, D. B., Hyslop, P. A., Schraufstätter, I. U., and Cochrane, C. G.,** Alterations in adenosine triphosphate and energy charge in cultured endothelial and P388D$_1$ cells after oxidant injury, *J. Clin. Invest.,* 76, 1471, 1985.
23. **Schraufstätter, I. U., Hinshaw, D. B., Hyslop, P. A., Spragg, R. G., and Cochrane, C. G.,** Oxidant injury of cells. DNA strand-breaks activate polyadenosine diphosphate-ribose polymerase and lead to depletion of nicotinamide adenine dinucleotide, *J. Clin. Invest.,* 77, 1312, 1986.
24. **Schraufstätter, I. U., Hinshaw, D. B., Hyslop, P. A., Spragg, R. G., and Cochrane, C. G.,** Glutathione cycle activity and pyridine nucleotide levels in oxidant-induced injury of cells, *J. Clin. Invest.,* 76, 1131, 1985.
25. **Berger, N. A., Sikorski, G. W., Petzold, S. J., and Kurohara, K. K.,** Association of poly(adenosine diphosphoribose) synthesis with DNA damage and repair in normal human lymphocytes, *J. Clin. Invest.,* 63, 1164, 1979.
26. **Berger, N. A.,** Poly(ADP-ribose) in the cellular response to DNA damage, *Radiat., Res.,* 101, 4, 1985.

Chapter 29

BIOCHEMICAL AND FUNCTIONAL ASPECTS OF OXYGEN-MEDIATED INJURY TO VASCULAR ENDOTHELIUM

Bruce A. Freeman, Robert M. Jackson, Sadis Matalon, and Susan M. Harding

TABLE OF CONTENTS

I. INTRODUCTION

Vascular injury, due to excess production of reactive oxygen species, occurs secondary to diverse phenomena including trauma, acute inflammation, sepsis, tissue ischemia-reperfusion, oxygen toxicity, and exposure to xenobiotics capable of redox cycling. Vascular lining cells, specifically the endothelium, can be both critical targets and sources of reactive oxygen species which can then yield secondary messengers which activate inflammatory cells to exacerbate initial oxidant insults. Thus, the endothelium is an extremely critical site of oxidant injury since damage can be induced by diverse etiologic factors, resulting in losses of both microvascular metabolic function and barrier properties.

It has been clear for many years that pulmonary oxygen toxicity results in permeability-type pulmonary edema, indicating damage to the pulmonary vasculature. Since the discovery of the endothelium as a differentiated cellular unit having specific functional and structural roles in the maintenance of lung fluid balance, much recent interest has focused on the way in which the pulmonary endothelium responds to hyperoxia. In recent years the endothelium has been shown to have a primary role in oxygen toxicity, because endothelial cells themselves are capable of producing and releasing highly reactive, partially reduced products of oxygen metabolism.

II. ULTRASTRUCTURAL OBSERVATIONS

The importance of the endothelium in oxygen toxicity began to be appreciated when detailed studies of alveolar capillary structure revealed that capillary endothelial cells were very sensitive to the toxic effects of oxygen. Morphological changes in the air-blood barrier of young rats (age 44 to 50 days) exposed to 100% oxygen at 1 atm, were characterized using quantitative ultrastructural techniques.[1] No detectable changes occurred for the first 24 hr. However, between 48 and 72 hr, interstitial edema began to collect around large blood vessels. By 72 hr of exposure, morphometric analysis revealed that 65% of all capillaries were obliterated. While the surface area of the alveolar epithelium remained constant, the capillary endothelial surface and the capillary blood volume fell after 72 hr to about 50% of their control values. This occurred in association with destruction of capillaries within the interalveolar septa and with thickening of the air-blood barrier. The cytoplasm in marginal regions of capillary endothelial cells swelled and was void of organelles after 2 days, and after 3 days a large fraction of the endothelium showed drastic changes. In late stages of oxygen toxicity, the cytoplasm became dark, and damaged cells partially detached from basement membranes. In some areas, complete cytolysis of endothelial cells was observed with capillaries bounded only by the basement membrane in direct contact with blood. Remarkably few morphological changes were observed in Type I and II alveolar epithelial cells, even while extensive destruction of the capillaries occurred. From this study it became obvious that in early stages of oxygen toxicity, interstitial edema, suggesting endothelial dysfunction, developed while the architecture and fine structure of the lung were still seemingly unchanged. Based on such early observations, endothelial cells and alveolar capillaries were regarded as the site of the most intense interaction between oxygen from alveolar air and those additional inflammatory factors supplied by the capillary blood. Functional studies to be detailed later will show the alveolar epithelium is also sensitive to oxygen injury.

Similar changes including swelling and vacuolization of the endothelium occurred in the lungs of mice exposed to 90% oxygen.[2] In addition to these pathologic changes, thymidine labeling showed no change in the labeling of alveolar epithelial cells. Yamamoto studied the Type II alveolar epithelium and capillary endothelial cells of the blood-gas barrier in rats exposed to 80 or 100% O_2.[3] They found swelling of capillary endothelial cells along with mitochondrial alterations including loss of matrix density and loss of cristae. The endoplasmic reticulum became swollen and contained electron-dense material, while free lysosomes were present thoughout the cytoplasm. Even in rats adapted to hyperoxia by prior exposure to 85% O_2, swelling of capillary endothelial cells to place by 14 days with total loss of endoplasmic reticulum by 21 days of O_2 exposure. Recently, the relative sensitivity of the endothelium of the lung to 85 or 100% oxygen was confirmed in rats.[4] These studies involved an extensive morphometric analysis which allowed quantitation of the distribution of cell types and the absolute volumes of various tissue components and surface areas. The total amount of alveolar epithelial surface area did not change during oxygen exposure, but the capillary endothelial surface area decreased by 45%. Total lung cell numbers increased while both the total number of endothelial cells and their percent as total lung cells decreased. The remaining endothelial cells were twice as large as the controls, resulting in only a small change in the total endothelial cell volume. The total number of endothelial cells decreased by 45% and capillary surface area decreased by the same amount. The decreased volume of the capillary bed resulted in decreased calculated diffusion capacity and an increase in the effective diffusion distance for oxygen. Thus, while endothelial cells are destroyed early during adaptive oxygen exposure, some of the cells remain and become resistant to hyperoxia.

III. FUNCTIONAL OBSERVATIONS

Animals with healthy lungs exposed to 100% O_2 at 1 atm develop arterial hypoxemia[5,6] or progressive uncompensated respiratory acidosis.[11,12] Death from normobaric O_2 toxicity in all animals is secondary to massive protein-rich pulmonary edema, resulting from damage to both the alveolar epithelium[6] and the capillary endothelium. Electron-microscopic examination of the blood-gas barrier of rats exposed to 95% O_2 for 72 hr revealed progressive pulmonary capillary destruction, leading to a 50% decrease in capillary area.[13] In this section, a number of studies that have used different techniques to quantify oxygen induced injury to the pulmonary microvasculature will be examined.

A common way of assessing fluid and solute transport across the pulmonary microvasculature is to measure the pulmonary lymph flow and its protein content. In sheep, pulmonary lymph may be obtained by cannulation of the caudal mediastinal lymph node.[14] This animal model has been used in many studies investigating the effects of different agents in the pulmonary circulation.

Bressack et al.[15] exposed six lambs to 100% O_2 until death and measured pulmonary and systemic vascular pressures, cardiac output, lymph flow, and albumin lymph to plasma ratios (L/P) at different intervals of exposure. All vascular pressures and cardiac output remained at control values throughout oxygen exposure. Lymph flows doubled after 120 hr in O_2, while (L/P) remained unchanged. At that time, the lambs had developed severe hypercapnia. From these data, the authors concluded that the increased production of partially reduced oxygen species damaged the microvascular membrane and increased the size of the transendothelial junctions. In the absence of any hemodynamic changes, this will result in higher albumin and solvent fluxes, producing noncardiogenic pulmonary edema. These changes occurred only shortly before death and after the onset of hypercapnia. Other have reported similar findings in adult sheep exposed to 100% O_2, the only difference being that significant increase in lymph flow was observed after 72 hr in O_2.[11,16]

Using a similar preparation, it was found that continuous exposure of sheep to 100% O_2 for 66 to 84 hr caused a smaller increase in the lymph/plasma (L/P) ratio for endogenous albumin than for uncharged dextrans with the same effective molecular radius.[17] It was concluded that charge, in addition to size, may play a factor in the movement of large macromolecules from the blood to the lymph. Indeed, the transendothelial channels seem to have positive charges lining their walls.[18] Whether exposure to hyperoxia may alter the charge distribution, in addition to the size of these channels, is not known at the present time.

A potential complication in the interpretation of these results is that some of the lymph entering the caudal mediastinal lymph node is systemic in origin.[19] This source of error may be especially relevant in these experiments, since there is a large increase in blood flow to the diaphragm of sheep exposed to 100% O_2 for 60 to 80 hr.[12] To circumvent this problem, the time course of the oxygen-induced damage to the pulmonary microvasculature was assessed by measuring the pulmonary filtration coefficient of isolated perfused lung.[5] An increase in filtration coefficient is a sensitive index of early capillary damage since it reflects an increase in either the number or the size of the different pathways for solute and solvent transport across the pulmonary microvasculature. Rabbits were exposed to 100% O_2 for 24, 48, or 66 hr. At the end of these periods, the animals were anesthetized and their heart and lungs were removed from the chest and suspended by the trachea from a linear force transducer. They were then perfused with a peristaltic pump at a constant flow with a solution of normal Ringer's bicarbonate containing 5% bovine serum albumin and equilibrated with 95% O_2 and 5% CO_2. The perfusion pressure and weight of the preparation were measured continuously.

After a steady weight was reached, the flow was increased rapidly, resulting in increases of the perfusion pressure and lung weight. The latter has been shown to be due to vascular distention, which is essentially complete by 30 sec, and to fluid flux from the vascular to the extravascular space.[20] The values of the filtration coefficient were calculated from the following formula:

$$K_f = (dw/dt) \times (1.3/(dP_a)) \qquad (1)$$

where dw/dt is the rate of weight gain in the preparation measured between 120 and 150 sec from the time the arterial pressure was increased; 1.3 is the value of the ratio of the arterial to venous resistance of the pulmonary circulation; and the dP_a is the change in perfusion pressure after the flow was increased.

The mean value of the filtration coefficient (± 1 SEM) in air-breathing rabbits was 0.036 $\pm$ 0.002 mℓ/(min $\times$ torr $\times$ g dry lung). After 48 and 66 hr in 100% O_2, it increased by 58 and 114%, respectively. These results indicate that oxygen damage to the pulmonary microvasculature is progressive and occurs concurrently with the increase in alveolar permeability to solute but before the compromise of the gas exchange.[5,6]

A decrease in the metabolic activity of the pulmonary endothelium has also been used as an index of oxygen-induced damage to the pulmonary circulation. Exposure of rats to 100% oxygen for 18 hr decreased the clearance of serotonin by the pulmonary circulation.[7] After 24 hr in hyperoxia there was significant decrease in the conversion of angiotensin I to angiotension II by canine pulmonary endothelium.[8] Angiotensin-converting enzyme activity decreased to 77% of its control value after 16 hr of exposure of conscious rabbits to 100% O_2, and to 66% of control value after 40 hr in 100% O_2. At the same time intervals, the authors found a significant depression in the clearance of ^{14}C-5-hydroxytryptamine by the pulmonary circulation. No capillary damage accountable for these lesions could be identified by light- or electron-microscopic examination of lung tissue after 20 hr in hyperoxia. Hyperoxia also decreased the plasma membrane fluidity of pulmonary artery and aortic endothelial cells in culture; the time course of this decrease paralleled the decrease in amine transport by these cells.[10]

In summary, exposure of mammals to 100% O_2 causes functional damage to the lung microvasculature which precedes the appearance of morphological changes or the onset of pulmonary edema.

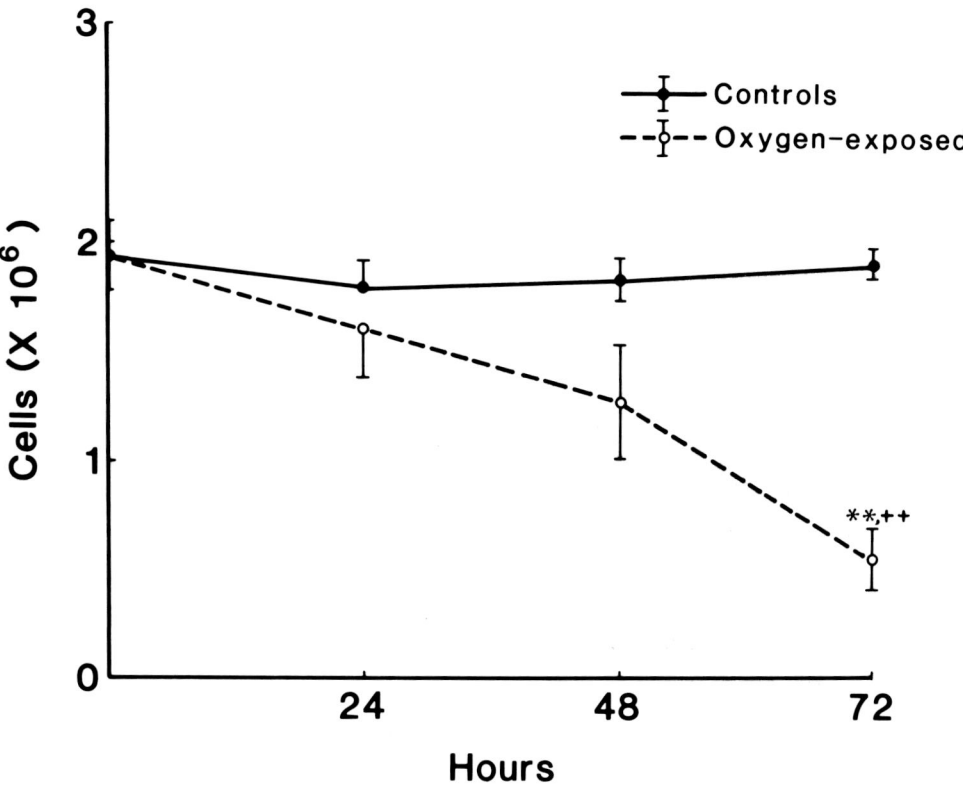

FIGURE 1. Changes in endothelial monolayer cell numbers occurring during incubation in air/5% CO_2 (controls) or 95% O_2/5% CO_2 (oxygen exposed). The number of cells in control monolayers did not change compared to the overall mean at 0 hr of 1,941,700 ± 162,500 cells per T25 flask. However, cell numbers in the oxygen-exposed monolayers decreased progressively to 560,000 ± 137,500 at 72 hr. Each point represents the mean of four to six monolayers with standard errors indicated. ** = $p < 0.01$, oxygen-exposed compared to air control at 72 hr; unpaired *t* test. + + = $p < 0.01$, 72-hr oxygen-exposed compared to 0-hr; ANOVA and Dunnett's *t* test.

IV. STRUCTURAL CHANGES OF CULTURED ENDOTHELIUM

Cultured endothelial cells from systemic or pulmonary arteries have been used as a tool to study the mechanism of hyperoxia-induced cytotoxic effects in the endothelium. The time course of changes in indices reflecting cell loss and cell damage during exposure of cultured endothelium to 95% O_2 has been studied (Figure 1). Endothelial cells from the thoracic aortas of freshly killed pigs were exposed to >95% O_2 for up to 5 days. After 5 days of O_2 exposure, the cultured endothelium had large fields containing cellular debris, granular material, and pyknotic nuclei along with strongly vacuolized cells and some apparently normal endothelial cells. DNA and protein content of the cultured vessels decreased with time in oxygen. Lactate dehydrogenase release (LDH) was also used to assess cell damage. Release of this intracellular enzyme into culture medium increased markedly between 3 to 5 days in high oxygen (Figure 2). Incorporation of ³H-thymidine into the cells, an indicator of DNA synthesis, also decreased significantly with time in oxygen. This appeared to be the most sensitive oxygen-related effect in cultured endothelial cells and occurred by only 24 hr of exposure. Similar changes in cell number and cell structure were detected in calf aortic endothelial cells during 42-hr oxygen exposures.[21] During exposure to 95% oxygen, the cells became larger and flatter and developed increased granularity. Vesicle and cysts

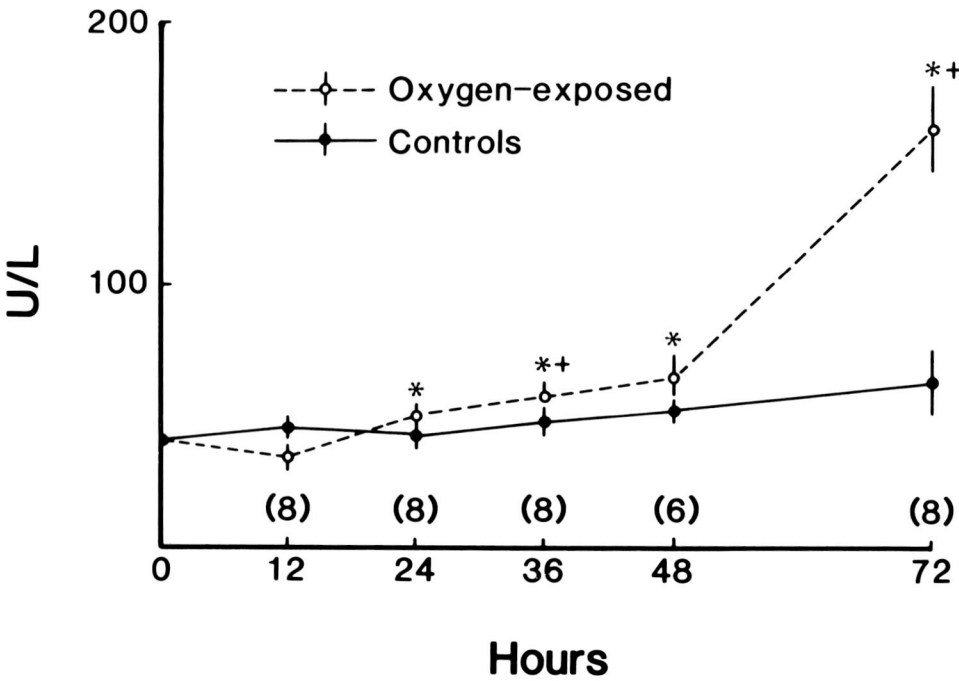

FIGURE 2.   Changes in cell medium LDH concentration during incubation in air/5% CO_2 (solid line and circles) or 95% O_2/5% O_2 (broken line and open circles). The medium LDH concentration was assayed at times ranging from 0 to 72 hr. The overall control mean LDH was 41.29 ± 1.84 units/liter (U/L); this represented the activity present in culture medium containing 20% heat-inactivated fetal bovine serum. The LDH concentration in cells exposed to air/5% CO_2 (controls) did not change significantly throughout the 72-hr period, indicating the relative absence of cytolysis. However, the LDH concentration in cells exposed to 95% O_2/5% CO_2 (oxygen-exposed) exceeded that in controls at 24, 36, 48, and 72 hr. The LDH concentration in oxygen-exposed cells also significantly exceeded its own baseline (0 hr) at 36 and 72 hr. * = $p < 0.05$, controls compared to matched oxygen-exposed; unpaired *t* test. + = $p < 0.05$, medium LDH compared to initial (0 hr) concentration; ANOVA and Dunnett's t test.

formed within the cells as did electron-dense inclusions. These morphologic changes preceded cell death as assessed by the ability to exclude vital dyes. Seriously injured cells eventually rounded up and floated off of the culture flask. Ultrastructural examination revealed that prominent abnormalities occurred in the mitochondria during cell death. Ultrastructural changes occurring in bovine pulmonary endothelial cells in culture during exposure to 80% oxygen are comparable to those in the lungs of intact animals. Lee et al. performed a detailed ultrastructural study of changes in bovine pulmonary artery endothelial cells exposed to hyperoxia in vitro.[22] They found that it required 48 hr of O_2 exposure for evidence of injury to become visible. The cells which were exposed to 80% O_2/5% CO_2/ 15% N_2 became enlarged and showed vacuolization with increased lysosomes between 24 to 48 hr. In contrast to previous studies, however, the appearance of mitochondria in oxygen-exposed cells remained unchanged from air-exposed controls. These morphologic changes were more marked in cells which were preconfluent when compared to those that were postconfluent. While the overall tempo and type of cytologic damage appears comparable in cultured endothelium, mitochondrial structure in cultured cells appears more resistant to hyperoxic damage than in vivo.

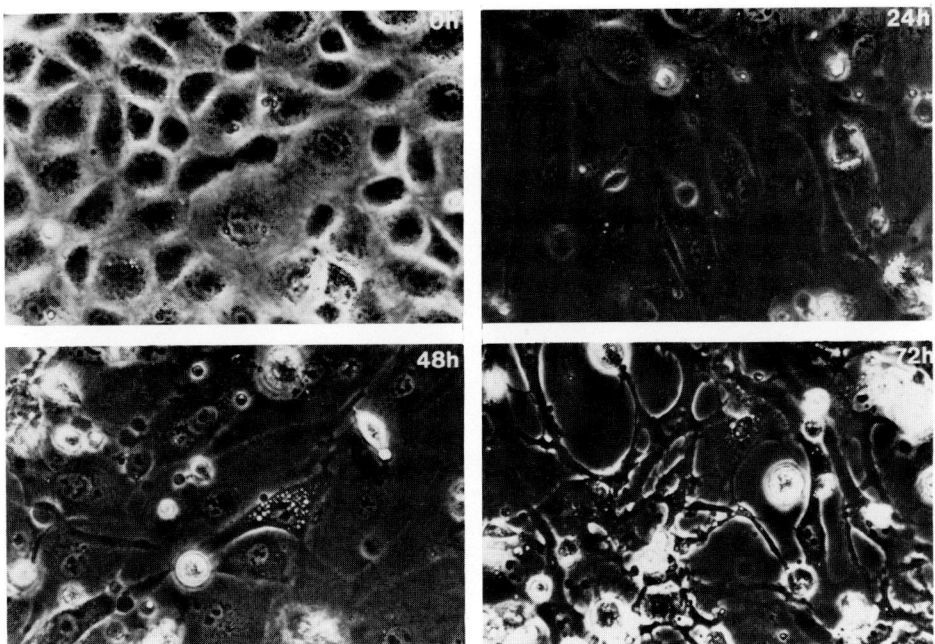

FIGURE 3. Morphological changes during incubation in 95% O_2/5% CO_2 for 72 hr. Postconfluent endothelial monolayers were exposed to hyperoxia, and the photographs are representative examples of the cytopathological changes observed at 0 hr (control) and after 24, 48, and 72 hr. Prior to exposure (0 hr), the cells formed a characteristic monolayer of squamous, epithelioid cells. Progressive deformation and cytolysis occurred during the exposure to hyperoxia. After 24 hr some enlargement and early vacuolization was present, which increased markedly by 48 hr. At the end of 72 hr in oxygen, the cells were grossly distorted, displaying irregular stellate shapes, and many cells had detached from the culture dish. (Phase contrast optics, magnification $\times$ 126.)

We have performed similar studies on cultures of bovine pulmonary artery cells.[23] Typically, postconfluent cells undergo cytolysis between 48 to 72 hr of exposure to hyperoxia. These cytolytic changes are preceded by definite changes in cell shape, with normally cuboidal cells assuming a more elongated appearance. These changes are illustrated in a series of phase contrast photographs in Figure 3.

The cytoskeleton of endothelium plays an important role in maintenance of conformation and junction formation. It has been shown that disruption of microfilaments with cytochalasin B and D results in increased endothelial cell monolayer permeability to albumin.[24,25] Thus, endothelial cytoskeletal integrity is linked to pulmonary microvascular permeability. The cytoskeleton consists of three distinct systems of filaments including microfilaments, microtubules,[26] and intermediate filament. Examination of bovine pulmonary artery endothelial cells near confluence by staining microfilament F-actin with the fluorescent stain NBD-phallacidin[27] showed well organized microfilaments under normoxic (20% O_2) conditions (Figure 4). When endothelial cell monolayers were exposed to 95% O_2 for up to 84 hr, there was disruption of microfilaments exhibited by clumping of the fluorescent stain prior to cytolysis or "rounding up" (Figure 5).

V. BIOCHEMICAL MEDIATORS OF OXYGEN INJURY

Biochemical studies have shown that cellular oxygen toxicity is initiated by intracellular overproduction of reactive species of oxygen, including O_2^-, H_2O_2, and $OH\cdot$. It is energetically unfavorable for oxygen to combine with additional spin-paired electrons, thus accounting

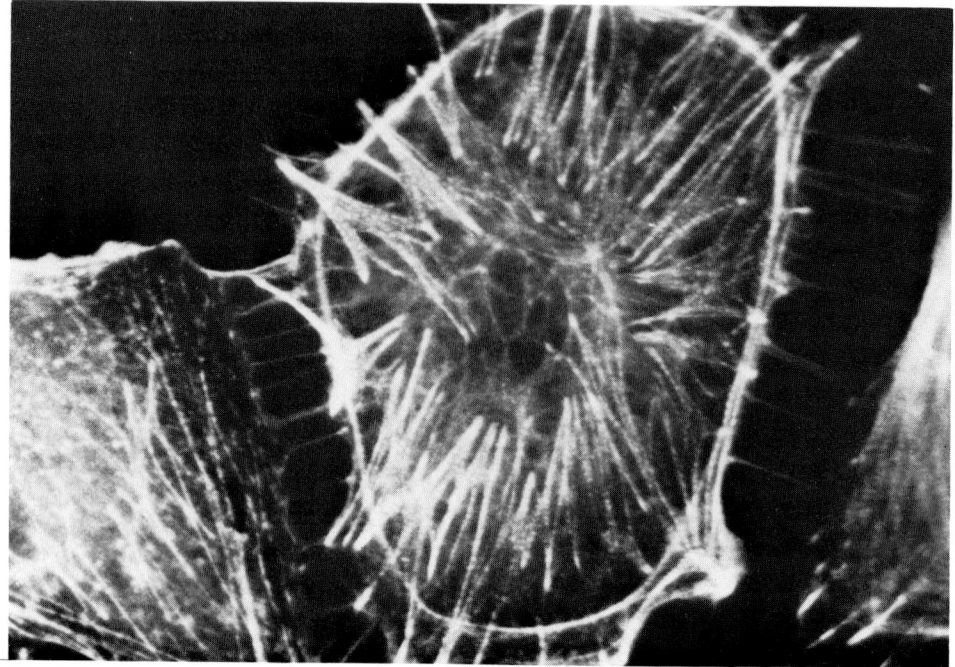

FIGURE 4.   Cytoskeleton stress fiber organization of bovine pulmonary artery endothelial cells cultured under normoxic conditions (20% O_2). Preconfluent monolayers were stained with NBD-phallacidin, a fluorescent marker specific for cellular filamentous (F)-actin, after being fixed with acetone at $-20°C$ for 10 min. These microfilaments of F-actin demonstrated well-organized forming stress fibers.

for the relatively low reactivity of oxygen as an oxidant in the absence of catalysis. The thermodynamic barrier to acceptance of electrons by oxygen can be bypassed by one electron reductions sequentially yielding O_2^-, H_2O_2, OH· plus H_2O, and $2H_2O$ (Figure 6). Mitochondrial cytochrome c oxidase binds oxygen to H_2O and transfers electrons to oxygen with no detectable release of reactive partially reduced species of oxygen. About 95% of the oxygen consumed by most mammalian cells can be accounted for by the catalytic reduction of oxygen to H_2O via this mitochondrial enzyme.

Hyperoxia increases the cyanide-resistant (noncytochrome-c oxidase) oxygen consumption of lung tissue slices, lung homogenates, lung mitochondria, and isolated and cultured cells.[27-31] This implies generally increased rates of tissue partial reduction of oxygen to toxic intermediates. The cyanide-resistant respiration of cells also includes the more benign oxidation of substrates including amino acids, nucleotides, and lipids. These processes do not involve radical intermediates in many cases, and are quantitatively minor compared to the contribution of oxygen radical generation to the net tissue cyanide-resistant respiration rate.[29,30] Endothelial cells are unique in their capacity to generate reactive oxygen species, determined from polarographic measurements of cellular cyanide-resistant respiration. The nonmitochondrial oxygen consumption of endothelial cells was 19% of total oxygen consumption, compared with other cell or organ preparations, which had 5 to 8% of total oxygen consumption as cyanide resistant.[31,32] This suggests that endothelial cells very actively utilize oxygen in biosynthetic and oxidative processes which involve endogenous rates of production of significant quantities of partially reduced oxygen species, i.e., O_2^-, H_2O_2, and OH·.

A. Cytosol

Soluble cytoplasmic molecules are quantitatively important sources of intracellular pro-

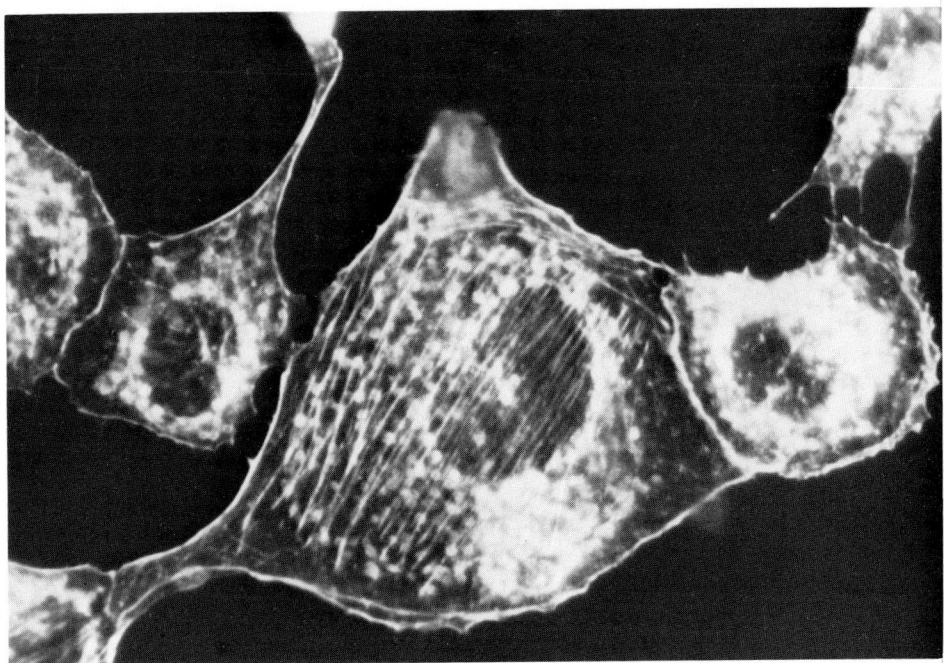

FIGURE 5.   Stress fiber organization of cultured bovine pulmonary artery endothelial cells after 84 hr exposure to 95% CO_2. The cells were fixed, extracted and stained with NBD-phallacidin using the same techniques as described in Figure 4. In hyperoxia, there is disruption of stress fibers. There is a reduction in stress fiber number, as well as clumping of the fluorescent stain in the cytoplasm. Morphological changes, including irregular stellate cell shapes are evident, which may affect microvascular permeability to changing endothelial surface area or intercellular junctions.

duction of reactive oxygen species. These molecules include thiols, hydroquinones, catecholamines, and flavins. In each instance, O_2^- is the primary radical generated by these substances.[33] Hydrogen peroxide is a secondary product of O_2^- formation, via spontaneous or enzymatically catalyzed dismutation of O_2^-. Cytoplasmic enzymes can also generate O_2^- and H_2O_2 as byproducts of catalytic cycling. Xanthine oxidase, aldehyde oxidase, tryptophan dioxygenase, flavoprotein dehydrogenases, peroxisomal oxidases, and hemoglobin are significant sources of O_2^- and H_2O_2. Free radical production by these sources almost universally increases directly as a function of oxygen tension.

B. Organelle
Membrane-bound proteins are also significant sources of cellular oxygen radical production. These sources exist in the nucleus, endoplasmic reticulum, mitochondria, and plasma membrane. Free radicals generated by these sources probably account for many of the early morphological changes observed in subcellular organelles of oxygen-exposed lungs.[34] For example, it would be reasonable to expect that relatively high local concentrations of reactive oxygen species would react near the site of production since their inherent reactivity ensures that diffusion distances in biological milieu will be very short. Thus, the mitochondrial swelling and dilatation of the inner mitochondrial membrane observed in early stages of oxygen toxicity may be due to overproduction of reactive oxygen species by electron transport components located on the inner mitochondrial membrane. We have seen that the production of O_2^- by submitochondrial particles derived from lung increases directly as a function of oxygen tension.[30]

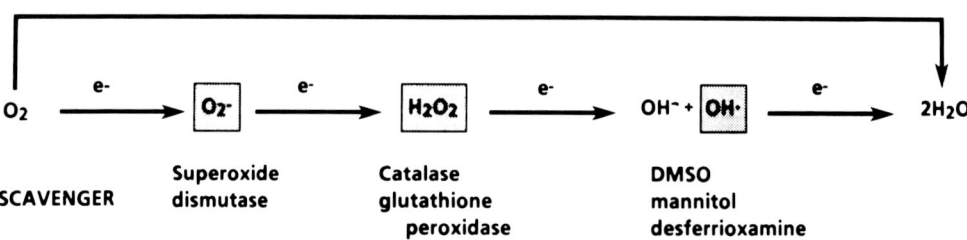

FIGURE 6.   Partial reduction of oxygen to reactive species. Oxygen is partially reduced by cellular metabolic processes or phagocytic cells to yield the reactive species O_2^-, H_2O_2 and $OH\cdot$. Cells have evolved a variety of antioxidant defenses designed to maintain low steady-state concentrations of toxic oxygen species via small molecule and enzymatic scavenging mechanisms.

The nuclear membrane and the endoplasmic reticulum contain a number of components capable of O_2^- and H_2O_2 generation. These include cytochrome P_{450}, which generates O_2^- by autooxidation and yields H_2O_2 by dissociation of peroxycytochrome P_{450}. Cytochrome B_5 can also autooxidize, yielding O_2^- and H_2O_2.[35,36] In the case of nuclear-membrane-derived reactive oxygen species, mechanism of gene expression which regulate cell differentiation and growth will be at risk. Nuclei isolated from lung and maintained under hyperoxic conditions generate O_2^- at twice the rate of nuclei incubated in air-saturated buffers.[37] Microsomes derived from endoplasmic reticulum generate O_2^- at a rate which increases three- to fivefold during hyperoxia.[38] Extrapolation of in vitro observations of mitochondrial and microsomal H_2O_2 generation to what they could maximally contribute in whole lung tissue showed the H_2O_2 production rate was 2.9 (mitochondria) and 19.7 (microsomes) $nmol \cdot min^{-1} \cdot g \, lung^{-1}$.

C. Membrane

Most, if not all mitochondrial H_2O_2 is derived from dismutation of O_2^- generated intramitochondrially. Lung mitochondrial O_2^- does not escape the superoxide dismutase of this organelle to reach cytosolic spaces. Hydrogen peroxide can diffuse extramitochondrially, and this rate increases as a function of oxygen tension. Thus, partial reduction of oxygen by mitochondria can damage other cytoplasmic components. Reactive oxygen species produced by the nuclear membrane in endoplasmic reticulum can undergo both intraorganelle reactions or diffuse directly into the cytosol.

Endothelial cell plasma-membrane-derived free radical production has not been well defined. Cyclooxygenase-mediated oxidation of arachidonate has been proposed to yield both a carbon and an oxygen-centered free radical intermediate. The oxygen-centered radical formed during the breakdown of the hydroperoxide of PGG_2, initially proposed to be $OH\cdot$ has recently been suggested to be a cyclooxygenase-hemoprotein radical, distinct from the oxygen-centered radical $OH\cdot$.[39,40] Free radical species are also involved in the conversion of prostaglandin endoperoxide (PGH_2) to thromboxanes, which is inhibited by the radical scavengers methional and nordihydroguiaretic acid.[41] It is interesting to note that the oxygen sensitive capillary endothelial cell is a major site of pulmonary eicosanoid synthesis. Because of the proximity of plasma-membrane-located free radical sources to cell and organelle surfaces, membrane-derived reactive oxygen species can potentially affect intracellular membrane and extracellular compartments, depending on the solubility and diffusion distances of the reactive molecules. The high and indiscriminate reactivity of $OH\cdot$ in biological milieu suggests that it will react within a few molecular radii of its site of generation. Superoxide is less reactive than $OH\cdot$ but still exerts significant toxicity, in part because it may be able to diffuse to critical targets distal to its site of generation. Hydrogen peroxide is uncharged

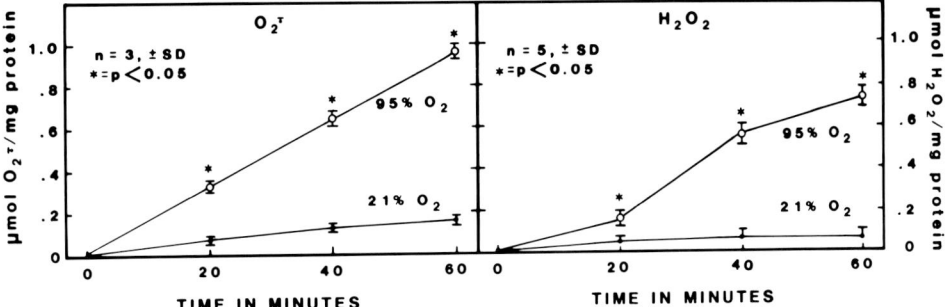

FIGURE 7.   Extracellular release of O_2^- and H_2O_2 by endothelium. Equilibration of cell culture medium with 95% oxygen enhanced both O_2^- H_2O_2 release by porcine thoracic aortic endothelial cells in monolayer culture. The enhanced rate of release of reactive oxygen species occurred immediately after increasing medium oxygen tension. Comparing rates of O_2^- and H_2O_2 release, it is apparent that about half of the extracellular H_2O_2 detected can be accounted for by spontaneous dismutation of O_2^-. Each point represent the mean of five monolayers with standard deviations indicated. * = $p < 0.05$, oxygen-exposed compared to air control; ANOVA and Dunnett's *t* test.

and less reactive than OH· and O_2^-. If H_2O_2 escapes cellular defense mechanisms, it is more likely to diffuse across organelle or plasma membranes to exert toxic effects distal to sites of generation than O_2^- or OH·. Thus, reactivity and toxicity are not necessarily directly linked.

Endothelial production of reduced oxygen species essentially obeys the Law of Mass Action. Many cellular components are available to reduce oxygen, and when the oxygen concentration increases, so does its rate of partial reduction. Experimental observations in general support the first-order nature of lung O_2^- and H_2O_2 production, with respect to oxygen concentration.[28-31,37] Other factors which would modify rates of lung cell oxygen radical production include enzyme inhibition or induction, cofactor availability, the concentration of substrates available for radical-producing enzymes and the presence of drugs or xenobiotics which are metabolized via free radical intermediates.

D. Intact Cells

Examination of monolayers of cultured pulmonary artery endothelium shows that both O_2^- and H_2O_2 are released extracellularly (Figure 7). Extracellular O_2^- release was quantified by measurement of vanadate-mediated cooxidation of NADH by O_2^-. Hydrogen peroxide release was monitored by fluorescence detection of horseradish-peroxidase-mediated oxidation of p-hydroxyphenylacetic acid. Endothelial cells and their medium equilibrated with 95% oxygen showed a significant increase in rates O_2^- and H_2O_2 release. The O_2^- release by endothelial cells was inhibited by stilbene sulfonate derivates which block the membrane anion channel. Additionally, 12-fold enhancement of intracellular catalase specific activity following treatment of endothelial cells with liposome-entrapped catalase resulted in complete inhibition of extracellular H_2O_2 release. Presumably, hyperoxia enhanced intracellular rates of O_2^- and H_2O_2 production, which then gained access to extracellular spaces by diffusion across endothelial plasma membranes since no cell lysis was detected in the 30-min oxygen metabolite assay periods. The egress of O_2^- and H_2O_2 from vascular lining cells exposes serum components to reactive oxygen species. This phenomenon can explain secondary effects of endothelial oxidant production, including modification of low-density lipoprotein to a cytotoxic species[42] and formation of an albumin-bound lipid neutrophil chemoattractant.[43] The dilemma of pulmonary oxygen toxicity, which results in primary free radical injury to lung cells and secondary injury due to infiltration of neutrophils, is an expected consequence of oxygen therapy. The observation that enhanced intracellular antioxidant enzyme activity prevents both hyperoxic injury to cells and animals[44,45] and attenuation of the extracellular release of reactive oxygen species which could mediate inflammatory responses shows that pharmacologic modification of oxygen toxicity is possible.

E. Modification of Defense Mechanisms

The importance of superoxide in mediating hyperoxic injury to endothelium was implied by prevention of development of biochemical endpoints of endothelial injury following provision of a specific scavenger for O_2^-, superoxide dismutase. Because hyperoxic-induced endothelial injury is derived from overproduction of intracellular reactive oxygen species, it was important to enhance intercellular antioxidant activity to protect from oxygen toxicity.[44] Antioxidant enzymes are not membrane permeable and would not be expected to scavenge intercellular reactive oxygen species. Because liposomes can entrap a wide variety of macromolecules that are normally membrane impermeable and can transfer entrapped macromolecules across plasma membranes by endocytosis or fusion, they were used as vectors for delivery of superoxide dismutase into endothelial monolayers. This approach has been used to render both cell and animal models resistant to oxygen-mediated toxicity. Liposome-mediated delivery of superoxide dismutase or catalase to cultured cells can result in up to a 100-fold increase in cellular specific activity.[45] The half-life of liposome-delivered catalase is shorter than that of superoxide dismutase (SOD), presumably because of the sensitivity of catalase to proteases compared with SOD. The predominant mechanism of liposome uptake by cells is via endocytosis and to a lesser extent via membrane fusion. More efficient techniques for enhancing target orgen or cell specificity of liposome atachment and cytosolic transfer of active liposome contents are being devised by a number of groups, so that liposome-mediated macromolecule transfer can have more widespread pharmacological application. By protecting intracellular sites at risk from oxidant stress, it may be possible to employ antioxidant enzyme-containing vectors in both mechanistic studies and for pharmacological modification of oxidant injury to endothelium.

VI. IN VITRO DEFINITION OF OXYGEN-INDUCED ENDOTHELIAL INJURY

The concentration of environmental oxygen controls cell growth in culture. The growth of human lung fibroblasts is inhibited by high O_2 tension, and no increase in their number can be observed at 80% oxygen.[46] In addition, hamster embryo cells have been shown to grow faster in 1 to 20% oxygen when compared to 50 to 97% oxygen.[47] Thus, high concentrations of oxygen appear to have an overall inhibitory effect on cell growth rates in culture.

Analogous decreases in endothelial cell growth have been assessed by ^{3}H-thymidine incorporation and changes in cell number of several groups of investigators. Synthesis of DNA is inhibited in endothelial cells from the lungs of mice exposed to 40 to 100% O_2, while other cell types are not affected.[48] This inhibition of DNA synthesis lasts for about 5 days and returns to control levels upon termination of oxygen exposure. Similar decreases in ^{3}H-thymidine incorporation have been observed in lung explants taken from animals after 4 to 6 days of exposure to oxygen.[49] An increase of ^{3}H-thymidine incorporation into lung endothelium occurs after oxygen-induced lung injury. This indicates both that the lung microvasculature is sensitive to oxygen toxicity, and that it is capable of repair and regeneration in this model in which mice were exposed to 90% oxygen for 6 days.[50]

The morphologic changes which occurred during the time course of oxygen toxicity in cultured endothelium have been described in detail. However, it is physiologically important to assess damage prior to structural derangement. For this reason, multiple techniques have been developed which capitalized on the normal metabolic functions of the endothelium to test early sublethal cell injury. In many cases, these correlate well with studies of changes of endothelial metabolism in intact animals exposed to high oxygen, as do changes in cell number and presence of lysed and detached endothelial cells. Techniques more sophisticated than the classical vital dye exclusion have allowed very early detection of changes in endothelial metabolism during oxygen exposure. Housset and colleagues have studied DNA content of cell monolayers and correlated this with changes in DNA synthesis assessed by ^{3}H-thymidine incorporation.[51] There is normally some cell turnover in postconfluent culture, so the rate of ^{3}H-thymidine incorporation serves as a baseline indicator of DNA synthesis. This index is significantly decreased by 24 hr of oxygen exposure and appears to be completely inhibited by 5 days of oxygen exposure. It is thus a sensitive measure of the ability of cells to take up a labeled precursor and to synthesize DNA for replication.

The most commonly used indicator of cytolysis is release of the intracellular enzyme LDH. This is easily measured using spectrophotometric assays, but it is relatively insensitive and requires a significant loss of cells for its detection. Cells in stable monolayer culture typically release 2 to 2.5% of the total monolayer LDH content per day. This release increases slightly after 24 hr, and the increase of release becomes precipitous between 2 to 5 days of oxygen exposure. Rubin and coworkers were able to demonstrate LDH release after only 24 to 48 hr of oxygen exposure.[52] Block et al. found a similar release of LDH from both porcine pulmonary arterial and aortic endothelial cells in monolayer culture when exposed to 95% O_2.[53] Some of the increased LDH release may however be accounted for by enzyme content in cells which have released from the endothelial monolayer. Problems with this technique include the possibility that LDH is labile in culture medium and that the LDH content of detached cells is different than that of those attached to the monolayer. Either of these possibilities would lead to artifacts that underestimate the extent of cytolysis in culture.

More sensitive and specific tests of endothelial cell damage reflect basic aspects of their metabolism. One standard technique is to label endothelial monolayers by incubating them with approximately 10 μCi/mℓ[51] Cr (Figure 8). This results in a sepcific activity of approximately 2×10^4 CPM per 5×10^5 endothelial cells.[54] Cytotoxicity is then assessed by release of the radiolabel into medium. [51]Cr binds nonspecifically to cellular proteins, and release of the label follows the kinetics of release of large cytoplasmic molecules. Spontaneous leakage of the label is usually minimal, and 85 to 95% is released when cells are completely disrupted.

Other more sensitive markers of in vitro cell injury have also been investigated. Andreoli and coworkers radiolabeled endothelial cell monolayers with trace amounts of 2-deoxy-^{3}H-glucose.[55] This assay was able to detect significant endothelial cell damage at much lower concentrations of stimulus than the ^{51}Cr release. It appears that release of this label occurs from cells which have not been irreversibly damaged, because viable cells (which were able to exclude trypan blue) released the label following oxidant stress.

Rubin examined the ATP content of endothelial monolayers to assess the metabolic state of cells exposed to hyperoxia.[52] Cell ATP was measured using the luciferin-luciferase assay. Total ATP content was reduced per endothelial cell monolayer but did not change when corrected for cellular DNA. However, changes in the ATP content of endothelial cell monolayers occurred in control cells as well. Thus, this indicator may reflect culture conditions other than the PO_2.

Vascular endothelium is the major site of 5-hydroxytryptamine uptake and metabolism. Platelets and other cells of the vessel wall may remove serotonin from the blood but do not metabolize it. Small et al. found that cells obtained from bovine aorta which morphologically resembled endothelium and were Factor VIII positive, produced 5-hydroxyindole acetic acid during incubation with 5-hydroxytryptamine.[56] Both imipramine and iproniazid inhibited 5-hydroxyindole acetic acid production, suggesting that the metabolism required transcellular uptake and oxidative deamination via monoamine oxidase. These conditions mirrored those found in the pulmonary vascular endothelium by Junod.[57] Bovine aortic endothelial cells

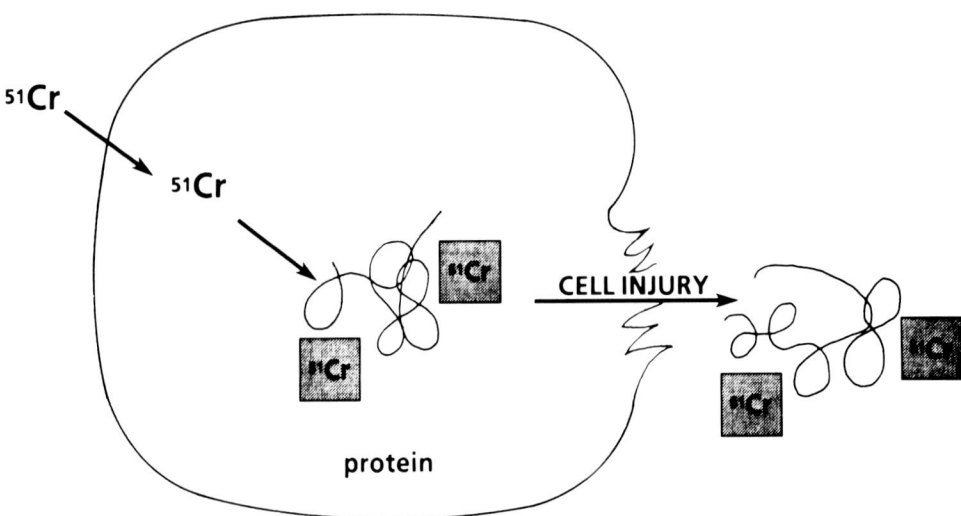

FIGURE 8.   Measurement of cell injury by release of ^{51}Cr label. Cells are incubated for 2 to 12 hr in 10 μCi/mℓ added to culture medium. The cells retain ^{51}Cr which is bound to proteins chromatographing as 10,000-dalton complexes or larger. Upon loss of cell membrane permeability, the ^{51}Cr-bound protein is released into culture medium in a manner proportional to release of intracellular LDH activity.

produced 39 ± 7.5 nmol 5-hydroxyindole acetic acid per 10^9 cells per hour when incubated in the presence of 4 μM ^{14}C-5-hydroxytryptamine. Such observations have now been extended to show 5-hydroxytryptamine uptake, and metabolism is regarded as a specific endothelial function whose impairment correlates with impaired metabolism of the endothelium. Block and Stalcup studied the specific effects of high O_2 tension on endothelial cells cultured from the thoracic aortas of calves.[21] These observations followed initial reports that high partial pressures of oxygen depressed the uptake of serotonin in isolated perfused rat lungs, and depression of 5-hydroxytryptamine uptake was suggested as a sensitive index of direct oxygen-induced endothelial cell injury.[58] These investigators incubated endothelial cells with ^{14}C-5-hydroxytryptamine and measured the uptake of 5-hydroxytryptamine by the disappearance of the labeled substrate in cultured medium. Uptake of 5-hydroxytryptamine by endothelial cells exposed to 95% O_2 for 20 to 42 hr was significantly depressed, compared with uptake by control cells maintained in 14% O_2. There was some recovery of 5-hydroxytryptamine uptake when hyperoxic-exposed monolayers were placed in medium equilibrated with 14% O_2. The mechanism by which hyperoxia interferes with 5-hydroxytryptamine uptake in endothelial cells is unknown. It may interfere with transcellular transport of 5-hydroxytryptamine or may depress intracellular metabolism of the amine. In conclusion, high PO_2 directly affects the metabolic activity of endothelial cells as manifested by the depression of 5-hydroxytryptamine in O_2-exposed cells.

Cultured endothelial cells produce appreciable quantities of cyclooxygenase metabolites. Prostacyclin is the major arachidonic acid metabolite of cultured endothelium, although smaller amounts of $PGF_{2\alpha}$, PGE_2, and thromoxane A_2 have been reported to be released by stimulated postconfluent monolayers.[58] In studies examining the effect of oxygen of this production, it is necessary to recognize that culture conditions can affect both arachidonic acid metabolism and cellular response to oxygen stress. To control for these factors, both control and experimental monolayers must be exactly matched for passage number, time after confluence and medium composition so that intracellular and extracellular antioxidants do not differ.[59] Diverse stimuli influence arachidonic acid metabolism in cultured endothelial cells.[60] Prostacyclin release typically parallels cytolysis and is stimulated by gamma irra-

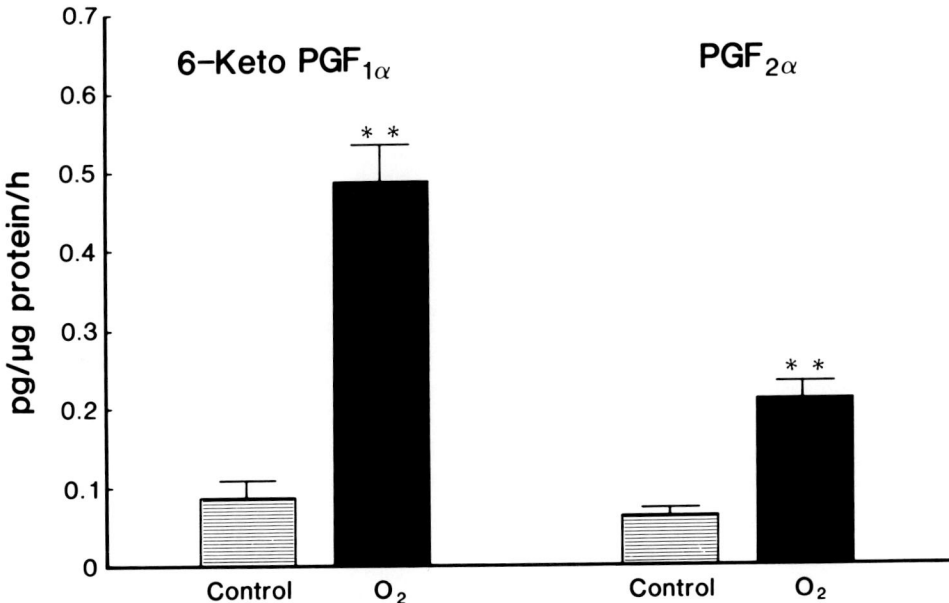

FIGURE 9. Apparent net production of 6-keto $PGF_{1\alpha}$ and $PGF_{2\alpha}$ during 72-hr incubation of endothelial monolayers (eight in each group) in air/5% CO_2 (control) or 95% O_2/5% CO_2 (O_2). The height of the bar represents the mean production rate with one standard error indicated. These data represent only the metabolites produced by endothelial cells, because background levels of 6-keto $PGF_{1\alpha}$ and $PGF_{2\alpha}$ in medium containing fetal bovine serum were subtracted prior to calculation. Nonstimulated control cells produced small amounts of both 6-keto $PGF_{1\alpha}$ (0.089 $\pm$ 0.013 pg/μg protein per hour and $PGF_{2\alpha}$ (050 + 0.008 pg per microgram of protein per hour. During hyperoxic exposure, production of both 6-keto $PGF_{1\alpha}$ (0.493 $\pm$ 0.187 pg/μg of protein per hour) and $PGF_{2\alpha}$ (0.202 $\pm$ 0.019 pg/μg of protein per hour) increased substantially by approximately five- and twofold, respectively. * = p < 0.01, control compared to O_2 exposed; unpaired t test.

diation. Accumulation of prostacyclin and prostacyclin production stimulated by exogenous arachidonic were correlated with cell detachment and a release of lactate dehydrogenase activity in endothelial cells exposed to 0.5 to 5 Gy of ^{60}Co gamma rays.[61]

We found that bovine pulmonary artery endothelial cells incubated in 95% oxygen released the prostacyclin metabolite 6-ketoprostaglandin $F_{1\alpha}$ and prostaglandin $F_{2\alpha}$ but no detectable thromboxane metabolite, thromboxane B_2. Those cells which were oxygen exposed released significantly more of 6-ketoprostaglandin $F_{1\alpha}$ and prostaglandin $F_{2\alpha}$ than did control cells when apparent net production rates over the entire 72-hr period were compared. These observations are illustrated in Figure 9.

Hydrogen peroxide from neutrophils can stimulate endothelial cell prostacyclin production.[62] However, preincubation of endothelial cells with high concentrations of H_2O_2 (around 35 μM) causes concentration-dependent inhibition of prostacyclin synthesis; exogenous H_2O_2 at high concentration appears to inhibit cyclooxygenase, but not prostacyclin synthetase or phospholipase A_2.[63] Some evidence exists for lipoxygenase pathways in endothelium. Chopped porcine pulmonary arteries, when stimulated by calcium ionophore, produce a substance that cochromatographs with synthetic leukotriene B_4.[64] Recently, preliminary evidence has been presented that aortic endothelial cells produce 12-hydroxy-5,8,10,14-eicosatetraenoic acid and dihydroxyeicosatetraenoic acid, the release of which can be inhibited by nordihydroguiaretic acid but not indomethacin.[65] Umbilical vein endothelial cells produce both 15-hydroxyeicosatetraenoic acid and 8,5,15-dihydroxyeicosatetraenoic acid molecules, implying a 15-lipoxygenase pathway and a system that can synthesize 14,15-LTA_4, a precursor of LTB_4.[66]

Because of their location in the pulmonary microvasculature, lung endothelial cells are in intimate contact with activated neutrophils destined to sites of lung inflammation. The mechanisms by which neutrophils are attracted into the lung are not completely described. However, the endothelium itself may be capable of elaborating neutrophil chemoattractants. For example, Mercandetti et al. studied conditioned media from confluent monolayers of endothelial cells taken from human umbilical veins.[67] They found that most of the monolayers elicited chemoattractant activity for neutrophils using the Boyden chamber assay technique. Chemoattractant activity in the medium increased with time and appeared in two molecular-weight fractions following gel filtration column chromatography. A migration-enhancing protein of approximately of 35,000 daltons was identified along with a lipid of approximately 1500 daltons. These findings provided some evidence that endothelial cells per se were capable of recruiting neutrophils to a vascular surface. Although no studies have specifically examined the effect of hyperoxia on neutrophil chemoattractant production, cultured bovine aortic and pulmonary artery endothelial cells release chemoattractant activity into culture medium when incubated with thiourea, a substance that causes increased permeability pulmonary edema in animals.[68] This occurred without cell death, and only minimal ^{51}Cr release was detectable during the incubation period. This activity appeared to be specifically derived from endothelial cells because pulmonary vascular smooth muscle cells and lung fibroblasts did not show increased neutrophil chemoattractant activity after incubation with thiourea. The material had both chemokinetic and chemotatic properties and was extractable into organic solvents. There is some indirect evidence that the chemotaxin may be a lipoxygenase product. Meclophenamate, a cyclooxygenase inhibitor, did not inhibit the production of this chemoattractant, whereas 5,8,11,14-eicosatetraeonic acid, an inhibitor of both cyclooxygenase and lipoxygenase, completely abolished generation of chemoattractants. No further studies have currently been done to characterize the chemoattractant.

In addition to releasing the compounds which may attract neutrophils, the endothelium becomes more adhesive to neutrophils after oxygen exposure. Bowman et al. demonstrated that a greater number of polymorphonuclear leukocytes (PMN) adhered to endothelial cells that had been exposed to hyperoxia for 28 to 48 hr than to control cells.[69] This may be an important mechanism, because the proximity of PMN to the endothelium appears to be critical for PMN-mediated endothelial cell damage. Stimulated neutrophils are able to injure endothelial cells in culture, and neutrophil-mediated cytotoxicity appears to be due to hydrogen peroxide.[70] Endothelial cells exposed to hyperoxia become more susceptible to neutrophil-mediated damage, and those complex interactions occur between the physicochemical environment and circulating inflammatory cells.[71] Activated neutrophils also release reactive oxygen species, which appears to trigger endothelial membrane phospholipase activation and cause prostacyclin secretion.[62]

ACKNOWLEDGMENTS

The authors acknowledge The Health Effects Institute, NIH RO1 NS 21782 and RO1 NS 24275 (BAF), RO1 HL31197 (SM), and the Veteran's Administration Research Service and American Lung Association (RMJ).

The editorial skills of Ms. Yvonne Lambott are also gratefully acknowledged.

REFERENCES

1. **Kistler, G., Caldwell, P., and Weibel, E.,** Development of fine structural damage to alveolar and capillary lining cells in oxygen-poisoned rat lungs, *J. Cell. Biol.,* 33, 605, 1967.
2. **Bowden, D., Adamson, I., and Wyatt, J.,** Reaction of the lung cells to a high concentration of oxygen, *Arch Pathol.,* 86, 671, 1968.
3. **Yamamoto, E., Wittner, M., and Rosenbaum, R.,** Resistance and susceptibility to oxygen toxicity by cell types of the gas-blood barrier of rat lung, *Am. J. Pathol.,* 59, 409, 1970.
4. **Crapo, J., Peters-Bolden, M., Marsh-Salin, J., and Shelburne, J.,** Pathologic changes in the lungs of oxygen-adapted rats, *Lab. Invest.,* 39, 640, 1978.
5. **Matalon, S. and Cesar, M. A.,** Effects of 100% oxygen breathing on the capillary filtration coefficient in rabbit lungs, *Microvasc. Res.,* 29, 70, 1985.
6. **Matalon, S. and Egan, E. A.,** Effects of 100% O_2 breathing on permeability of alveolar epithelium to solute, *J. Appl. Physiol.,* 859, 863, 1981.
7. **Block, E. R. and Fisher, A. B.,** Depression of serotonin clearance by rat lungs during oxygen exposure, *J. Appl. Physiol.,* 42, 33, 1977.
8. **Harabin, A. L., Homer, L. D., and Bradley, M. E.,** Pulmonary oxygen toxicity in awake dogs: metabolic and physiologic effects, *J. Appl. Physiol.,* 57, 1480, 1984.
9. **Dobuler, K. J., Catravas, J. D., and Gillis, C. N.,** Early detection of oxygen-induced lung injury in conscious rabbits. Reduced in vivo activity of angiotensin converting enzyme and removal of 5-hydroxy-tryptamine, *Am. Rev. Respir. Dis.,* 126, 534, 1982.
10. **Block, E. R., Patel, J. M., Angelides, K. J., Sheridan, N. P. and Garg, L. C.,** Hyperoxia reduces plasma membrane fluidity: a mechanism for endothelial cell dysfunction, *J. Appl. Physiol.,* 60, 826, 1986.
11. **Erdmann, J., III, Huttemeier, P. C., Landolt, C., and Zapol, W. M.,** Pure oxygen breathing increases sheep lung microvascular permeability, *Anesthesiology,* 58, 153, 1983.
12. **Matalon, S., Nesarajah, E., and Farhi, L. E.,** Pulmonary and circulatory changes in conscious sheep exposed to 100% O_2 at 1 ATA, *J. Appl. Physiol.,* 53, 110, 1982.
13. **Kistler, G. S., Caldwell, P. R. B., and Weibel, E. R.,** Development of fine structural damage to alveolar and capillary lining cells in oxygen-poisoned rat lungs, *J. Cell Biol.,* 32, 605, 1967.
14. **Staub, N. C., Bland, R. D., Brigham, K. L., Demling, R. H., Erdmann, A. J., III, and Woolverton, W. C.,** Preparation of chronic lymph fistulas in sheep, *J. Surg. Res.,* 19, 315, 1975.
15. **Bressack, M. S., McMillan, D. D., and Bland, R. D.,** Pulmonary oxygen toxicity: increased microvascular permeability to proteins in unanesthetized lambs, *Lymphology,* 12, 133, 1979.
16. **Newman, J. H., Loyd, J. E., English, D. K., Olgetree, M. L., Fulkerson, W. J., and Brigham, K. L.,** Effects of 100% O_2 oxygen on lung vascular function in awake sheep, *J. Appl. Physiol.,* 54, 1379, 1983.
17. **Hansen-Flaschen, J. H., Lanken, P. N., Pietra, G. G., Sampson, P. M., Johns, L., and Fishman, A. P.,** Effects of 100% O_2 on passage of uncharged dextrans from blood to lung lymph, *J. Appl. Physiol.,* 60, 1797, 1986.
18. **Simionescu, N., Simionescu, M., and Palade, G. E.,** Differentiated microdomains on the luminal surface of the capillary endothelium. I. Preferential distribution of anionic sites, *J. Cell Biol.,* 90, 605, 1981.
19. **Drake, R., Adair, T., Traber, D., and Gabel, J.,** Contamination of caudal mediastinal node efferent lymph in sheep, *Am. J. Physiol.,* 241 (Heart Circ. Physiol. 10), H354, 1981.
20. **Lunde, P. K. M. and Waaler, B. A.,** Transvascular fluid balance in the lung, *J. Physiol. (London),* 205, 1, 1969.
21. **Block, E. and Stalcup, S.,** Depression of serotonin uptake by cultured endothelial cells exposed to high O_2 tension, *J. Appl. Physiol.,* 59, 1212, 1981.
22. **Lee, S., Douglas, W., Deneke, S., and Fanburg, B.,** Ultrastructural changes in bovine pulmonary artery endothelial cells exposed to 80% O_2 in vitro, *In Vitro,* 19, 714, 1982.
23. **Jackson, R., Chandler, D., and Fulmer, J.,** Production of arachidonic acid metabolites by endothelial cells in hyperoxia, *J. Appl. Physiol.,* 61, 584, 1986.
24. **Shasby, D. M., Lind, S. E., Shasby, S. S., Goldsmith, J. C., and Hunninghake, G. W.,** Reversible oxidant-induced increases in albumin transfer across cultured endothelium: alterations in cell shape and calcium homeostasis, *Blood,* 65, 605, 1985.
25. **Shasby, M. D., Shasby, S. S., Sullivan, J. M., and Peach, M. J.,** Role of endothelial cell cytoskeleton in control of endothelial permeability, *Circ. Res.,* 51, 657, 1982.
26. **Weber, K. and Osborn, D. M.,** The molecules of the cell matrix, *Sci. Am.,* 253, 110, 1985.
27. **Barak, L. S. et al.,** Fluorescence staining of the active cytoskelton in living cells with 7-nitrobenz-2-oxa-1,3-diazole-phallacidin, *Proc. Natl. Acad. Sci. U.S.A.,* 77, 980, 1980.
28. **Freeman, B. A., Topolosky, M. K., and Crapo, J. D.,** Hyperoxia increases oxygen radical production in rat lung homogenates, *Arch. Biochem. Biophys.,* 216, 477, 1982.

29. **Freeman, B. A. and Crapo, J. D.,** Hyperoxia increases oxygen radical production in rat lungs and lung mitochondria, *J. Biol. Chem.,* 256, 10986, 1981.
30. **Turrens, J. F., Freeman, B. A., Levitt, J. G., and Crapo, J. D.,** The effect of hyperoxia on superoxide production by lung submitochondrial particles, *Arch. Biochem. Biophys.,* 217, 401, 1982.
31. **Beckman, J. S. and Freeman, B. A.,** Antioxidant enzymes as mechanistic probes of oxygen-dependent toxicity, in *Physiology of Oxygen Radicals,* Taylor, A. and Matalon, S., Eds., American Physiological Society, Bethesda, Md., 1986, 39.
32. **Crapo, J. D., Freeman, B. A., Barry, B. E., Turrens, J. F., and Young, S. L.,** Mechanisms of hyperoxic injury to the pulmonary microcirculation, *Physiologist,* 26, 170, 1983.
33. **Freeman, B. A. and Tanswell, A. K.,** Biochemical and cellular aspects of pulmonary oxygen toxicity, *Adv. Free Radical Biol. Med.,* 1, 133, 1985.
34. **Crapo, J. D., Barry, B. E., Foscue, H. A., and Shelburne, J.,** Structural and biochemical changes in rat lungs occurring during exposures to lethal and adaptive doses of oxygen, *Am. Rev. Respir. Dis.,* 122, 123, 1980.
35. **Estabrook, R. W. and Werringloer, J.,** Cytochrome P_{450}: its role in oxygen activation for drug metabolism, in *Drug Metabolism Concepts,* Donald, M. J. et al., Eds., American Chemical Society, Washington, D.C., 1976, 1.
36. **Archakov, A. I., Bachmanova, B. I., Isotaov, M. V., and Kuznetsova, C. P.,** Reduction of microsomal haemoproteins by superoxide radical generated by the NADPH-specific flavoprotein, in *Microsomes, Drug Oxidations and Chemical Carcinogenesis,* Estabrook, R. W., Ed., Academic Press, New York, 1980, 289.
37. **Yusa, T., Crapo, J. D., and Freeman, B. A.,** Hyperoxia enhances lung and liver nuclear superoxide generation, *Biochim. Biophys. Acta,* 798, 167, 1984.
38. **Freeman, B. A. and Crapo, J. D.,** Biology of disease: free radicals and tissue injury, *Lab. Invest.,* 47, 1, 1982.
39. **Egan, R. W., Paxton, J., and Kuehl, F. A.,** Mechanism for irreversible self-deactivation of prostaglandin synthetase, *J. Biol. Chem.,* 251, 7329, 1976.
40. **Kalyanaraman, B., Mason, R. P., Tainer, B., and Eling, T. E.,** The free radical formed during the hydroperoxide-mediated deactivation of ram seminal vesicles is hemoprotein-derived, *J. Biol. Chem.,* 257, 4764, 1982.
41. **Moncada, S., Bunting, S., Mullane, K., Thorogood, P., Vane, J. R., Raz, A., and Needleman, P.,** Imidazole: a selective inhibitor of thromboxane synthetase. *Prostaglandins,* 13, 611, 1977.
42. **Steinbrecher, U. P., Parthasarathy, S., Leake, D. S., and Steinberg, D.,** Modification of LDL by endothelial cells involves lipid peroxidation and degradation of LDL phospholipids, *Proc. Natl. Acad. Sci. U.S.A.,* 81, 3883, 1984.
43. **McCord, J. M. and Roy, R. S.,** The pathophysiology of superoxide: roles in inflammation and ischemia, *Can. J. Physiol. Pharmacol.,* 60, 1346, 1982.
44. **Freeman, B. A., Young, S. L., and Crapo, J. D.,** Liposome-mediated augmentation of superoxide dismutase in endothelial cells prevents oxygen injury, *J. Biol. Chem.,* 258, 12534, 1983.
45. **Freeman, B. A., Turrens, J. F., Mirza, Z., Crapo, J. D., and Young, S. L.,** Modulation of oxidant lung injury by using liposome-entrapped superoxide dismutase and catalase, *Fed. Proc.,* 44, 2591, 1985.
46. **Balin, A. K, Fisher, A., and Carter, D.,** Oxygen modulates growth of human cells at physiologic partial pressures, *J. Exp. Med.,* 160, 152, 1984.
47. **Goetz, I.,** Oxygen toxicity in normal and neoplastic hamster cells in culture, *In Vitro,* 11, 382, 1975.
48. **Evans, M. and Hackney, J.,** Cell proliferation in lungs of mice exposed to elevated concentrations of oxygen, *Aerospace Med.,* 6, 620, 1972.
49. **Adamson, I. and Bowden, D.,** Pulmonary injury and repair. Organ culture studies of murine lung after oxygen, *Arch Pathol. Lab. Med.,* 100, 640, 1976.
50. **Bowden, D. and Adamson, I.,** Endothelial regeneration as a marker of the differential vascular response in oxygen-induced pulmonary edema, *Lab. Invest.,* 30, 350, 1974.
51. **Housset, B., Ody, C., Rubin, D., Elemer, G., and Junod, A.,** Oxygen toxicity in cultured aortic endothelium: selenium-induced partial protective effect, *J. Appl. Physiol.,* 55, 343, 1983.
52. **Rubin, D., Housset, B., Jean-Mairet, Y., and Junod, A.,** Effects of hyperoxia on biochemical indexes of pig aortic endothelial function, *In Vitro,* 19, 625, 1983.
53. **Block, E. R., Patel, J., and Sheridan, N.,** Effect of oxygen and endotoxin on lactate dehydrogenase release, 5-hydroxytryptamine uptake, and antioxidant enzyme activities in endothelial cells, *J. Cell. Physiol.,* 122, 240, 1985.
54. **Moldow, C. and Jacob, H.,** Endothelial culture, neutrophil or enzymic generation of free radicals: in vitro methods for the study of endothelial injury, *Meth. Enzymol.,* 105, 378, 1984.
55. **Andreoli, W., Baehner, R., and Bergstein, J.,** In vitro detection of endothelial cell damage using 2-deoxy-d-^{3}H-leucine, ^{3}H-adenine, and lactate dehydrogenase, *J. Lab. Clin. Med.,* 106, 253, 1985.
56. **Small, R., Macarak, E., and Fisher, A.,** Production of 5-hydroxyindole acetic acid from serotonin by cultured endothelial cells, *J. Cell. Physiol.,* 90, 225, 1977.

57. **Junod, A. F.**, Uptake, metabolism and efflux of ^{14}C-5-hydroxytryptamine in isolated, perfused rat lungs, *J. Pharmacol. Exp. Ther.*, 183, 341, 1974.
58. **Ager, A., Gordon, J., Moncada, S., Pearson, J., Salmon, J., and Trevethick, M.**, Effects of isolation and culture on prostaglandin synthesis by porcine aortic endothelial and smooth muscle cells, *J. Cell. Physiol.*, 110, 9, 1982.
59. **Bishop, C. T., Mirza, Z., Crapo, J., and Freeman, B.**, Free radical damage to cultured porcine aortic endothelial cells and lung fibroblasts: modulation by culture conditions, *In Vitro*, 21, 229, 1985.
60. **Sacks, T., Moldow, C., Craddock, P., Bowers, T., and Jacob, H.**, Oxygen radicals mediate endothelial cell damage by complement stimulated granulocytes. An in vitro model of immune vascular damage, *J. Clin. Invest.*, 1161, 1978.
61. **Rubin, D., Drab, E., Ts'ao, C., Gardner, D., and Ward, W.**, Prostacyclin synthesis in irradiated endothelial cells cultured from bovine aorta, *J. Appl. Physiol.*, 58, 592, 1985.
62. **Harlan, A. and Collaban, K. S.**, Role of hydrogen peroxide in the neutrophil-mediated release of prostacyclin from cultured endothelial cells, *J. Clin. Invest.*, 74, 442, 1984.
63. **Whorton, A. R., Montgomery, M. E., and Kent, R. S.**, Effect of hydrogen peroxide on prostaglandin production and cellular integrity in cultured porcine aortic endothelial cells, *J. Clin. Invest.*, 76, 295, 1985.
64. **Piper, P. J. and Galton, S. A.**, Generation of leukotriene B$_4$ from porcine pulmonary artery, *Prostaglandins*, 8, 3, 1974.
65. **Kuhn, H., Ponicke, K., Halle, W., Weisner, R., Schew, T., and Forster, W.**, Metabolism of 1-^{14}C-arachidonic acid by cultured calf aortic endothelial cells: evidence for presence of a lipoxygenase pathway, *Prostaglandins Leukotrienes Med.*, 17, 291, 1985.
66. **Hopkins, N., Oglesby, T., Bundy, G., and Gosman, R.**, Biosynthesis and metabolism of 5-hydroperoxy-5,8,11,13 eicosatetraenoic acid by human umbilical vein endothelial cells, *J. Biol. Chem.*, 259, 14348, 1984.
67. **Mercandetti, A. and Colmerauer, M.**, Cultured human endothelial cells elaborate neutrophil chemoattractants, *J. Lab. Clin. Med.*, 104, 370, 1984.
68. **O'Brien, R. F., Seta, M., Mararski, J., Center, D., and Rounds, S.**, Thiourea causes endothelial cells in tissue culture to produce neutrophil chemoattractant activity, *Am. Rev. Respir. Dis.*, 130, 103, 1984.
69. **Bowman, C. M., Butler, E., and Repine, J.**, Hyperoxia damages cultured endothelial cells causing increased neutrophil adherance, *Am. Rev. Respir. Dis.*, 128, 469, 1983.
70. **Weiss, S. J., Young, J., LoBuglio, A., Slivka, A., and Nimeh, F.**, Role of hydrogen peroxide in neutrophil-mediated destruction of cultured endothelial cells, *J. Clin. Invest.*, 68, 714, 1981.
71. **Sutton, N. and Simon, L. M.**, Lung cell oxidant injury. Enhancement of polymorphonuclear leukocyte-mediated cytotoxicity in lung cells exposed to sustained in vitro hyperoxia, *J. Clin. Invest.*, 70, 342, 1982.

Chapter 30

PHAGOCYTIC PROPERTIES OF ENDOTHELIAL CELLS

Una S. Ryan

TABLE OF CONTENTS

I. HISTORICAL SURVEY

The term "endothelium" was introduced in 1865 by His[1], and in 1886 Wyssokowitsch (see Reference 2) reported that endothelium of peripheral vessels may phagocytize microorganisms in those organs where the velocity of blood is decreased, e.g., spleen, bone marrow, lungs, and kidneys. As reviewed by Altschul in 1954,[2] the literature in the two decades after 1900 argued the issue of phagocytosis by endothelial cells back and forth. Some authors claimed that bacteria entered the endothelium "in its agonal state;" others asserted that living endothelium is capable of phagocytosis. Altschul discusses at length whether phagocytosis is physiologic or pathologic without coming to any definite conclusion. However, the frequency of the early reports of phagocytosis is greatest for endothelium of the lungs and kidney after those of the reticuloendothelial system and lymph vessels.

Altschul further reviews data involving experimental endocarditis lenta in which endocardial cells as well as aortic endothelial cells were found to contain streptococci. In such cases, there was a distortion of the endothelial cells and their nuclei. Hammerschmidt[3] injected *Corynebacterium murisepticum* into mice and found excessive storage of bacilli in endothelium. Interestingly, Domagk and colleagues[4-6] report that repeated injections of staphylococci into rats resulted in anaphylactic reactions with swelling of capillary endothelium and phagocytosis in the lungs. Early after intravenous injection, the pulmonary endothelium, especially of the capillaries, was filled with gram-positive cocci. The same authors reported that injection of microorganisms into the renal artery caused proliferation of capillary endothelium in the glomeruli and a great swelling of its nuclei. After repeated injections the endothelium reacted with mitoses. Thus, even the early literature linked endothelial phagocytosis with increased division (*vide infra*). Rosenthal[7] reported that experimentally introduced cocci are taken up by endothelial cells which then emigrate into surrounding tissues, thus suggesting that phagocytosis may induce migration in endothelial cells.

Although the concept that phagocytosis is an "activation" response of endothelium was introduced independently;[11] in fact Dietrich[13] had already suggested that endothelium of brain and skin can be "activated" and turn phagocytic. McClung (see Reference 14) included inflammation among such activating processes. Although it is never explicitly stated as such, some of the earlier literature could be interpreted as implying that phagocytic responses are inducible rather than constitutive properties of endothelium. For example, Downey[8] reports that splenectomy enhances the selective phagocytosis of pulmonary endothelium for carbon particles.

It also seems that loss of phagocytic properties may accompany differentiation. There are reports[9,10] that in the early embryo, endothelial cells are phagocytic but that in many vessels this property is gradually lost during development, while on the other hand, it increases in the endothelium of organs such as liver and spleen.

Although endothelial cells are not considered "professional phagocytes", there are several reports of its phagocytic activity in the recent literature, many of these involving in vitro experiments.

A variety of models, including organ cultures of human heart valves and endothelial cell cultures, have been used to study the initial pathobiology of acute infective bacterial endocarditis caused by *Staphylococcus aureus*. Gould et al.,[15] using both gram-positive and -negative bacteria, reported that those microorganisms most frequently associated with endocarditis showed the greatest in vitro adherence. Peterson et al.[16] showed selective adherence by a single strain, but no overall difference in adherence between gram-positive and -negative bacteria. Cooper et al.[17] indicated that there is an initial insult (toxin or other agent) which traumatizes the normal endothelial cell. Staphylococci then attach, followed by invasion and destruction of the endothelium. Hamill et al.[18] have reported that phagocytosis of *S. aureus* by endothelial cells, followed by intracellular persistence, may be an important postadherence event in the pathogenesis and pathophysiology of endovascular infections.

In addition to *S. aureus*,[11,12,18] a wide variety of organisms have been reported to be phagocytized by endothelial cells, including *Histoplasma capsulatum*[19,20] and *Streptococcus pyogenes*.[21] There is an extensive literature on the presence of *Mycobacterium leprae* in endothelium in vivo and in vitro.[22-26] Moreover, uptake has been reported for inert particles such as polystyrene or latex spheres[11,12,27,28] bovine serum albumin-gold conjugates,[29] colloidal carbon particles,[30] elastin particles,[31] fat,[32] India Ink,[10,22] and benzidine dyes.[9]

An interesting role of endothelium in enhancing phagocytosis and killing of Staphylococci by leukocytes has also been reported.[34] The supporting effect of endothelial cells was not due to opsonization of the bacteria by immunoglobulin or complement from the endothelial cell surface, nor to coating with fibronectin.

Recently, we have been able to show that uptake of bacteria by endothelial cells is selective,[35,36] stimulates an oxidative burst,[37] results in unmasking of Fc receptors on the endothelial surface,[11,38,39] and causes an increase in the rates of division, migration, and propensity for further phagocytosis.[35] These consequences of phagocytosis have been grouped together as "activation" responses of endothelial cells.[12]

II. PHAGOCYTOSIS BY PULMONARY ENDOTHELIAL CELLS

Several years ago, quite by accident, we "rediscovered" that pulmonary endothelium is capable of ingesting polystyrene beads. Following the development of techniques for culture of endothelium on 100- to 150 μ microcarriers[40-42] and for isolating endothelial cells from both arterial and venous microvessels using 40- to 80 μm microcarriers,[43-45] we felt that a similar perfusion of blood-free lungs with 5- to 10-μm beads should yield a harvest of capillary endothelial cells. Contrary to expectation, no capillary endothelial cells were collected on the beads. Instead, the capillary endothelial cells were found to have phagocytized the beads, and many beads had made their way from the vasculature to the interstitum or even to the airways. We set out to determine some of the characteristics of phagocytosis of pulmonary endothelial cells in situ and in culture.[11,46]

A. Time Course of Uptake and Size of Particulates

Monolayers of bovine pulmonary artery endothelial cells in culture when exposed to unopsonized polystyrene microspheres, fixed-swelled red cells, cholesterol crystals, or heat-killed *Staphylococcus aureus,* ingest the particles in a time-dependent manner (see Figures 1 and 2). The cells continue to take up particles over 72 hr until they become engorged. When the particles are added to stationary confluent monolayers they are ingested, but eventually particles may be returned to the medium and float free without apparent reuptake. However, if these beads are retrieved from the medium and offered to fresh cultures, uptake occurs with kinetics similar to those in the first culture, indicating that phagocytosis per se does not render the beads incapable of further binding or phagocytosis. If the particles are added to preconfluent endothelial cell cultures (or wounded monolayers; see below) the dividing, migrating cells continue to phagocytize beads. Cells containing particles are capable of both division and migration and at division apportion the ingested beads between the daughter cells.

Pulmonary endothelial cells seem to be able to phagocytize particles from 0.1 to approximately 20 μm in diameter. Particles larger than about 30 μm in diameter are treated as microcarriers, and endothelial cells will form confluent monolayers around them, either after seeding of cells onto beads in suspension (i.e., in roller bottles or spinner flasks[40-42]), or if the microcarrier beads are introduced onto confluent monolayers. As described previously, endothelial cells from microcarriers covered with confluent monolayers will migrate onto

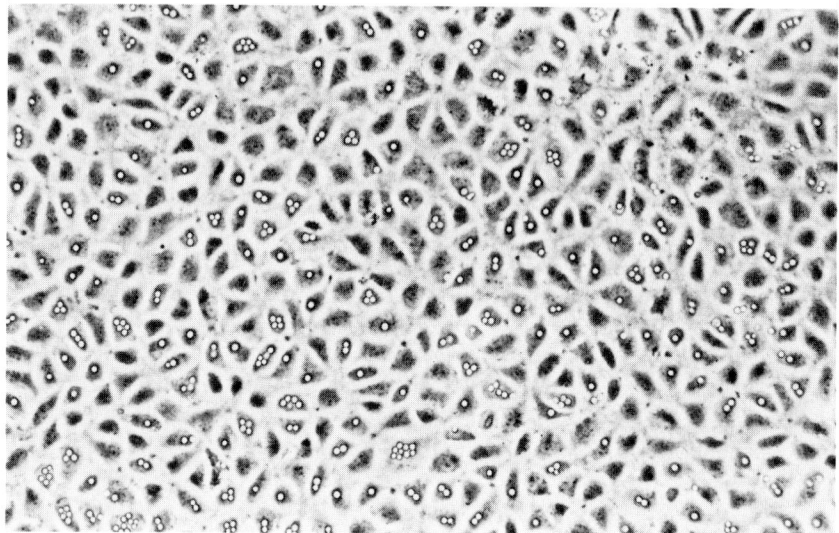

FIGURE 1.   Phase contrast micrograph showing a monolayer of endothelial cells that had been exposed to unopsonized polystyrene beads, 5- to 10 μm in diameter. Many of the cells contain phagocytized beads. (Magnification × 160.)

flasks, grafts, filters, or fresh microcarriers.[40,47] Particles of approximately 25 μμ in diameter appear to represent the cross-over size and can be regarded as covered by, or partially ingested by, two endothelial cells.

B. Mechanism of Uptake as Judged by Electron Microscopy and Immunocytochemistry

Particles and microorganisms adhere to endothelial cells within minutes. Scanning electron microscopy indicates that the endothelial surface becomes ruffled and flaps of cell membrane enfold the adherent particles which are then drawn within the cells (Figure 3). Video time-lapse microscopy reveals that long pseudopod projections can be extended by migrating endothelial cells which engulf particles and internalize them. Thus it is clear that endothelial cells possess active mechanisms for bringing outlying particles towards the center of the cell where they become sequestered within lysosomes. Using antibodies to vimentin (provided by Dr. Gluckhova, Cardiology Institute, Moscow) we have shown that phagocytized particles are surrounded by a web of vimentin, that the projections and extensions of the cell contain vimentin cables and the particles that become parked within the main body of the cell are enclosed by a basketwork of vimentin (Plate 1).* As reported by others,[18] cytochalasin treatment inhibits phagocytosis, indicating an involvement of microfilaments.

There is some indication that repeated phagocytic events occur at certain points on the cell surface since more than one particle is frequently engulfed by one set of membrane extensions resulting in the formation of membrane-bounded vacuoles containing multiple particles (Figure 4).

C. Selectivity of Uptake of Living Bacteria

Sieving of particulates is one of the primary functions of the pulmonary capillary bed. There are a number of mechanisms for disposing of trapped material, including removal by macrophages and digestion by extracellularly released products of a variety of cells. However, endothelial phagocytosis may be a second line of defense.

---

*   Plate 1 appears after page 52.

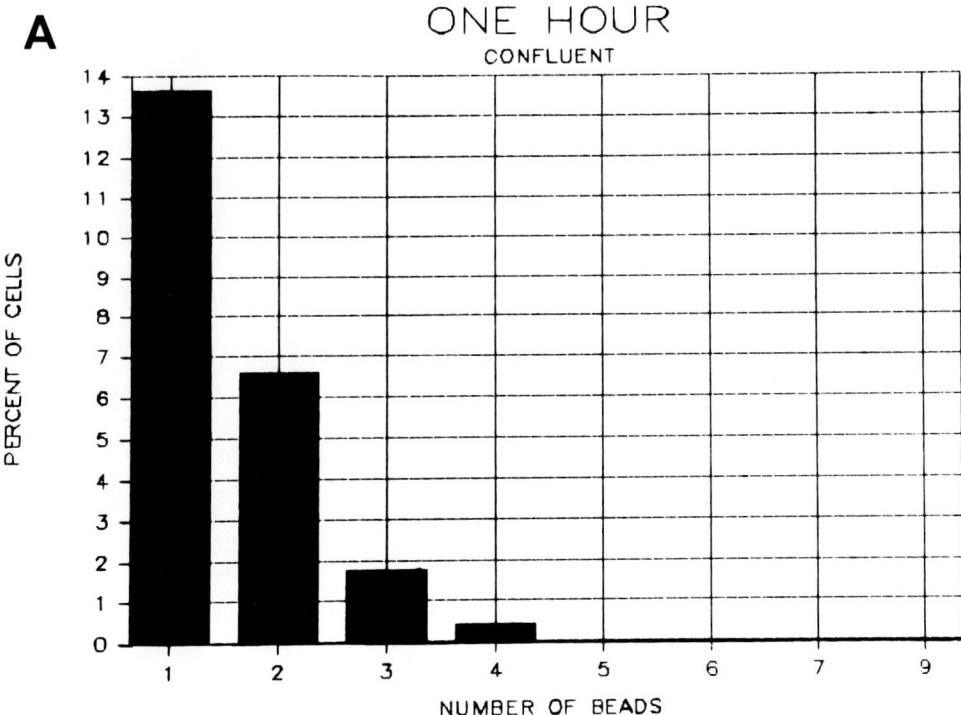

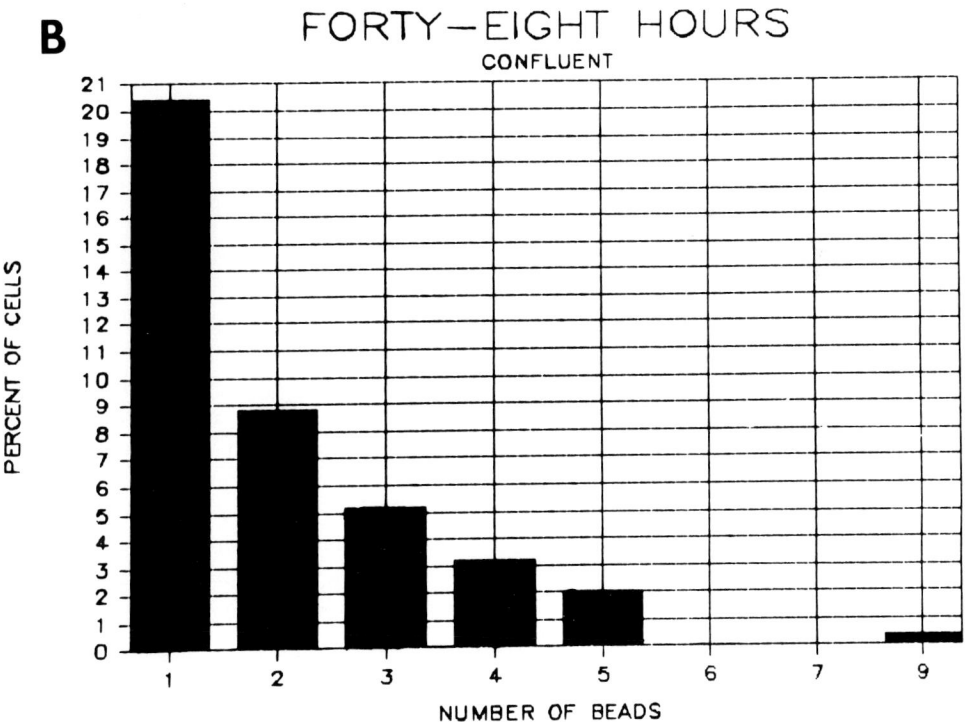

FIGURE 2.   Graphs showing time-dependent uptake of unopsonized 5- to 10-μm beads by bovine pulmonary endothelial monolayers. (A) 1-hr incubation; (B) 48-hr incubation.

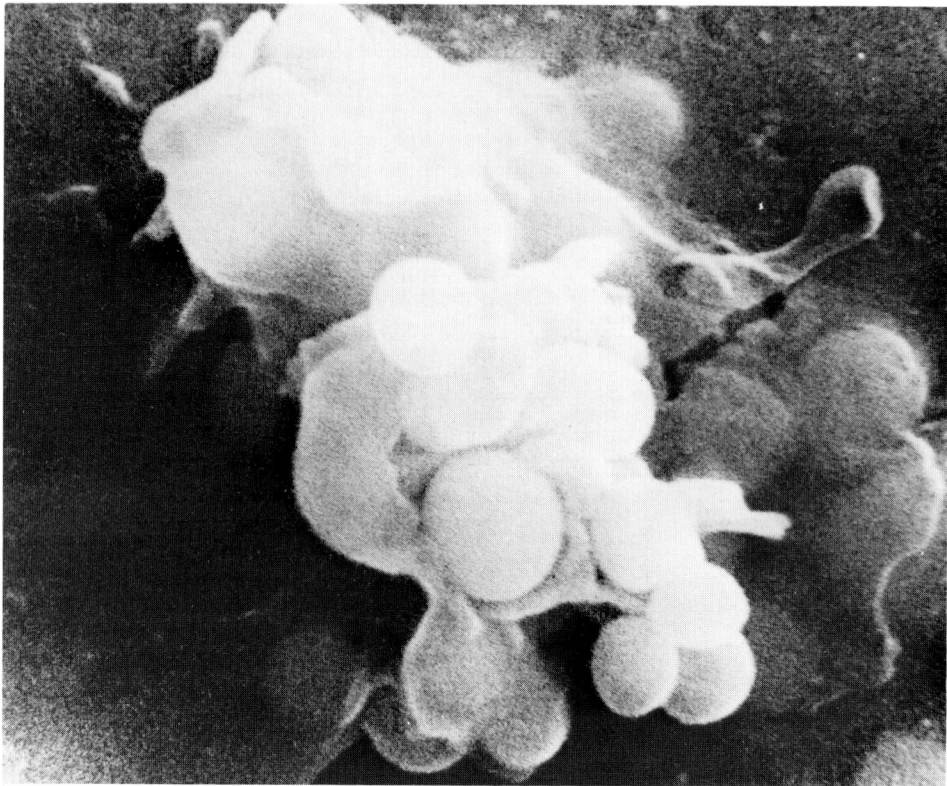

FIGURE 3.   Scanning electron micrograph showing pulmonary endothelial cell that had been exposed to *S. aureus*. As binding occurs, the endothelial cell membrane begins to ruffle and flaps of cell membrane engulf the bacteria and internalize them. (Magnification × 16,490.)

Despite the apparently indiscriminate uptake of particles and dead bacteria, endothelial cells show great selectivity in phagocytosis of living bacteria and can distinguish between wild type and mutant strains of *Salmonella minnesota*. As shown in Figure 5, monolayers of pulmonary endothelial cells exposed to wild (S) and mutant (Re) types of *S. minnesota* for 15 min at a concentration of 100 bacteria per cell showed far greater binding and uptake of the Re than the S form (83 vs. 11% with >3 bacteria per cell).[35] These results were confirmed by scanning electron microscopy.[36] Confluent endothelial cultures were grown on glass coverslips and treated with Re or S type *S. minnesota* or served as untreated controls. The cells were washed extensively to remove unreacted bacteria. Samples were prepared for scanning electron microscopy and conductively coated using the OTOTO method.[48] This allowed a softer image and aided localization of engulfed bacteria which appeared as rod-shaped protrusions beneath the cell surface. At 30 min of incubation there was a significant difference between S vs. Re types bound and phagocytized. The Re showed a higher degree of binding while the S form showed negligible association with endothelial cells (see Figure 6). Scanning electron microscopy indicates that although the bound bacteria lie flat on the endothelial cell surface, during the process of phagocytosis the Re mutant appears to be ingested end-on (Figure 7). The endothelial cell surface responds to the presence of bound bacilli by a ruffling similar to that seen during engulfment cocci or spherical particles (Figures 3 and 7).

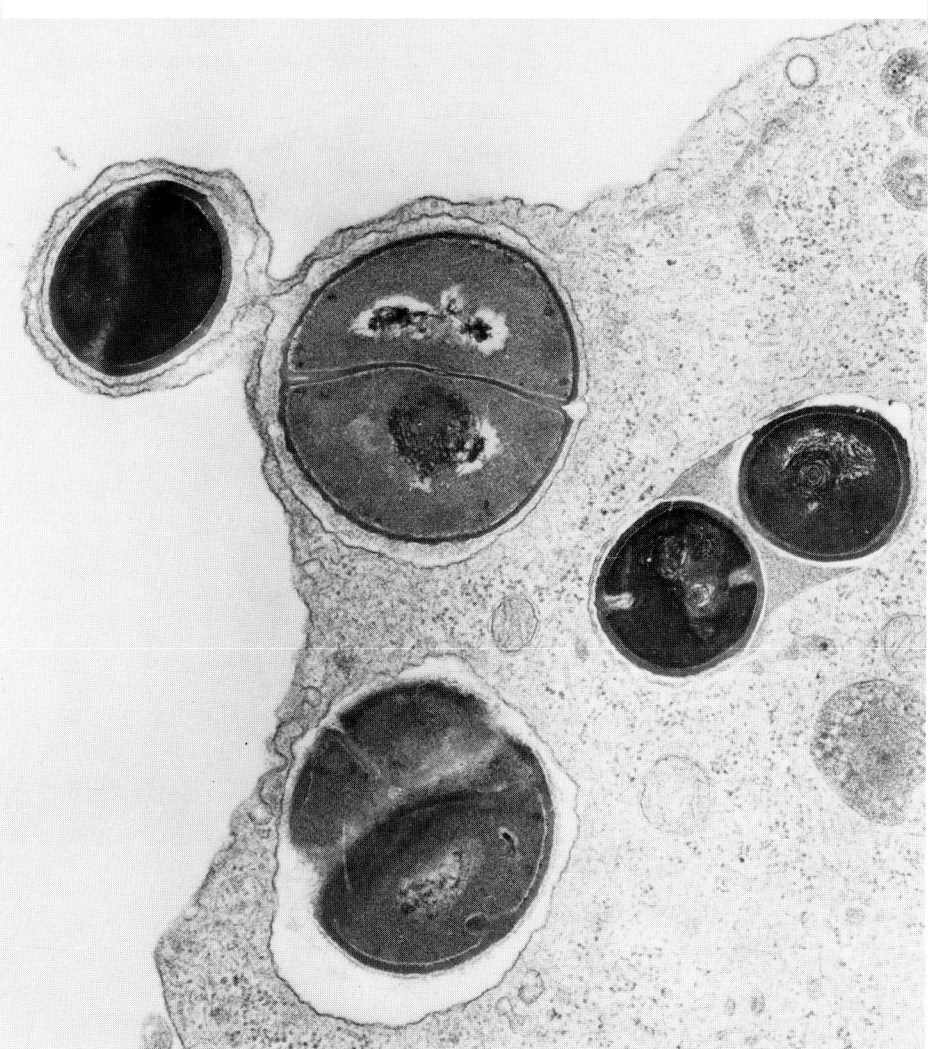

FIGURE 4. Transmission electron micrograph showing ingestion of *S. aureus* by a pulmonary endothelial cell. As described in Figure 3, the bacteria are enveloped by membrane folds and sequestered in membrane bounded vesicles within the cell. (Magnification × 36,400.)

D. Endothelial Cell Responses to Phagocytosis

There are a number of consequences of phagocytosis that affect the endothelial cell as well as the ingested microorganism. Endothelial cells do not normally express receptors for the Fc (constant fraction) of the IgG antibody or for C3b (fraction of the third component of complement).[38] However, certain types of injury, such as viral infection, white cell lysates, endotoxin, and antibodies to endothelial surface enzymes in the presence of complement, cause an unmasking of Fc receptors and, to a lesser extent, C3b receptors on the endothelial cell surface.[11,39,49] We have found that phagocytosis of large numbers of particles results in the expression of Fc receptors by endothelial cells (Figure 8).[11] This would have the effect of rendering the endothelial surface adhesive for immune complexes and would promote phagocytosis where opsonization with IgG serves as the recognition factor.

Macrophages and polymorphonuclear leukocytes respond to phagocytosis by producing an "oxidative burst". This is a coordinated sequence of events initiated by an increase in

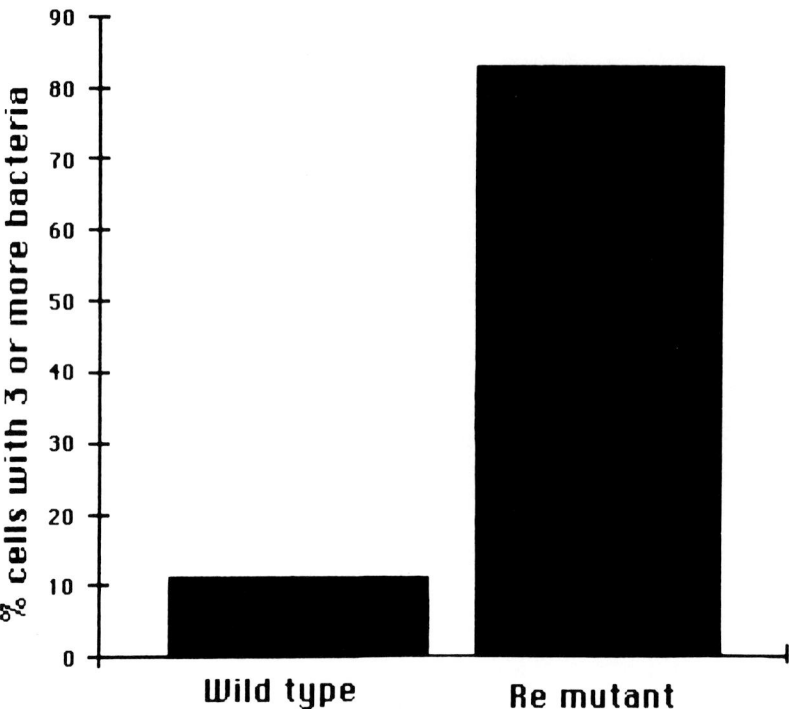

FIGURE 5.   Uptake of *S. minnesota* wild (S) and mutant (Re) types by pulmonary endothelial cells. Bacteria were added at a concentration of 100 per endothelial cell for 15 min in the absence of serum. The cells containing more than three bacteria were scored as positive. The wild type is only ingested to a small degree, while over 80% of the cells phagocytize the Re mutant. (From Ryan, U. S., Ed., Pulmonary endothelium in Health and Disease, Marcel Dekker, New York, 32, 3, 1987. With permission.)

oxygen uptake followed by one-electron reduction of oxygen to superoxide anions (O_2^-) using NADPH and NADH as the electron donor and catalyzed by a NAD(P)H oxidase (see References 50 and 51). O_2^- is subsequently converted to hydrogen peroxide (H_2O_2) by spontaneous or enzyme-mediated dismutation.[52] Further reactions, some of which are catalyzed by myeloperoxidase, produce additional singlet oxygen and OH radical. These highly reactive oxygen compounds react preferentially with unsaturated double bonds in the cell and have a microbicidal function. In the process, derivatives of dioxetane dione may arise which are rich in energy and can give their energy to molecules of the cell that are capable of fluorescence. These are then converted to the electronically excited state and the energy is emitted as light. The chemiluminescence that occurs naturally during phagocytosis is only very slight.[53] This natural chemiluminescence can be potentiated by a factor of 10^3 if a chemiluminescent compound such as luminol or ʌcigenin is added to the system.[54]

We examined whether endothelial cells are capaɔle of producing an oxidative burst using bioluminescence methods and by measurement of O_2^- and H_2O_2 released. O_2^- released into

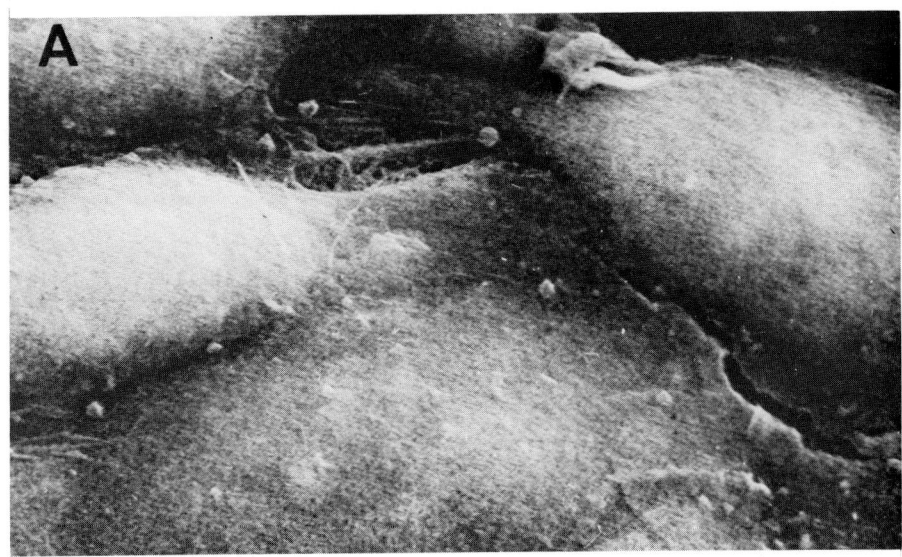

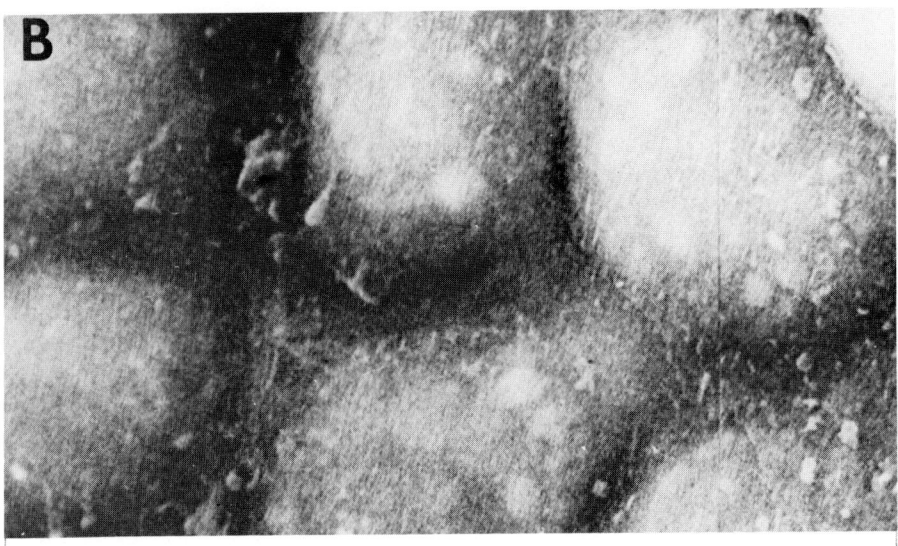

FIGURE 6. Scanning electron micrographs showing binding of *S. aureus* to endothelial cells; (A) shows a control monolayer (magnification × 3000); (B) shows negligible binding of S (wild type) (magnification × 2400); and (C) shows significant binding by Re mutant (magnification × 2000).

the medium can be assayed using the superoxide dismutase inhibitable reduction of ferricytochrome C, determined by measuring its absorbance at 550 nm.[55] In addition, semiautomated microassays for the measurement of both O_2^- and H_2O_2 can be used.[56] The assays are performed directly in 96-well flat-bottom tissue-culture plates. The endothelial cells and bacteria are covered with ferricytochrome C to measure O_2^- by increase in absorbance at 550 nm. Measurement of H_2O_2 is based on the horseradish-peroxidase-dependent oxidation of phenol red which is assayed by its increased absorbance at 600 nm. Both measurements are made directly using a BioRad® Model 2550 EIA reader.

We have found that cultures of bovine pulmonary endothelial cells respond to binding

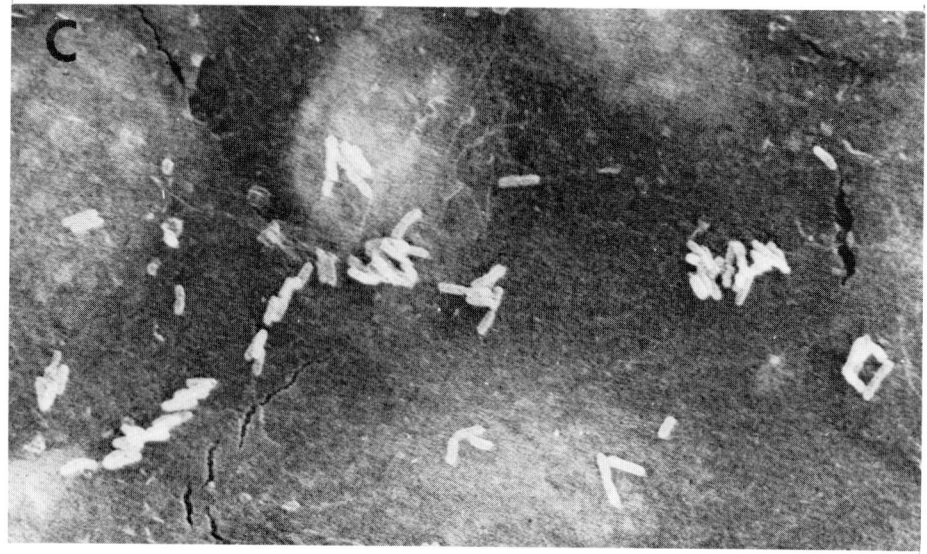

FIGURE 6C

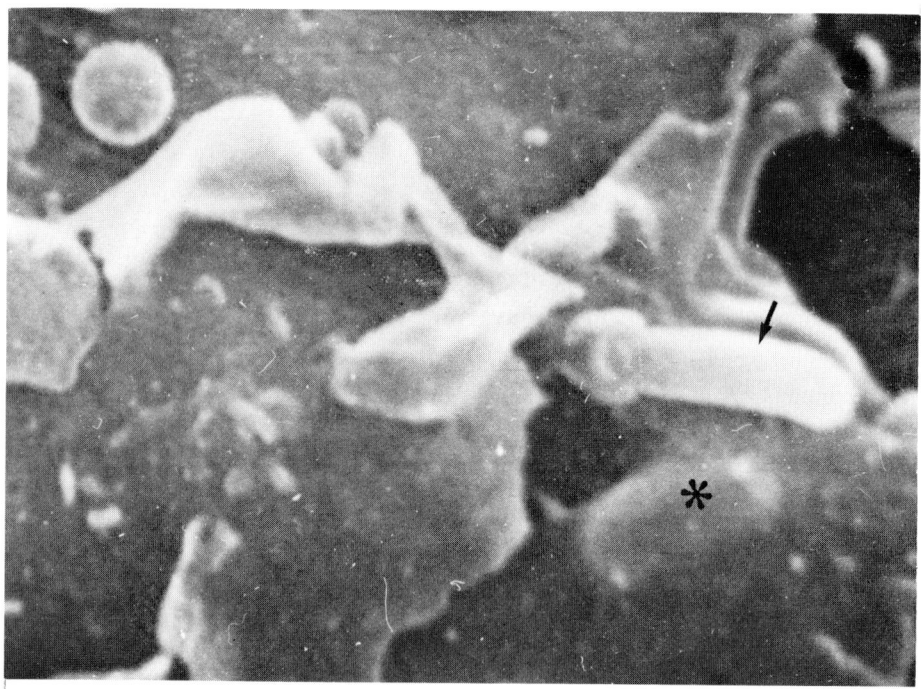

FIGURE 7. High-power scanning electron micrograph showing ingestion of Re mutant, the bacterium is engulfed end-on (arrow). A phagocytized bacterium can be seen beneath the cell surface (asterisk). (Magnificaton × 1,960.)

and ingestion of *S. minnesota* by producing an oxidative burst.[35,37] The generation of O_2^- by endothelial cells has been reported previously.[57-59] Since superoxide anion release can be triggered by phorbol ester (PMA) and calcium ionophore (A23187),[37] it seems that this response, like other activation responses of endothelial cells,[12] may involve activation of protein kinase C and mobilization of Ca^{++}.[60]

FIGURE 8.   Scanning electron micrograph showing expression of Fc receptors on the endothelial surface after ingestion of 5-µm beads. IgG coated erythrocytes (EA7S) were bound to the surface (arrows), indicating the presence of Fc receptors. Phagocytized beads can be seen within the cell (asterisks). Bound but not yet internalized beads are also shown (arrowheads). (Magnification × 2800.)

Having demonstrated an oxidative burst as a consequence of bacterial phagocytosis we next investigated the microbicidal effectiveness of endothelial cells. Endothelial cell monolayers were lysed 30 min after ingestion of S or Re types of *S. minnesota,* followed by quantitation of colony forming units in the lysates. A far greater number of living bacteria were released from cells incubated with Re than S type (2.3 × 10^6 vs. 1 × 10^6 bacteria per milliliter of lysate).[35] Since endothelial cells preferentially bind and ingest the Re mutant, these data suggest that endothelial cells phagocytize the Re mutant and harbor it intracellularly but are not efficient at killing it despite the generation of reactive oxygen species. Thus for some strains of bacteria, endothelium may provide an environment protected from antibiotics.

Clearly, phagocytic interactions between bacteria and endothelial cells involve a struggle for survival by both parties. In studies of the effects of intracellular *Staphylococcus aureus* on endothelial cells,[61] both intracellular *S. aureus* and heat-labile factors in *S. aureus* culture supernatants were shown to be cytotoxic for endothelial cells, as measured by loss of ^{3}H-adenine. The ability to damage endothelial cells appears to be due to intracellular production of alpha hemolysin, since mutants that produce no hemolysin do not damage endothelial cells.

In a study of rabbit corneal endothelium after phagocytosis of polystyrene spheres, the endothelium degenerated and detached from Descemet's membrane.[27] Measurement of extralysosomal release of acid phosphatase, *N*-acetyl-β-*D*- glucosaminidase and β-glucuronidase, suggested that released lysosomal enzymes could have accounted for the degeneration of the corneal endothelium.

In vivo, mechanisms which result in lysis of endothelial cells would have the effect of releasing viable bacteria back into the bloodstream and may in part explain recurrent infections that are difficult to treat with antibiotic therapy alone.

E. Behavioral Outcomes of Phagocytosis by Endothelial Cells

In stationary, confluent monolayers, endothelial cells neither divide nor migrate. However, if the monolayer is "wounded" by removal of cells with a rubber policeman, cells will migrate out from the area of preserved confluence and divide to fill the wound.[62] We have studied the affects of phagocytosis using this wounded monolayer assay coupled with a computer program for analysis of cell kinetics.[63] If the monolayer is allowed to ingest beads prior to wounding, the rate of migration of the cells is increased. If the cells phagocytize additional beads en route, the rate of migration increases per bead ingested up to four beads (Figure 9). Similarly, the interval between successive mitoses, or interdivision time, decreases in the cells that have phagocytized beads and the increase in division rate is proportional to the number of beads ingested, up to four beads (Figure 10). At division, the phagocytized beads are apportioned between the daughter cells. The daughter cells appear to "remember" the state of activation of the parent and continue to migrate and divide at rates characteristic of the parent.

Not only are migration and division increased but the propensity to phagocytize other beads also increases with each ingested bead up to four (Figure 11).

III. CONCLUDING COMMENTS

The increased migration, division, and phagocytosis shown by cells undergoing phagocytosis can be regarded as activation phenomena. Endothelial cells respond to a wide variety of stimuli that induce activation responses.[12] The manifestions of activation are as diverse as the signals; however, there is some indication that the mechanisms of signal transduction are quite conservative and include mobilization of Ca^{++}[60] and activation of protein kinase C.[64] Many of the stimuli that lead to activation responses induce shape changes in the endothelial cells.

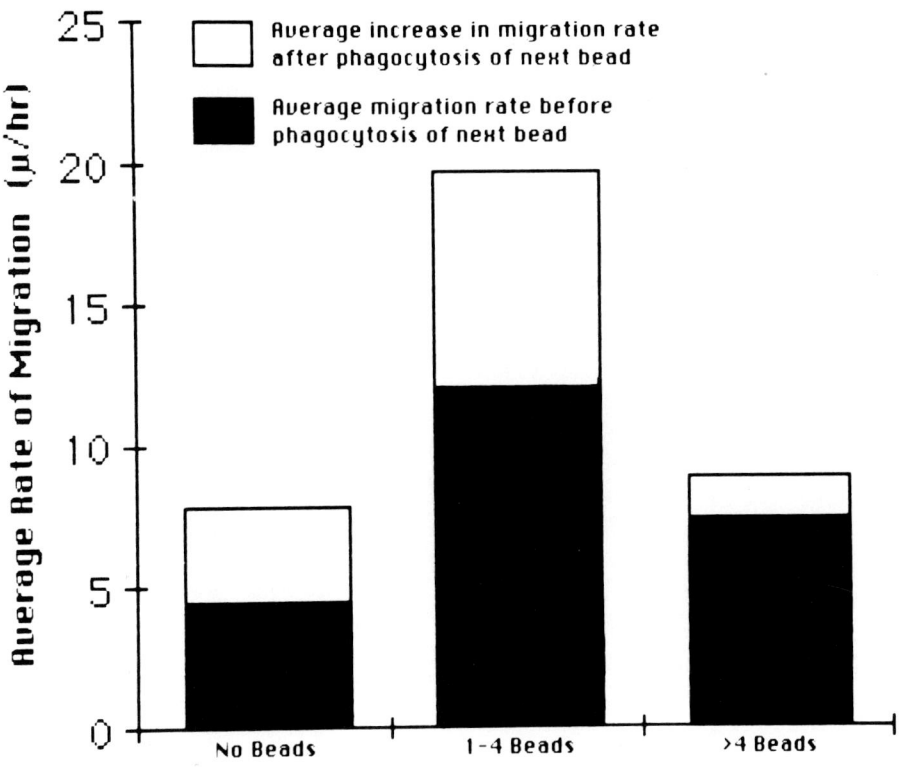

FIGURE 9.   Graph showing the increase in migration rate caused by ingestion of each additional bead. The increase is greatest for uptake of one through four extra beads; thereafter the increase is less. The average rate of migration of cells containing up to four beads is more than four times that of cells containing no beads. (From Ryan, U. S., Ed., *Pulmonary Endothelium in Health and Disease,* Marcel Dekker, New York, 32, 3, 1987. With permission.)

Phagocytosis appears to be both a stimulus for and an expression of endothelial cell activation in that ingestion of particles induces activation responses such as unmasking of Fc receptors, generation of superoxide anion, mitosis, increased migration, and further phagocytic activity.

It remains to be seen whether other aspects of endothelial activation such as the induction of procoagulant activity[65] or release of endothelium dependent relaxing factor (EDRF)[60] are accompanied by changes in behavior kinetics similar to those seen in the phagocytic response.

For the moment, it is interesting to note that despite their separate origins, endothelial cells and macrophages share a number of functional similarities. Upon appropriate stimulation endothelial cells can be induced to express Class II histocompatibility antigens,[66] present antigen, express Fc receptors, engage in phagocytosis, and produce toxic oxygen radicals. Both cell types possess LDL receptors[67] and can elaborate PDGF-like molecules.[68] Endothelial cells not only respond to macrophage-derived molecules such as interleukin-1 (IL-1), but can accumulate IL-1 mRNA and secrete biologically active IL-1 protein product in response to endotoxin and tumor necrosis factor.[69] Thus, some of the effects of bacterial endotoxin on endothelium may be direct (including procoagulant activity and increased leukocyte adhesion) or may be mediated and amplified by IL-1 produced by the endothelial

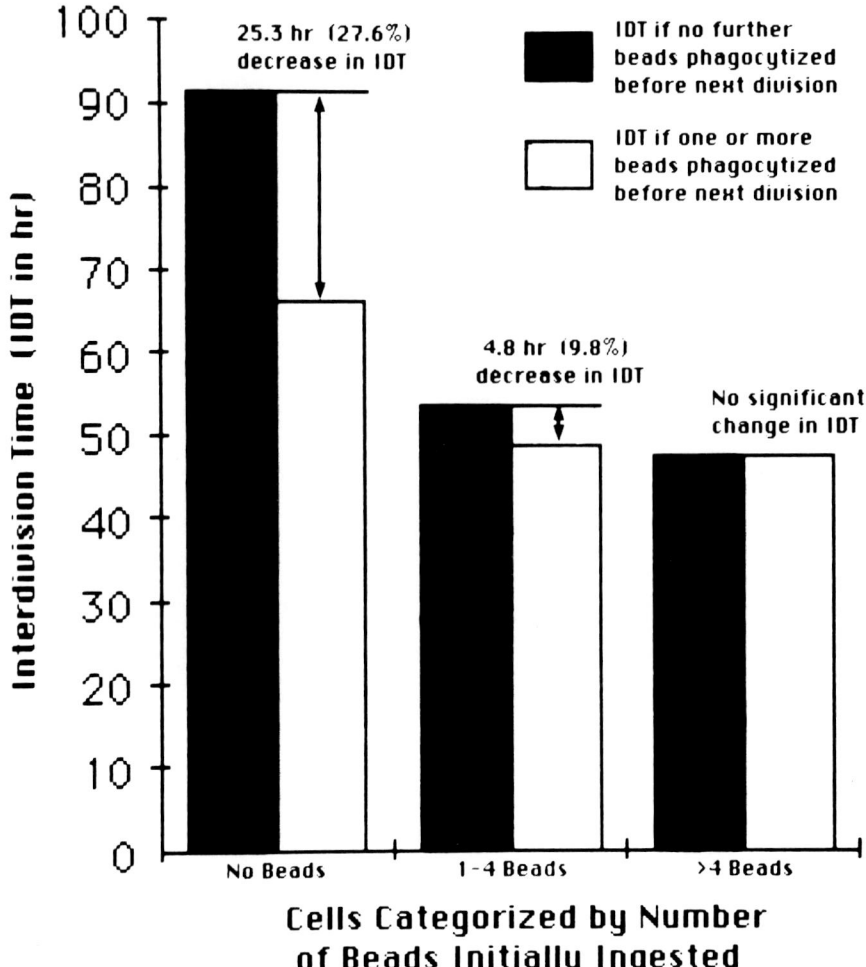

FIGURE 10.   Graph showing decrease in interdivision time (IDT) as a result of phagocytosis. As is the case for acceleration of migration rate (Figure 10), the maximum effect is seen up to four ingested beads. The greatest acceleration in mitotic rate is seen after ingestion of the first bead. It is significant that cells not associated with beads before or after wounding have IDTs that, on average, are nearly double the IDTs of cells that phagocytize at least one bead during the course of the experiment (92 vs. 52 hr). (From Ryan, U. S., Ed., *Pulmonary Endothelium in Health and Diseases,* Marcel Dekker, New York, 32, 3, 1987. With permission.)

cells themselves. Endogenous IL-1 production by endothelial cells could provide an early warning system to signal local tissue invasion or injury. Clearly, phagocytic responses are an aspect of the active participation of endothelial cells in the host response to injury.

ACKNOWLEDGMENTS

It is a pleasure to thank Carlos Diaz-Silveira for typing the manuscript. The work was supported by grants HL21568 and HL33064 from NHLBI and a grant from the Council for Tobacco Research-USA.

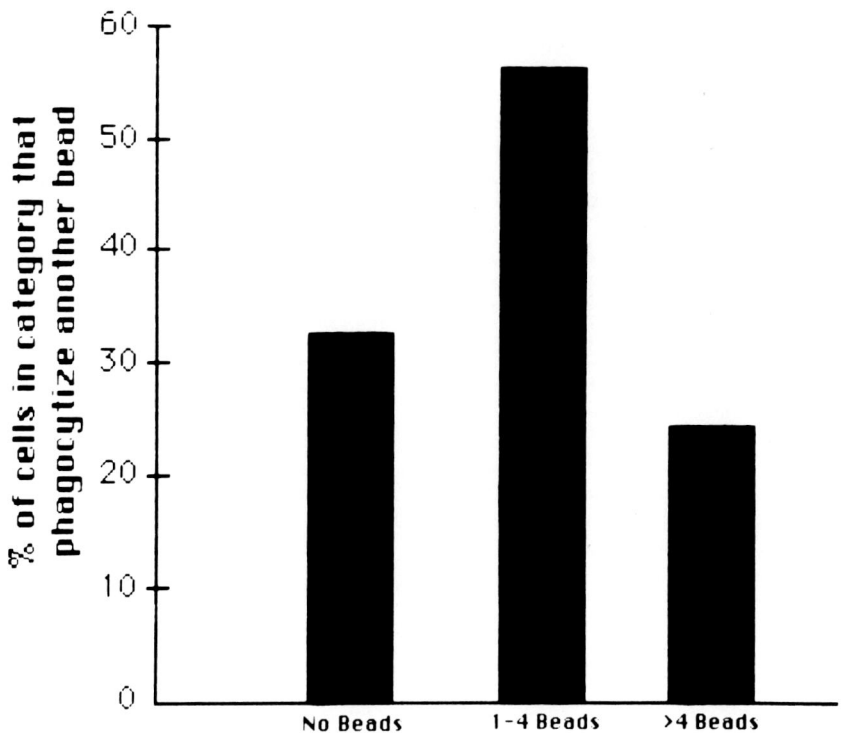

Cells Categorized by Number
of Beads Initially Ingested

FIGURE 11. Graph showing that the propensity of endothelial cells to phagocytize another bead is increased in those cells that have previously ingested one to four beads.

REFERENCES

1. **His, W.,** *Die Häute und Höhlen des Körpers,* Schweighhauserische Universitätsbuchdruckerei, Basel, 1865.
2. **Altschul, R.,** *Endothelium,* Macmillan, New York, 1954.
3. **Hammerschmidt, J.,** Die Rolle Des Gefässendothels bei septischen Prozessen, *Z. Immunitaetsforsch,* 54, 205, 1928.
4. **Domagk, G.,** Über die bedeutung der Endothelien für die Abwehr von Infektionserregern und die Entstehung des Amyloids, *Klin. Wochenschr.,* 3, 1338, 1924.
5. **Domagk, G.,** Bei der Eklampsie auftretende Endothelveränderungen und ihre Bedeutung, *Klin. Wochenschr.,* 4, 1011, 1925.
6. **Domagk, G. and Neuhaus, C.,** Die experimentelle Glomerulonephritis, *Arch. Pathol., Anat.,* 264, 522, 1927.
7. **Rosenthal, W.,** Phagozytose durch Endothelzellen, *Z. Immunitaetsforsch.,* 31, 372, 1921.
8. **Downey, H.,** Further studies on the reactions of blood-and tissue-cells to acid colloidal dyes, *Anat. Rec.,* 12, 429, 1917.
9. **Wislocki, G. B.,** The staining of amphibian larvae with benzidine dyes with especial reference to the behavior of the lymphatic endothelium, *Am. J. Physiol.,* 42, 124, 1916.
10. **Beard, J. W. and Beard, L. A.,** The phagocytic activity of endothelium in the embryo chick, *Am. J. Anat.,* 40, 295, 1927.
11. **Ryan, U. S., Ryan, J. W., and Crutchley, D. J.,** The pulmonary endothelial surface, *Fed. Proc.,* 44, 2603, 1985.
12. **Ryan, U. S.,** Endothelial cell activation responses, in *Pulmonary Endothelium in Health and Disease,* Ryan, U. S., Ed., Marcel Dekker, New York, 32, 3, 1987.

13. **Dietrich, A.,** *Thrombose,* J. Springer, Berlin, 1932.

14. **Altschul, R.,** *Selected Studies on Atherosclerosis,* Charles C Thomas, Springfield, Ill., 1950.

15. **Gould, K., Ramirez-Ronda, C. H., Holmes, R. K., and Sanford, J. P.,** Adherence of bacteria to heart valves in vitro, *J. Clin. Invest.,* 56, 1364, 1975.

16. **Peterson, L. R., Sinha, A. A., and Gruninger, R. P.,** Selective bacterial adherence to cardiac endothelial cells in tissue culture, *Cardiovasc. Res.,* 15, 404, 1981.

17. **Cooper, M. D., Jeffery, C., Gall, D. L., and Anderson, A. S.,** Scanning electron microscopy studies of staphylococcal adherence to heart valve endothelial cells in organ culture: an *in vitro* model of acute endocarditis, *Scanning Electron Microsc.,* III, 1231, 1985.

18. **Hamill, R. J., Vann, J. M., and Proctor, R. A.,** Phagocytosis of *Staphylococcus aureus* by cultured bovine aortic endothelial cells: model for postadherence events in endovascular infections, *Infect. Immun.,* 54, 833, 1986.

19. **Bedrosian, N., Gatchel, S. L., and Tewari, R. P.,** Interactions of *Histoplasma capsulatum* (HC) with murine lung endothelial cells (EC) in culture, 26th Intersci. Conf. Antimicrob. Agents. Chemother., p. 250, 1986. Abstract (830)

20. **Gatchel, S. L., Shigematsu, M. L., Maciag, T., and Tewari, R. P.,** Interactions of *Histoplasma capsulatum* (HC) with human endothelial cells (EC) in culture, *Am. Soc. Microbiol.,* F-15, 400, 1986.

21. **Chaudhary, S., Alred, T., Gatchel, S. L., Maciag, T., and Tewari, R. P.,** Interaction of *Streptococcus pyogenes* with umbilical vein endothelial cells (HUVEC), *Am. Soc. Microbiol.,* D-39, 72, 1986.

22. **Fite, G. L.,** The vascular lesions of leprosy, *Int. J. Lepr.,* 9, 193, 1941.

23. **Coruh, G. and McDougall, A. C.,** Untreated lepromatous leprosy: histopathological findings in cutaneous blood vessels, *Int. J. Lepr.,* 47, 500, 1979.

24. **Turkel, S. B., Van Hale, H. M., and Rea, T. H.,** Ultrastructure of the dermal microvasculature in leprosy, *Int. J. Lepr.,* 50, 164, 1982.

25. **Damavandi, N. M. and Mehta, L.,** *In vitro* assessment of endothelial cell response to *Mycobacterium leprae, Int. J. Lepr.,* 54, 135, 1986.

26. **Burchard, G. and Bierther, M.,** An electron microscopic study of the small cutaneous vessels in lepromatous leprosy, *Int. J. Lepr.,* 53, 70, 1985.

27. **Hara, S., Ishiguro, S., and Mizuno, K.,** Phagocytosis of polystyrene spheres in the rabbit corneal endothelium: contribution of lysosomal enzymes to the endothelial degeneration, *Opthalmol. Visual Sci.,* 26, 1631, 1985.

28. **Tripathi, R. C. and Tripathi, B. J.,** Human trabecular endothelium, corneal endothelium, keratocytes, and scleral fibroblasts in primary cell culture. A comparative study of growth characteristics, morphology, and phagocytic activity by light and scanning electron microscopy, *Exp. Eye Res.,* 35, 611, 1982.

29. **Geoffroy, J. S. and Becker, R. P.,** Endocytosis by endothelial phagocytes: uptake of bovine serum albumin-gold conjugates in bone marrow, *J. Ultrastruct. Res.,* 89, 223, 1984.

30. **Oghiso, Y. and Matsuoka, O.,** Time dependent changes of microscopic localization of intravenously administered colloidal carbon particles in mouse lymph nodes, *J. Toxicol. Sci.,* 8, 291, 1983.

31. **Podor, T. J. and Sorgente, N.,** Aortic endothelial cell phagocytosis and elastolytic activity *in vitro:* elevated elastin degradation is associated with cell migration following monolayer injury, *J. Cell. Physiol.,* in press.

32. **Clark, E. R. and Clark, E. L.,** A study of the reaction of lymphatic endothelium and of leucocytes, in the tadpole's tail, toward injected fat, *Am. J. Anat.,* 21, 421, 1917.

33. **McJunkin, F. A.,** The phagocytic activity of vascular endothelium of granulation tissue, *Am. J. Pathol.,* 4, 587, 1928.

34. **Vandenbroucke-Grauls, C. M. J. E., Thijssen, H. M. W. M., and Verhoef, J.,** Phagocytosis of Staphylococci by human polymorphonuclear leukocytes is enhanced in the presence of endothelial cells, *Infect. Immun.,* 50, 250, 1985.

35. **Ryan, U. S., Goodwin, J. D., Curless, B. L., and Schultz, D. R.,** Endothelial cell responses to bacterial ingestion, *Fed. Proc.,* 46, 971, 1987.

36. **Gonzalez, E. T., Goodwin, J. D., and Ryan, U. S.,** Binding and phagocytosis of S. versus Re types of live *Salmonella minnesota* by bovine endothelial cell monolayers, *Fed. Proc.,* 46, 662, 1987.

37. **Ryan, U. S. and Vann, J. M.,** Endothelial cells: a source and target of oxidant damage, in *Proceedings of the 4th International Congress on Oxygen Radicals,* Simic, M. G., Taylor, K. A., and Ward, A. F., Eds., 1988.

38. **Ryan, U. S., Schultz, D. R., Del Vecchio, P., and Ryan, J. W.,** Endothelial cells of bovine pulmonary artery lack receptors for C3b and for the Fc portion of immunoglobulin G, *Science,* 208, 748, 1980.

39. **Ryan, U. S., Schultz, D. R., and Ryan, J. W.,** Fc and C3b receptors on pulmonary endothelial cells: induction by injury, *Science,* 214, 557, 1981.

40. **Ryan, U. S., Mortara, M., and Whitaker, C.,** Methods for microcarrier culture of bovine pulmonary artery endothelial cells avoiding the use of enzymes, *Tissue Cell,* 12, 619, 1980.

41. **Ryan, U. S.,** Culture of pulmonary endothelial cells on microcarrier beads, in *Biology of the Endothelial Cell,* Jaffe, E. A., Ed., Martinus Nijhoff, The Hague, 1984, chap. 4.

42. **Ryan, U. S. and Maxwell, G.,** Microcarrier cultures of endothelial cells, *J. Tissue Cult. Meth.,* 10, 7, 1986.
43. **Ryan, U. S., White, L., Lopez, M., and Ryan, J. W.,** Use of microcarriers to isolate and culture pulmonary microvascular endothelium, *Tissue Cell,* 14, 597, 1982.
44. **Ryan, U. S. and Ryan, J. W.,** Inflammatory mediators, contraction and endothelial cells, in *Progress in Microcirculation Research,* Vol. II, Courtice, F. C., Garlick, D. G., and Perry, M. A., Eds., Committee in Postgraduate Medical Education, University of New South Wales, Sydney, 1984, 424.
45. **Ryan, U. S. and White, L.,** Microvascular endothelium, isolation with microcarriers: arterial, venous, *J. Tissue Cult. Meth.,* 10, 9, 1986.
46. **Ryan, U. S., Schultz, D. R., Goodwin, J. W., Vann., J. M., Selvaraj, M. P., and Hart, M. A.,** Role of C1Q in phagocytosis of S. Minnesota by pulmonary endothelial cells, submitted.
47. **Ryan, U. S. and Olazabal, B.,** Endothelial seeding of filters, grafts and tubes, *J. Tissue Cult. Meth.,* 10, 61, 1986.
48. **Ryan, U. S. and Hart, M. A.,** Electron microscopy of endothelial cells in culture. II. Scanning electron microscopy and OTOTO impregnation method, *J. Tissue Cult. Meth.,* 10, 35, 1986.
49. **Ryan, U. S.,** Metabolic activity of pulmonary endothelium: modulation of structure and function, *Annu. Rev. Physiol.,* 48, 263, 1986.
50. **Babior, B. M.,** Oxygen-dependent microbial killing by phagocytes (first of two parts), *N. Engl. J. Med.,* 298, 659, 1978.
51. **Johnston, R. B.,** Oxygen metabolism and the microbicidal activity of macrophages, *Fed. Proc.,* 37, 2759, 1978.
52. **Root, R. K., Metcalf, J. A., Oshino, N., and Chance, B.,** H_2O_2 Release from human granulocytes during phagocytosis, *J. Clin. Invest.,* 55, 945, 1975.
53. **Allen, R. C.,** Biochemiexcitation: chemiluminescence and the study of biological oxygenation reactions, in *Chemical and Biological Generation of Excited States,* Adam, W. and Cilento, G., Eds., Academic Press, New York, 1982, 309.
54. **Allen, R. C. and Loose, L. D.,** Phagocytic activation of a Luminol-dependent chemiluminescence in rabbit alveolar and peritoneal macrophages, *Biochem. Biophys. Res. Commun.,* 69, 245, 1976.
55. **Babior, B. M., Kipnes, R. S., and Curnutte, J. T.,** Biological defense mechanisms. The production by leukocytes of superoxide, a potential bactericidal agent, *J. Clin. Invest.,* 52, 741, 1973.
56. **Pick, E. and Mizel, D.,** Rapid microassays for the measurement of superoxide and hydrogen peroxide production by macrophages in culture using an automatic enzyme immunoassay reader, *J. Immunol. Meth.,* 46, 211, 1981.
57. **Rosen, G. M. and Freeman, B. A.,** Detection of superoxide generated by endothelial cells, *Proc. Natl. Acad. Sci. U.S.A.,* 81, 7269, 1984.
58. **Matsubara, T. and Ziff, M.,** Superoxide anion release by human endothelial cells: synergism between a phorbol ester and a calcium ionophore, *J. Cell. Physiol.,* 127, 207, 1986.
59. **Matsubara, T. and Ziff, M.,** Increased superoxide anion release from human endothelial cells in response to cytokines, *J. Immunol.,* 137, 3295, 1986.
60. **Ryan, U. S., Johns, A., and van Breemen, C.,** Role of calcium in Receptor-mediated endothelial cell responses, *Chest,* 1988.
61. **Vann, J. M., Proctor, R. A., and Ryan, U. S.,** Mechanism of the cytotoxic effect of intracellular *S. aureus* on bovine endothelial cells, *Fed. Proc.,* 46, 1400, 1987.
62. **Ryan, U. S., Absher, M., Olazbal, B. M., Brown, L. M., and Ryan, J. W.,** Proliferation of pulmonary endothelial cells: time-lapse cinematography of growth to confluence and restitution of monolayer after wounding, *Tissue Cell,* 14, 637, 1982.
63. **Ryan, U. S. and Mayfield, L. J.,** Video analysis of endothelial cell genealogy, migration and division, *J. Tissue Cult. Meth.,* 10, 55, 1986.
64. **Grigorian, G. Y. and Ryan, U. S.,** Platelet-activating factor effects on bovine pulmonary artery endothelial cells, *Circ. Res.,* 61, 389, 1987.
65. Chapter 10, Volume I, of this book.
66. Chapter 27, Volume II, of this book.
67. Chapter 12, Volume I, of this book.
68. Chapter 16, Volume II, of this book.
69. **Libby, P., Ordovas, J. M., Auger, K. R., Robbins, A. H., Birinyi, L. K., and Dinarello, C. A.,** Endotoxin and tumor necrosis factor induce interleukin-1 gene expression in adult human vascular endothelial cells, *Am. J. Pathol.,* 124, 179, 1986.

Chapter 31

ENDOTHELIUM-DERIVED RELAXING FACTOR

Anthony Johns, Raouf A. Khalil, Una S. Ryan, and Cornelis Van Breemen

TABLE OF CONTENTS

I. THE ENDOTHELIUM-DERIVED RELAXING FACTOR (EDRF)

The paradox that acetylcholine, the potent dilator of intravascular beds, often produced contractions in isolated blood vessels was resolved by the discovery that the vasodilator action of acetylcholine on the isolated blood vessel was completely dependent on the presence of an intact endothelium.[1-3] This led to a number of investigations into the mechanisms of action of a variety of vasodilator agents, many of which have now been shown to produce relaxation of blood vessels only in the presence of an intact endotheium (Table 1). The characteristics of endothelium-mediated vasodilation induced by acetylcholine have been studied most extensively and shown to be mediated by muscarinic receptors sensitive to atropine blockade.[2,3,7-10] The importance of the endothelium in controlling blood-vessel tone is emphasized by experiments showing that removal of the endothelium increases the sensitivity of blood vessels to a variety of vasoconstrictors,[2,6,7,11] suggesting that not only is there a release of a vasodilator from the endothelium in response to a stimulating agent but also there is a background release of this substance under resting conditions.

It is now generally accepted that the hypothesis that best explains endothelium-dependent vasodilation is that endothelium releases a relaxing factor which was termed the endothelium-derived relaxing factor (EDRF) by Furchgott and co-workers.[2,3] The secretion of a diffusible relaxing factor has been elegantly shown by his group using experiments where a deendothelialized blood vessel (the recipient strip) is mounted with its intimal surface in close proximity to the endothelial surface of a donor strip in a so-called sandwich mount,[12,13] and also in experiments involving cultured endothelial cells grown on microcarrier beads.[14-16] These later experiments have shown that although the endothelial cells seemed to have lost their muscarinic receptors during culture, they are still capable of releasing EDRF upon stimulation with other vasodilators. Identification of EDRF has been greatly hampered by its very short half-life of about 6 sec.[14,15] A number of studies have determined that the half-life of EDRF can be prolonged with cupric ions and by superoxide dismutase which suggests that it is a free radical.[3,4,15] Inhibition of acetylcholine-induced relaxation with the potent free radical scavenger hydroquinone is consistent with this speculation.[2,3] Experiments on bovine pulmonary artery showed that both acetylcholine-induced EDRF release in preparations with intact endothelium and nitric oxide radical in deendothelialized preparations caused comparable relaxation and cGMP accumulation, both of which were inhibited by ferrous ion. The remarkably similar properties of EDRF and nitric oxide radical suggested EDRF to be a radical species chemically related to nitric oxide.[17] Other studies provided evidence that EDRF appeared to be a negatively charged substance with hydrophilic properties since it was retained on anion exchange columns and not on hydrophobic extraction columns.[14] Other workers have suggested that EDRF has a carbonyl group in or near its active site.[18] Early studies have ruled out the possibility that EDRF was a prostaglandin, AMP, or adenosine,[2,18] but the potent lipoxygenase inhibitors (ETYA and NDGA) have been shown to inhibit the endothelium-dependent relaxation of blood vessels, suggesting that the enzyme lipoxygenase could be involved in the production of EDRF.[2-6,19-29]

Inhibition of the synthesis, release, or action of EDRF has been studied extensively by a number of investigators and the results are summarized in Table 2.

II. MECHANISMS INVOLVED IN THE RELEASE OF EDRF

Since A23187 is one of the most potent releasers of EDRF, it was hypothesized that an increase of the intracellular free calcium concentration could have a key role in the synthesis and/or release of EDRF.[3] Calcium is known to be a cofactor for many enzyme activities including the enzyme phospholipase A_2,[30] which is possibly one of the early key steps in a series of intracellular events that eventually result in the synthesis of EDRF.[3] Therefore, a

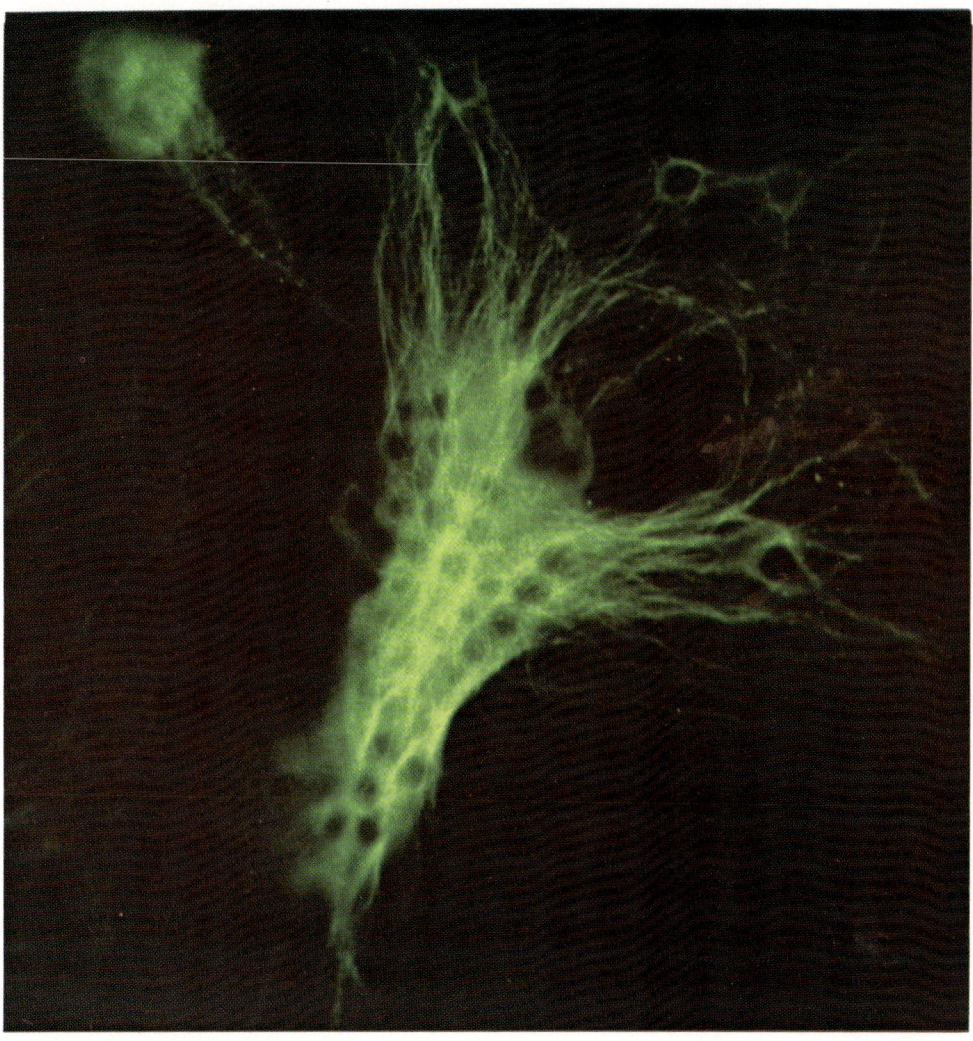

PLATE 1, Chapter 30. Immunofluorescence localization using monoclonal antibodies to vimentin showing extensions of the cell containing cytoskeletal vimentin filaments. Outlying phagocytized beads are surrounded by vimentin filaments as are the phagocytized beads packed in the center of the cell. (Magnification × 1050.)

Table 1
THE VARIOUS VASODILATORS CLASSIFIED ACCORDING TO
THEIR MECHANISM OF ACTION

Vasodilators thought to require intact endothelium		Vasodilators thought to act directly on smooth muscle	
Vasodilator	Ref.	Vasodilator	Ref.
A23187	4, 6, 19, 20, 24, 27	Adenosine	7, 22, 53
Acetylcholine	1—3, 7—10	Glyceryl trinitrate	2
ADP	4, 7, 19, 22, 53	Isoproterenol	2
Arachidonic acid	2, 6, 26, 54	Papaverine	8, 9
ATP	4, 7, 19, 22, 53	Prostaglandin E_2	8, 21
Bradykinin	8, 19, 21, 53, 55—60	Prostaglandin I_2	8, 21
Calcitonin gene-related peptide	68, 69	Sodium azide	2
Histamine	10, 20	Sodium nitrite	2
Hydralazine	61	Sodium nitroprusside	20
Melittin	39		
Oxytocin	67		
Serotonin	62, 63		
Substance P	19, 25, 64		
Thrombin	23, 43, 65		
Trypsin	66		
Vasopressin	67		

casual link between the increase in intracellular free calcium concentration and the release of EDRF was postulated.[2,12,30-33]

It has been of interest to determine whether agents modifying the release of EDRF do so by altering calcium fluxes into the endothelial cells.[31] Initially the source of calcium was considered to be mainly extracellular, based on studies showing that the release of EDRF was completely inhibited by calcium-free medium.[31,34-36] This was further supported by the results of Singer and Peach[37] who showed that slow calcium channel blockers (verapamil and nifedipine) partially inhibited the release of EDRF. It has also been shown that EDRF could be released from unstimulated aortic preparations, indicating a continuous background release of EDRF.[15,18,38] It has been reported that the relaxant effect of acetylcholine upon calcium replenishment is greater than that to acetylcholine in the continuous presence of calcium, suggesting that both the stimulated and the background release of EDRF were calcium dependent. Inhibition of the background release of EDRF may lead to a buildup of EDRF within endothelial cells which results in a large release of EDRF on the restoration of calcium.[31] In addition it was found that the increased tissue cGMP levels induced by acetylcholine were dramatically reduced after removal of calcium.[32]

These studies have, however, been challenged by a number of reports showing that the relaxation produced by arachidonate and melittin was not altered in calcium free solutions, suggesting that the archidonate- and melittin-induced release of EDRF did not depend on the presence of extracellular calcium.[35,39] It has also been shown that intracellular calcium transients induced in endothelial cells by ATP, thrombin, and bradykinin are only slightly inhibited by removal of extracellular calcium,[33,40,41] and that the increase in cGMP in smooth muscle due to EDRF release from endothelial cells can still occur in the presence of calcium channel blockers.[32] The latter observation would suggest that the endothelial cells do not contain voltage-sensitive calcium channels, which is further supported by the ineffectiveness of the voltage-operated calcium-channel activator Bay K 8644 in releasing EDRF from endothelial cells or raising cGMP levels in smooth muscle cells.[38]

Table 2
COMPOUNDS THAT INHIBIT THE RELEASE OR
ACTION OF ENDOTHELIUM-DEPENDENT
VASODILATORS AND THEIR POSSIBLE
MECHANISM OF ACTION

Agent	Possible mechanism	Ref.
Amiloride and dichlorobenzamil	Blockade of sodium/calcium exchange	76
Anoxia	Possibly release of endothelium-derived contracting factor	2,6,24,70
5,8,11,14-Eicosatetrayonic acid (EYTA)	Inhibition of lipoxygenase enzyme	2,3,6,20—26
Fe^{2+}	Chemical ateration of EDRF or scavenging free radical	15,17
Hemoglobin	Either by scavenging EDRF or by inhibition of cGMP formation	12
Hydroquinone	Short exposure possibly by scavenging free radicals, but long exposure perhaps by removing endothelial cells	3,4,10,12
Methylene blue	Inhibiting formation of cGMP in smooth muscle	48, 73—75
Nordihydroguiaretic acid (NDGA)	Inhibition of lipoxygenase enzyme	19,26,27, 29,71
p-Bromophenacyl-bromide (BPB)	Inhibition of phospholipase A_2 and possibly due to loss of damage of endothelial cells	6,19,27
Quinacrine	Possibly by blockade of receptor-operated channel	3,20,21,23, 25,28,72
SKF 525A	Inhibition of a cytochrome P-450 pathway possibly involved in EDRF synthesis	35

However, extracellular Ca^{2+} is undoubtedly crucial to endothelial stimulus-secretion coupling (see Figure 1). Three separate mechanisms may be involved: (1) an unregulated Ca^{2+} leak into the cells due to a finite Ca^{2+} permeability of all biological membranes containing lipids and proteins may be responsible for the background release of EDRF; (2) agonist-stimulated Ca^{2+} influx through receptor-operated cation channels; and (3) extracellular Ca^{2+} is necessary for the refilling of intracellular Ca^{2+} stores. The latter mechanism is suggested by our finding that bradykinin, ATP, and thrombin, agents which release Ca^{2+} from cultured endothelial cells bathed in Ca^{2+} free solution, fail to release Ca^{2+} upon a second challenge. Under these conditions, Ca^{2+} appears not to be recycled intracellularly, but instead there is a requirement for the intracellular store to be refilled from the extracellular space. The mechanism of intracellular Ca^{2+} release is likely to be inositol,1,4,5-trisphosphate (IP_3)-induced activation of Ca^{2+} channels in the endoplasmic reticulum (ER) since bradykinin, a potent Ca^{2+} releaser,[16] increases IP_3 in cultured endothelial cells.[42] The relationship between the dose of bradykinin and the increase in IP_3 as well as the release of ^{45}Ca indicates an ED_{50} of 1×10^{-9} M, whereas the dose-response relationship related to EDRF release as measured by bioassay indicates an ED_{50} of 4×10^{-10} M bradykinin.[16,42] These differences would be reconciled if low concentrations of drugs that release EDRF could increase intracellular calcium by stimulation influx through receptor-operated calcium channels while higher agonist concentrations could, in addition, induce intracellular calcium release.[16]

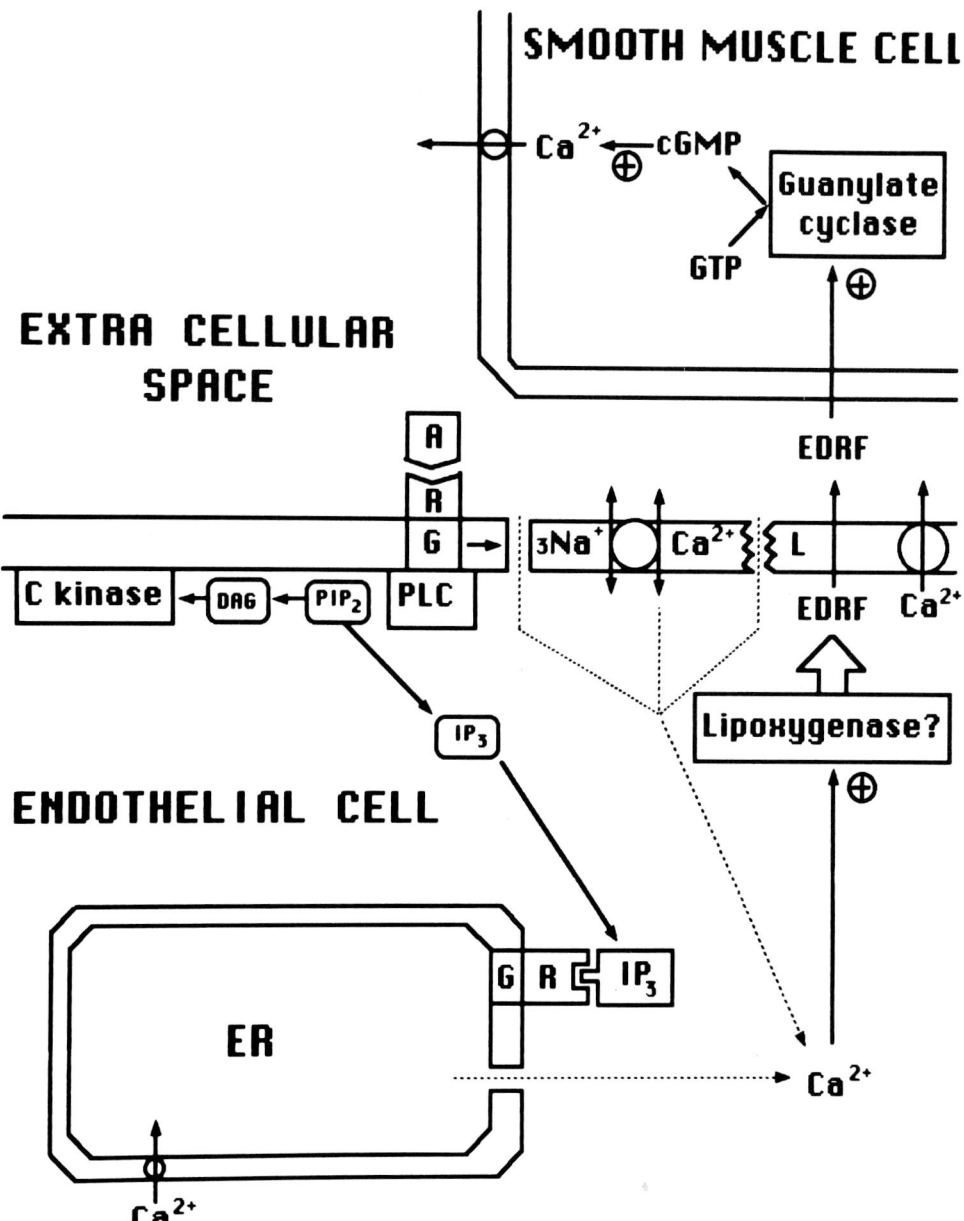

FIGURE 1. A diagrammatic representation of a hypothesis for Ca^{2+} activation of endothelial cells. Agonists (A) such as bradykinin, ATP and thrombin activate receptors (R) on the plasmalemma which through the mediation of G proteins (G) activate phospholipase C (PLC). This results in the hydrolysis of phosphatidyl inositol bisphosphate (PIP_2) to inositol 1,4,5-trisphosphate (IP_3) and diacylglycerol (DAG). The IP_3 liberates Ca^{2+} from the endoplasmic reticulum (ER). Receptor occupation also activates cation channels which supply Ca^{2+} directly from the extracellular space. A background Ca^{2+} leak (L) supplies Ca^{2+} when the cell is not activated. Ca activates the synthesis or release of EDRF, perhaps by stimulating lipoxygenase. EDRF diffuses out of the endothelial cell and into the smooth muscle cell, where it activates guanylate cyclase. The raised levels of cGMP in the smooth muscle activate Ca removal from the myoplasm to cause relaxation.

Experiments on both rat and rabbit aorta preparations showed that the relaxations elicited by acetylcholine and A23187 but not by sodium nitroprusside were functionally antagonized and abolished by the amiloride analog dichlorobenzamil, an inhibitor of calcium influx via sodium-calcium exchange, suggesting that sodium-calcium exchange may be an important mechanistic step in the release of EDRF.[36] However, since the available amiloride analogs, including dichlorobenzamil, are known to possess additional pharmacological actions, verification of the role of sodium-calcium exchange in EDRF release awaits the development of more specific antagonists.

III. MECHANISMS INVOLVED IN THE RELAXATION INDUCED BY EDRF

It has been shown that the contractions produced by increasing K^+ in the Krebs solution are less sensitive to relaxation by acetylcholine than are equivalent contractions produced by norepinephrine in the case both of rabbit aorta[12,5] and canine arteries.[6,8,43,44] However, the fact that there is still significant relaxation in a completely depolarizing solution indicates that hyperpolarization of the smooth muscle cells is not required for endothelium-dependent relaxation by acetylcholine.[12]

A number of agents that relax smooth muscle have been found to increase the activity of smooth muscle guanylate cyclase,[32,45-51] including arachidonate hydroperoxides,[46,47] the phospholipase A_2 activator melittin,[39] and free radicals, particularly nitric oxide and the hydroxyl radical.[48] It has also been demonstrated in smooth muscle that relaxation induced by nitric-oxide-producing agents such as organic nitrites, nitroglycerin, and nitroprusside all involve stimulation of soluble guanylate cyclase and increased cellular levels of cGMP.[48] In addition, it has been found that treatment of deendothelialized rabbit aorta preparations with 8-bromocyclic GMP reduced their contractile response to norepinephrine.[11] The effects of EDRF on smooth muscle have been shown to be inhibited by methylene blue which inhibits guanylate cyclase and in experiments using both rat and rabbit aorta preparations, it has been demonstrated that in the presence of norepinephrine the vasodilator agents acetylcholine, histamine, and A23187 increase cGMP levels by 20- to 40-fold only if the endothelium is intact, whereas glyceryl trinitrate had the same effect on cGMP levels in both intact and deendothelialized tissues.[20,52]

Some studies, however, relate the relaxant effect of EDRF to reduction of calcium influx. In rabbit aorta preparations, acetylcholine reduced the norepinephrine-induced calicum influx only when the endothelium was present. This acetylcholine effect is probably mediated by EDRF release since it is blocked by agents which inactivate EDRF like dithiothreitol, potassium borohydride, phenylhydrazine, or phenidone.[53] However, the effect of EDRF on calcium fluxes has yet to be clarified.

REFERENCES

1. **Furchgott, R. F. and Zawadzki, J. V.,** Acetylcholine relaxes arterial smooth muscle by releasing a relaxing substance from endothelial cells, *Fed. Proc.,* 39, 581, 1980.
2. **Furchgott, R. F. and Zawadzki, J. V.,** The obligatory role of endothelial cells in the relaxation of arterial smooth muscle by acetylcholine, *Nature (London),* 288, 373, 1980.
3. **Furchgott, R. F., Zawadzki, J. V., and Cherry, P. D.,** Role of endothelium in the vasodilator response to acetylcholine, in *Vasodilation,* Vanhoutte, P. and Leusen, I., Eds., Raven Press, New York, 1981, 49.
4. **Furchgott, R. F.,** The requirement for endothelial cells in the relaxation of arteries by acetylcholine and some other vasodilators, *Trends Pharmacol. Sci.,* 2, 173, 1981.
5. **Furchgott, R. F.,** Acetylcholine and blood vessel relaxation: complications and clarifications, in *Trends in Autonomic Pharmacology,* Vol. 2, Kalsner, S., Ed., Urban & Schwarzenberg, Baltimore, 497, 1982.

6. **Furchgott, R. F.**, Role of endothelium in responses of vascular smooth muscle, *Circ. Res.*, 53, 557, 1983.
7. **De Mey, J. G. and Vanhoutte, P. M.**, Role of the intima in cholinergic and purinergic relaxation of isolated canine femoral arteries, *J. Physiol.*, 316, 347, 1981.
8. **Chand, N. and Altura, B. M.**, Acetylcholine and bradykinin relax intrapulmonary arteries by acting on endothelial cells: role in lung vascular diseases, *Science*, 213, 1376, 1981.
9. **Lee, T. J.-F.**, Cholinergic mechanism in the large cat cerebral artery, *Circ. Res.*, 50, 870, 1982.
10. **Van de Voorde, J. and Leusen, I.**, Role of endothelium in the vasodilator response of rat thoracic aorta to histamine, *Eur. J. Pharmacol.*, 87, 113, 1983.
11. **Godfraind, T.**, EDRF and cyclic GMP control gating of receptor-operated calcium channels in vascular smooth muscle, *Eur. J. Pharmacol.*, 126, 341, 1986.
12. **Furchgott, R. F.**, The role of endothelium in the responses of vascular smooth muscle to drugs, *Annu. Rev. Pharmacol. Toxicol.*, 24, 175, 1984.
13. **Rubanyi, G. M. and Vanhoutte, P. M.**, Superoxide anions and hyperoxia inactivate endothelium-derived relaxing factor, *Am. J. Physiol.*, 250, H822, 1986.
14. **Cocks, T. M., Angus, J. A., Campbell, J. H., and Campbell, G. R.**, Release and properties of endothelial-derived relaxing factor (EDRF) from endothelial cells in culture, *J. Cell. Physiol.*, 123, 310, 1985.
15. **Grugglewski, R. J., Palmer, R. M. J., and Moncada, S.**, Superoxide anion is involved in the breakdown of endothelium-derived vascular relaxing factor, *Nature (London)*, 320, 454, 1986.
16. **Johns, A., Ryan, U. S., and van Breemen, C.**, unpublished observation.
17. **Ignarro, L. J., Byrns, R. E., and Wood, K. S.**, Pharmacological and biochemical properties of endothelium-derived relaxant factor (EDRF): evidence that EDRF is closely related to nitric oxide (NO) radical, *Circulation, Suppl. 2*, 74(4), 287, 1986.
18. **Griffith, T. M., Edwards, D. H., Lewis, M. J., Newby, A. C., and Henderson, A. H.**, The nature of endothelium-derived relaxant factor, *Nature (London)*, 308, 645, 1984.
19. **Furchgott, R. F., Cherry, P. D., and Zawadzki, J. V.**, Endothelium dependent relaxation of arteries by acetylcholine, bradykinin and other agents, in *Vascular Neuroeffector Mechanisms: 4th International Symposium*, Bevon, J., et al., Eds., Raven Press, New York, 1983, 37.
20. **Rapoport, R. M. and Murad, F.**, Agonist-induced endothelium-dependent relaxation in rat thoracic aorta may be mediated through cGMP, *Circ. Res.*, 52, 352, 1983.
21. **Cherry, P. D., Furchgott, R. F., Zawadzki, J. V., and Jothianandan, D.**, The role of endothelial cells in the relaxation of isolated arteries by bradykinin, *Proc. Natl. Acad. Sci. U.S.A.*, 79, 2106, 1982.
22. **Furchgott, R. F. and Zawadzki, J. V.**, ATP relaxes rabbit aortic smooth muscle by both an indirect action via endothelial cells and a direct action, *Pharmacologist*, 22, 271, 1980.
23. **DeMey, J. G., Claeys, M., and Vanhoutte, P. M.**, Endothelial-dependent inhibitory effects of acetylcholine, adenosine triphosphate, thrombin and arachidonic acid in the canine femoral artery, *J. Pharmacol. Exp. Ther.*, 222, 166, 1982.
24. **Zawadzki, J. V., Cherry, P. D., and Furchgott, R. F.**, Comparison of endothelium-dependent relaxation of rabbit aorta by A23187 and by acetylcholine, *Pharmacologist*, 22, 271, 1980.
25. **Zawadzki, J. V., Furchgott, R. F., and Cherry, P. D.**, The obligatory role of endothelial cells in the relaxation of arterial smooth muscle by substance P, *Fed. Proc.*, 40, 689, 1981.
26. **Singer, H. A. and Peach, M. J.**, Endothelium-dependent relaxation of rabbit aorta. I. Relaxation stimulated by archidonic acid (AA), *J. Pharmacol. Exp. Ther.*, 227, 790, 1983.
27. **Furchgott, R. F., Zawadzki, J. V., Jothianandan, D., and Cherry, P. D.**, Nordihydroguiaretic acid (NDGA) and α, p-dibromoacetophenone (BPB) inhibit endothelium-dependent relaxation of arteries by acetylcholine, bradykinin and A23187, *Fred. Proc.*, 1982.
28. **Singer, H. A. and Peach, M. J.**, Endothelium-dependent relaxation of rabbit aorta. II. Inhibition of relaxation stimulated by methacholine and A23187 with antagonists of arachidonic acid metabolism, *J. Pharmacol. Exp. Ther.*, 227, 796, 1983.
29. **Chand, N. and Altura, B. M.**, Inhibition of endothelial cell-dependent relaxations to acetylcholine and bradykinin by lipoxygenase inhibitors in canine isolated renal arteries, *Microcirculation*, 1, 211, 1981.
30. **Van der Bosch, H.**, Intracellular phospholipase A, *Biochem. Biophys. Acta*, 604, 191, 1980.
31. **Long, C. J. and Stone, T. W.**, The release of endothelium-derived relaxant factor is calcium dependent, *Blood Vessels*, 22, 205, 1985.
32. **Miller, R. C., Schoeffter, P., and Stoclet, J. C.**, Insensitivity of calcium-dependent endothelial stimulation in rat isolated aorta to the calcium entry blocker, flunarizine, *Br. J. Pharmacol.*, 85, 481, 1985.
33. **Lückhoff, A. and Busse, R.**, Increased free calcium in endothelial cells under stimulation with adenine nucleotides, *J. Cell. Physiol.*, 126, 414, 1986.
34. **Edwards, D. F., Griffith, T. M., Henderson, A. H., Lewis, M. G., and Newby, A. C.**, Production of endothelium derived relaxant factor is both ATP and calcium dependent, *Br. J. Pharmacol.*, 85, 344P, 1985.

35. **Dusting, G. J. and MacDonald, P. S.,** Endothelium-dependent vasodilation: role of beta-adrenoceptors, calcium and cytochrome P-450, *Blood Vessels,* 23(2), 66, 1986.
36. **Winguist, R. J., Bunting, P. B., and Schofield, T. L.,** Blockade of endothelium-dependent relaxation by the amiloride analog dichlorobenzamil: possible role of Na^+/Ca^{++} exchange in the release of endothelium-derived relaxant factor, *J. Pharmacol. Exp. Ther.,* 235(3), 644, 1985.
37. **Singer, H. A. and Peach, M. H.,** Calcium and endothelial-mediated vascular smooth muscle relaxation in rabbit aorta, *Hypertension,* 4 (Suppl. 2), 19, 1982.
38. **Spedding, M., Schini, V., Schoeffter, P., and Miller, R. C.,** Calcium channel activation does not increase release of endothelial-derived relaxant factors (EDRF) in rat aorta although tonic release of EDRF may modulate calcium channel activity in smooth muscle, *J. Cardiovasc. Pharmacol.,* 8, 1130, 1986.
39. **Loeb, A. L., Johns, R. A., and Peach, M. J.,** Phospholipase activation by melittin releases prostacyclin and an endothelium-derived relaxing factor from rabbit aorta, *Blood Vessels,* 23(2), 86, 1986.
40. **Hallam, T. J. and Pearson, J. D.,** Elevation of cytoplasmic calcium concentration in human umbilical vein endothelial cells by thrombin, *J. Physiol.,* 377, 131p, 1986.
41. **Hallam, T. J. and Pearson, J. D.,** ATP and bradykinin stimulate increased cytoplasmic calcium concentration in cultured piglet aortic endothelial cells, *Br. J. Pharmacol.,* 88, 247p, 1986.
42. **Derian, C. K. and Moskowitz, M. A.,** Phosphoinositide hydrolysis in endothelial cells and carotid artery segments. Bradykinin-2 receptor stimulation is calcium independent. *J. Biol. Chem.,* 261, 3831, 1986.
43. **De Mey, J. G. and Vanhoutte, P. M.,** Heterogeneous behavior of the canine arterial and venous wall, *Circ. Res.,* 51, 439, 1982.
44. **De Mey, J. G. and Vanhoutte, P. M.,** Interaction between Na^+, K^+ exchanges and the direct inhibitory effect of acetylcholine on canine femoral arteries, *Circ. Res.,* 46, 826, 1980.
45. **Rapoport, R. M., Draznin, M. D., and Murad, F.,** Endothelium-dependent vascular relaxation may be mediated through cyclic GMP-dependent protein phosphorylation, *Clin. Res.,* 31, 526A, 1983.
46. **Hidaka, H. and Asano, T.,** Stimulation of platelet guanylate cyclase by unsaturated fatty acid peroxides, *Proc. Natl. Acad. Sci. U.S.A.,* 74, 3657, 1977.
47. **Goldberg, N. D., Graff, G., Haddox, M. K., Stephenson, J. H., Glass, D. B., and Moser, M. E.,** Redox modulation of splenic cell soluble guanylate cyclase activity: activation by hydrophilic and hydrophobic antioxidants represented by ascorbic and dehydroascorbic acids, fatty acid hydroperoxides, and prostaglandin endoperoxides, *Adv. Cyclic Nucleotide Res.,* 9, 101, 1978.
48. **Murad, F., Arnold, W. P., Mittal, C. K., and Braughler, J. M.,** Properties and regulation of guanylate cyclase and some proposed functions for cyclic GMP, *Adv. Cyclic Nucleotide Res.,* 11, 175, 1979.
49. **Katsuki, S. and Murad, F.,** Regulation of adenosine cyclic 3′,5′-monophosphate levels and contractility in bovine tracheal smooth muscle, *Mol. Pharmacol.,* 13, 330, 1977.
50. **Böhme, E., Graf, H., and Schultz, G.,** Effects of sodium nitroprusside and other smooth muscle relaxants on cyclic GMP formation in smooth muscle and platelets, *Adv. Cyclic Nucleotide Res.,* 9, 131, 1978.
51. **Schultz, K. D., Böhme, E., Kreye, V. W., and Schultz, G.,** Relaxation of hormonally stimulated smooth muscular tissues by the 8-bromo derivative of cyclic GMP, *Naunyn Schmiedebergs Arch. Pharmakol.,* 301, 1, 1979.
52. **Furchgott, R. F. and Jothianandan, D.,** Relation of cyclic GMP levels to endothelium-dependent relaxation by acetylcholine in rabbit aorta, *Fed. Proc.,* 42, 619, 1983.
53. **Gordon, J. L. and Martin, W.,** Endothelium-dependent relaxation of the pig aorta: relationship to stimulation of Rb efflux from isolated endothelial cells, *Br. J. Pharmacol.,* 79, 531, 1983.
54. **Cherry, P. D., Furchgott, R. F., and Zawadzki, J. V.,** The endothelium-dependent relaxation of vascular smooth muscle by unsaturated fatty acids, *Fed. Proc.,* 42, 619, 1983.
55. **Altura, B. M. and Chand, N.,** Bradykinin-induced relaxation of renal and pulmonary arteries is dependent upon intact endothelial cell, *Br. J. Pharmacol.,* 74, 10, 1981.
56. **Cherry, P. D., Furchgott, R. F., and Zawadzki, J. V.,** The indirect nature of bradykinin relaxation of isolated arteries: endothelial dependent and independent components, *Fed. Proc.,* 40, 689, 1981.
57. **Aiken, J. W.,** Inhibitors of prostaglandin synthesis specifically antagonize bradykinin- and angiotensin-induced relaxations of the isolated celiac artery from rabbit, *Pharmacologist,* 1, 295, 1974.
58. **Blumberg, A. L., Denny, S. E., Marshall, G. R., and Needleman, P. A.,** Blood vessel-hormone interactions: angiotensin, bradykinin and prostaglandin, *Am. J. Physiol.,* 232, H305, 1977.
59. **Needleman, P., Marshall, G., and Sobel, B. E.,** Hormone interactions in the isolated rabbit heart: synthesis and coronary vasomotor effects of prostaglandins, angiotensin, and bradykinin, *Circ. Res.,* 37, 802, 1975.
60. **Toda, N.,** Actions of bradykinin on isolated cerebral and peripheral arteries, *Am. J. Physiol.,* 232, H267, 1977.
61. **Spokas, E. G., Folco, G., Quilley, J., Chandler, P., and McGiff, J. C.,** Endothelial mechanism in the vascular action of hydralazine, *Hypertension,* 5 (Suppl. 1), 1104, 1983.
62. **Cohen, R. A., Shepherd, J. T., and Vanhoutte, P. M.,** Inhibitory role of the endothelium in the response of isolated coronary arteries to platelets, *Science,* 221, 273, 1983.

63. **Cohen, R. A., Shepherd, J. T., and Vanhoutte, P. M.,** 5-hydroxytryptamine can mediate endothelium-dependent relaxation of coronary arteries, *Am. J. Physiol.,* 245, H1077, 1983.
64. **Zawadzki, J. V., Furchgott, R. F., and Cherry, P. D.,** Endothelium-dependent relaxation of arteries by octasubstance P, Kassinin and octacholecystokinin, *Fed. Proc.,* 42, 619, 1983.
65. **Ku, D.,** Coronary vascular reactivity after acute myocardial ischemia, *Science,* 218, 576, 1982.
66. **Burdet, R., Criscione, L., Powell, J., and Sippola, L.,** Role of the endothelium in vasodilator effects of acetylcholine, histamine and trypsin in resistance vessels, *Br. J. Pharmacol.,* 88, 266P, 1986.
67. **Katusic, Z. S., Sheperd, J. T., and Vanhoutte, P. M.,** Oxytocin causes endothelium-dependent relaxation of canine basilar arteries by activation V_1-vasopressinergic receptors, *J. Pharmacol. Exp. Ther.,* 236, 166, 1986.
68. **Al-Kazwini, S. J., Craig, R. K., Holman, J. J., and Marshall, I.,** Different potencies of calcitonin gene-related peptides (CGRP) on mesenteric vasculature constricted by noradrenaline or potassium, *Br. J. Pharmacol.,* 88, 439P, 1986.
69. **Sever, P., Hughes, A., Martin, G., and Thom, S.,** Calcitonin gene-related peptide: in vitro endothelium-dependent dilation of human arteries and in vitro vasodilation in the forearm vascular bed, *Blood Vessels,* 23(2), 99, 1986.
70. **Rubanyi, G. M. and Vanhoutte, P. M.,** Hypoxia releases a vasoconstrictor substance from the canine vascular endothelium, *J. Physiol.,* 364, 45, 1985.
71. **Flower, R. J.,** Drugs which inhibit prostaglandin biosynthesis, *Pharmacol. Rev.,* 26, 33, 1974.
72. **Flower, R. J. and Blackwell, G. J.,** The importance of phospholipase-A_2 in prostaglandin biosynthesis, *Biochem. Pharmacol.,* 25, 285, 1975.
73. **Gruetter, C. A., Kadowitz, P. J., and Ignarro, L. F.,** Methylene blue inhibits coronary arterial relaxation and guanylate cyclase activation by nitroglycerin, sodium nitrite and amyl nitrite, *Can. J. Physiol. Pharmacol.,* 59, 150, 1981.
74. **Gruetter, C. A., Gruetter, D. Y., Lyon, J. E., Kadowitz, P. J., and Ignarro, L. F.,** Relationship between cyclic guanosine 3':5'-monophosphate formation and relaxation of coronary arterial smooth muscle by glyceryl trinitrate, nitroprusside, nitrite and nitric oxide: effects of methylene blue and methemoglobin, *J. Pharmacol. Exp. Ther.,* 219, 181, 1981.
75. **Kukovetz, W. R., Pöch, G., and Holzmann, S.,** Cyclic nucleotides and relaxation of vascular smooth muscle, in *Vasodilation,* Vanhoutte, P. and Leusen, I., Eds., Raven Press, New York, 1981, 339.
76. **Schoefter, P. and Miller, R. C.,** Role of sodium-calcium exchange and effects of calcium entry blockers on endothelial mediated-responses in rat isolated aorta, *Mol. Pharmacol.,* 30, 53, 1986.

Chapter 32

ENDOTHELIUM-DERIVED VASOCONSTRICTOR FACTORS

Gabor M. Rubanyi

TABLE OF CONTENTS

I. INTRODUCTION

The vascular endothelium plays an important role in modulating the tone of underlying vascular smooth muscle. This role is achieved by (1) interposing a physical barrier between the vascular smooth muscle and vasoactive substances circulating in the blood; (2) extracting and metabolically degrading vasoactive substances (e.g., norepinephrine, serotonin, and kinins); (3) converting precursors (e.g., angiotensin I) into vasoactive products; and (4) synthesizing and releasing vasoactive mediators in response to various stimuli. In recent years it became evident that in addition to prostacyclin, endothelial cells synthesize and release a still unidentified relaxing substance (endothelium-derived relaxing factor or EDRF[1]), which mediates relaxations of isolated arteries in response to acetylcholine[2] and to a number of other vasodilators.[1,3,4] Substantial data are now available which demonstrate that in addition to mediating relaxations, endothelial cells can also facilitate contractile responses of the vascular smooth muscle. Although briefly described in previous reviews,[1,4] a comprehensive summary of this indirect mechanism of vascular smooth msuscle contraction was still lacking. This review therefore summarizes the experimental evidence for endothelium-mediated facilitatory responses in various blood vessels and attempts to speculate on the nature and mechanisms of action on vascular smooth muscle of endothelium-derived vasoconstrictor factors which mediate these responses.

II. FACILITATION OF VASCULAR SMOOTH MUSCLE CONTRACTION BY THE ENDOTHELIUM

A. Agonists

The concept of endothelium-mediated facilitation of vascular smooth muscle contraction was introduced on the basis of experimental evidence that removal of the endothelium depressed or prevented contractile responses of isolated blood-vessel preparations under certain conditions. In the femoral artery of the dog, removal of the endothelium reduced maximal contractile responses to high K^+ and norepinephrine.[5,6] Maximal responses to norepinephrine were also reduced in pulmonary, saphenous, and splenic arteries and in femoral, saphenous, and splenic veins of the dog.[6] A major argument against a (specific) facilitatory role of endothelial cells in these experiments was the possibility that the procedure of endothelium removal damaged the underlying vascular smooth muscle cells. However, several observations argued against this possibility: (1) sensitivity (ED_{50} value of the dose-response curves) to these agonists was not affected; (2) in some blood vessels (e.g., pulmonary vein) endothelium-removal did not reduce maximal responses to norepinephrine; and (3) contractile responses to several other agonists (e.g., acetylcholine) were enhanced rather than depressed by the procedure.[6] It was therefore postulated that in some blood-vessel preparations an endogenous vasoconstrictor is released from the intima, which contributes to the contractile response to high K^+ and to norepinephrine.[5,6] This hypothesis seemed to be confirmed by earlier observations made in helical strips of the rabbit aorta that the presence of the endothelium is essential for norepinephrine (released from glass beads placed on the intimal surface) to evoke contractions.[7] However, the considerably higher sensitivity to the contractile action of norepinephrine of smooth muscle cells nearer to the intima than of those nearer to the adventitia[8-11] and the lack of effect of endothelium removal on norepinephrine-induced contractions in strips or rings of rabbit aorta[2] argued against the possibility that norepinephrine triggers the release of a constrictor mediator from the intima of rabbit aortic strips. Increased extracellular concentrations of K^+ (3 to 20 mM) triggered rhythmic contractile activity in rings of canine basilar artery with endothelium, which was significantly reduced by removal of the endothelium.[12] That damage of vascular smooth muscle during mechanical rubbing of the intimal surface was not responsible for the depression of K^+-induced rhythmic activity was demonstrated by the occurrence of similar rhythmic activity in rings without endothelium after exposure to tetraethylammonium.[12]

In several systemic arteries of the dog and of other species the endothelium plays an essential role in mediating vasorelaxation induced by agonists like acetylcholine, the calcium ionophore A23187, thrombin, adenine nucleotides, arachidonic acid, and 5-hydroxytryptamine, and also by aggregating platelets.[1,3,4] In contrast, in systemic and pulmonary veins of the dog acetylcholine does not, or does only to a very moderate degree, trigger endothelium-mediated relaxations, and thrombin and arachidonic acid produce contractions, which are reduced or prevented by removal of the endothelium.[6] Acetylcholine, arachidonic acid, and the calcium ionophore A23187 evoke endothelium-mediated facilitation of basal myogenic tone in rings of basilar artery from the same species.[13]

These observations suggested that some agonists which stimulate the release of a relaxing mediator from endothelial cells of canine arteries, trigger the synthesis/release of a contractile mediator (or mediators) from the endothelium of canine veins and basilar arteries. Endothelium-dependent relaxations in response to acetylcholine are depressed in rings of thoracic aorta isolated from spontaneously hypertensive rats (SHR) compared to those from normotensive Wistar Kyoto rats (WKY).[14,15] Acetylcholine induces endothelium-dependent contractions in quiescent aortic rings from SHR, but not in those from WKY rats.[16] Although these contractions are evoked by concentrations (3×10^{-7} to 10^{-5} M) of acetylcholine higher than those required to induce endothelium-dependent relaxations, they could be inhibited by atropine, indicating that muscarinic receptors are involved in the endothelium-dependent contractile responses as well. These observations suggest that the reduced endothelium-dependent relaxations to acetylcholine in the SHR are probably not due to a decreased release of endothelium-derived relaxing factor but to the simultaneous release of endothelium-derived vasoconstrictor substance(s).[16] The depressed endothelium-dependent relaxations to adenosine diphosphate in the aorta of SHR compared to that of WKY was also explained by assuming the corelease of relaxing and facilitatory mediators.[17] The presence of endothelium facilitated contractions to aggregating platelets and 5-hydroxytryptamine in rings of thoracic aorta from SHR,[17] probably because these stimuli trigger the release of a vasoconstrictor mediator from the endothelium.

B. Stretch

Stretch applied to isolated canine basilar artery rings with endothelium caused development of active tension, which could not be observed in rings where the endothelium has been removed.[18] Stretch-induced endothelium-dependent contractions started within a few seconds after the onset of stretch and they lasted for at least 120 sec, resembling the time course of an autoregulatory response to changes in transmural pressure.[19] Rapid elevation of transmural pressure from near zero to 20 to 50 mmHg in perfused canine carotid artery segments evoked active contractions of the vessel wall, which could also be prevented by removal of the endothelium.[20] These observations suggest that the endothelium may play an important role in mediating stretch-induced vasoconstriction, either by reduced release of relaxing mediator(s) or by the synthesis and release of vasoconstrictor factor(s).[20]

C. Anoxia

Anoxia (induced by gassing organ chambers with 95% N_2 and 5% CO_2 gas mixture) facilitates contractile responses in isolated coronary,[21-23] femoral,[24] pulmonary, saphenous, splenic,[6] and basilar arteries,[25] and in pulmonary and splenic veins of the dog,[6] and in the coronary[26] and pulmonary artery of the pig.[27] Anoxic facilitation is observed with contractions evoked by several agonists including norepinephrine, KCl, $BaCl_2$, 5-hydroxytryptamine, prostaglandin $F_{2\alpha}$, uridine triphosphate, and ouabain.[21,23-27] In quiescent femoral artery preparations (which do not exhibit spontaneous myogenic tone) anoxia does not trigger con-

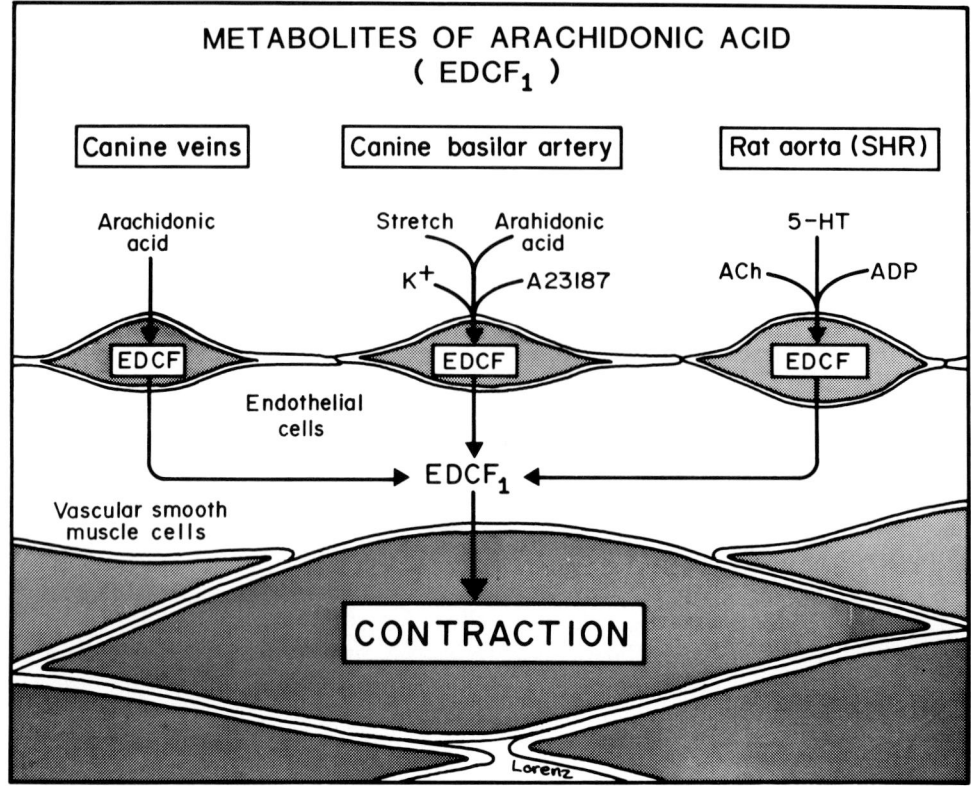

FIGURE 1. Schematic representation of various stimuli which evoke endothelium-dependent contractions in different blood vessels by the release of vasoconstrictor metabolites of arachidonic acid from the endothelium. To differentiate from other endothelium-derived vasoconstrictor factor (EDCF), this group of endothelial mediators is called $EDCF_1$.

tractions,[24] but in arterial preparations with considerable intrinsic myogenic tone (e.g., coronary and basilar arteries), anoxic facilitation can be observed in the absence of contractile agonists.[23,25] Removal of the endothelium reduces or reverses anoxic facilitation in most preparations studied,[6,23,24,27] demonstrating that the intimal layers of the blood vessel wall contribute to the anoxic potentiation, presumably by releasing vasoconstrictor mediator(s).[23]

III. NATURE OF ENDOTHELIUM-DERIVED VASOCONSTRICTOR FACTORS (EDCF)

A. Metabolites of Arachidonic Acid ($EDCF_1$)

The blood vessels in which various stimuli evoke endothelium-dependent contractions, presumably by the release of vasoconstrictor product(s) of arachidonic acid metabolism from endothelial cells, are illustrated in Figure 1. To differentiate from other endothelium-derived vasoconstrictor mediators, the term $EDCF_1$ is used to describe this group of endothelial factors.

1. Exogenous Arachidonic Acid

The primary source of vasoactive metabolites of arachidonic acid in the blood vessel wall is the intima.[28,29] All three known pathways (i.e., cyclooxygenase, lipoxygenase, and cytochrome P-450 monooxygenase) of arachidonic acid metabolism have been identified in vascular endothelial cells.[30-33] Thus arachidonic acid (administered exogenously or liberated from membrane phospholipids by phospholipases) can induce endothelium-mediated vascular responses by stimulating the synthesis/release of vasoactive intermediates or end products of any of these metabolic pathways. Pharmacological studies revealed that endothelium-dependent contractions to exogenous arachidonic acid may be mediated by cyclooooxygenase and lipoxygenase products of the fatty acid.

a. Cyclooxygenase

In most (but not all) systemic arteries of the dog, endothelium-dependent relaxations in response to arachidonic acid can be prevented by various inhibitors of cyclooxygenase, indicating that they are mediated by vasodilator prostanoids (most likely by prostacyclin).[29,34] However, prostacylin contracts the aorta and coronary artery of the pig, the aorta and pulmonary artery of the rabbit, the human umbilical artery, the femoral and pulmonary veins of the dog, and the aorta and vena cava of the rat,[2,16,35-40] indicating that in certain vascular beds, endothelium-dependent contraction to exogenous arachidonic acid may be mediated by the same endothelial product (e.g., prostacyclin) which mediates relaxation in other blood vessels. Indeed, endothelium-mediated contractions to exogenous arachidonic acid in canine femoral and pulmonary veins[40] and the rabbit aorta[41] could be blocked by different inhibitors of cyclooxygenase. However, the inhibitor of prostacyclin synthetase, tranylcypromine, did not affect the contractions induced by arachidonic acid in canine pulmonary and femoral veins, indicating that cyclooxygenase products other than prostacyclin may mediate the facilitatory responses in these preparations.[40] It is also unlikely, that thromboxane A_2 (a vasoconstrictor cyclooxygenase product) plays a major role, since two inhibitors of thromboxane synthetase (imidazole and BW 149H) did not reduce the contractions caused by arachidonic acid in canine veins.[40] Inhibition of the synthesis of prostacyclin or thromboxane A_2 could increase the production of other vasoconstrictor end products of cyclooxygenase, such as $PGF_{2\alpha}$. In blood vessels of the rat, the production of $PGF_{2\alpha}$ is greater in veins compared with the corresponding arteries.[42] If the same difference exists between arteries and veins in other species, $PGF_{2\alpha}$ may be a likely candidate as mediator of endothelium-dependent contraction in response to arachidonic acid in canine veins. Indomethacine prevented endothelium-dependent contractions to exogenous arachidonic acid in canine basilar arteries,[13] suggesting that as in canine veins, products of cyclooxygenase mediate the facilitatory response in this blood vessel as well.

b. Lipoxygenase

In the canine saphenous vein, high concentrations of arachidonic acid can be metabolized via the lipoxygenase pathway.[43] Unsaturated fatty acid hydroperoxide intermediates[44,45] and leukotrienes[46-48] can cause contraction of vascular smooth muscle. Two inhibitors of lipoxygenase (nordihydroguaiaretic acid and Bay G 6575) reduced the endothelium-dependent contractions to arachidonic acid in canine pulmonary veins.[40] Similarly, the leukotriene-receptor antagonist FPL 55712 reduced the contractions to 10^{-5} M arachidonic acid in this blood vessel.[40] These observations suggest that as in the saphenous vein,[43] arachidonic acid can be metabolized through the lipoxygenase pathway in canine pulmonary veins, and products of this pathway (most probably leukotrienes) may also contribute to arachidonic acid-induced endothelium-dependent contractions.

2. Acetylcholine

Endothelium-dependent contractions in response to acetycholine in quiescent rings of SHR aorta were abolished by the phospholipase A_2 inhibitor quinacrine,[16] suggesting that liberation of arachidonic acid from membrane phospholipids is an initial step in the production of endothelium-derived vasoconstrictor factor(s) in this blood vessel, and that the factor may be a vasoconstrictor metabolite of the fatty acid. Indomethacin and meclofenamate also abolished the endothelium-mediated contractions, indicating that product(s) of cyclooxygenase may be the contractile mediator(s).[16] Indeed, the leukotriene synthetase inhibitor diethylcarbamazine had no effect on the contractions, ruling out the potential involvement of the end products of lipoxygenase metabolism of arachidonic acid. Since inhibitors of prostacyclin synthetase (tranylcypromine) and thromboxane synthetase (imidazole) did not reduce the endothelium-dependent contractions evoked by acetylcholine, the vasoconstrictor mediator is not likely to be prostacyclin or thromboxane A_2.[16] Although other prostaglandins (PGD_2, PGE_1, PGE_2, $PGF_{2\alpha}$) caused contractions in aortic rings without endothelium, no significant differences were observed in the responsiveness between aortas of SHR and WKY.[16] Thus, the exact nature of the cyclooxygenase product involved in the endothelium-dependent contractions to acetylcholine in aortas of SHR is still uncertain.

3. Potassium

Endothelium-dependent rhythmic contractions triggered by increased extracellular K^+ concentration in canine basilar arterial rings could be inhibited by two inhibitors of cyclooxygenase, indomethacin and meclofenamate.[12] This suggests that similar to endothelium-mediated contractions to exogenous arachidonic acid in the same blood vessel (see Section III.A.1), K^+-induced rhythmic contractile activity may also be mediated by vasoconstrictor product(s) of cyclooxygenase synthetized and released by the endothelium. Since exogenous prostacyclin (10^{-8} M) inhibited and the prostacyclin synthetase inhibitor tranylcypromine facilitated the K^+-induced rhythmic contractions, it is unlikely that prostacyclin is involved in the contractile response.[12] In contrast, $PGF_{2\alpha}$ and PGE_2 caused concentration-dependent increases in the frequency and amplitude of K^+-induced rhythmic contractions. In addition, $PGF_2\alpha$ and PGE_2 also reversed the inhibitory effect of indomethacin.[12] These data suggest that increased production and release of these prostaglandins from endothelial cells may play an important role in mediating the contractile response to high K^+ in this blood vessel.

4. Stretch

Stretch-induced endothelium-dependent contractions of isolated canine basilar artery rings were significantly reduced by indomethacin, but they were not affected by the leukotriene synthetase inhibitor diethylcarbamazine.[18] Indomethacin did not affect contractions to high K^+, indicating that the effect of the drug is due to inhibition of cyclooxygenase and not of calcium entry. The inhibition of stretch-induced endothelium-dependent contractions by indomethacin is consistent with the postulation that acute increases in transmural pressure stimulate the synthesis of cyclooxygenase products in cerebral arteries.[49] Thus, the endothelium-dependent contractions to stretch in canine basilar arteries must be mediated, at least in part, by vasoconstrictor cyclooxygenase products produced and released from endothelial cells. However, the exact nature of the cyclooxygenase product(s) remains to be determined.

In contrast to stretch-induced repsonses in basilar artery rings, the endothelium-mediated contractions of perfused canine carotid artery segments in response to rapid increases in transmural pressure were not affected by indomethacin.[20] This then suggests that the contractile factor(s) produced in endothelial cells in response to stretch may be different in blood vessels from different anatomical origin.

B. Vasoconstrictor Polypeptide Factor Produced by Cultured Endothelial Cells (EDCF₂)

Increasing concentrations of the culture media from bovine aortic endothelial cells caused progressive contractions of rings isolated from porcine, bovine, and canine coronary arteries.[50] Vasoactivity of the media from endothelial cells in primary culture reaches a maximal potency on the 3rd or 4th day of incubation. Control culture media (i.e., incubated under similar conditions in the absence of endothelial cells) or media obtained from cultures of fibroblasts (human skin and mouse 3T3 cells) and vascular smooth muscle cells (rat aorta)

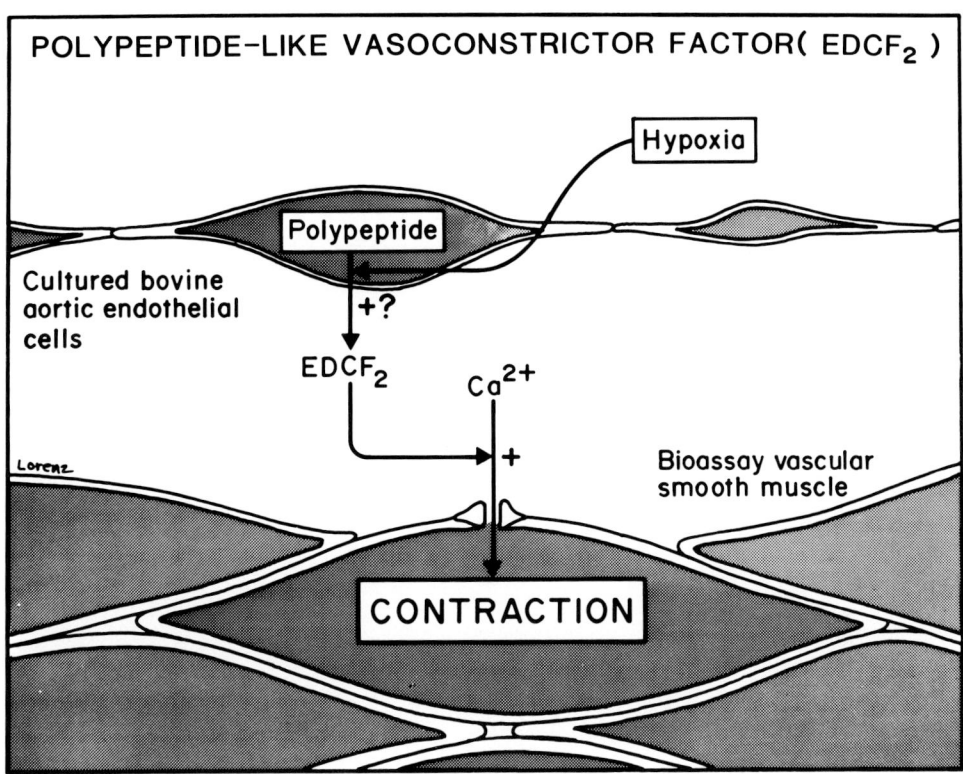

FIGURE 2. Polypeptide-like vasoconstrictor factor (EDCF$_2$) produced by unstimulated cultured bovine aortic endothelial cells. Hypoxia facilitates the basal rate of the production. The factor evokes contraction of bioassay coronary arteries, presumably by enhancing Ca^{2+} influx into vascular smooth muscle cells.

had no vasoconstrictor activity, demonstrating that endothelial cells specifically produce a vasoconstrictor factor. The vasoconstriction induced by the media from endothelial cells was unaffected by inhibitors of receptors (e.g., α- and β-adrenergic, muscarinic, serotoninergic, and histaminergic) for known endogenous vasoactive substances. The vasoconstrictor factor is not likely to be a product of arachidonic acid metabolism, since treatment of the cultured endothelial cells or the bioassay coronary artery preparations with indomethacin, meclofenamate, or phenidone had no effect on the vasoconstrictor activity of the media.

The action of the vasoconstrictor factor on bioassay tissues was abolished by several treatments (i.e., sodium duodecyl sulfate, trypsin, alkali, or acid hydrolysis) known to affect protein structure and/or conformation. It was therefore postulated that the factor is a polypeptide[50] (Figure 2). Dialysis and ultrafiltration studies indicated that the molecular weight of the factor is greater than 3500 but less than 12,000. Filtering the media on a column of Biogel P-10 and bioassaying the resulting fractions on a coronary artery ring demonstrated a single peak of vasoconstrictor activity eluted midway between pancreatic trypsin inhibitor (6500 mol wt) and cytochrome c (12,500 mol wt). These studies suggested, that cultured bovine aortic endothelial cells produce a potent coronary vasoconstrictor polypeptide with an average molecular weight of 8500 ± 1500, having approximately 60 to 70 amino acid residues.[50] Consistent with these findings are the recent observations that unstimulated cultured bovine aortic and pulmonary endothelial cells release a polypeptide-like factor, which contracts bovine pulmonary arteries.[51] Although cultured endothelial cells can produce the potent vasoconstrictor polypeptide angiotensin II,[52,53] the factor responsible for the vasoconstrictor activity of the media is not likely to be identical with it.[51]

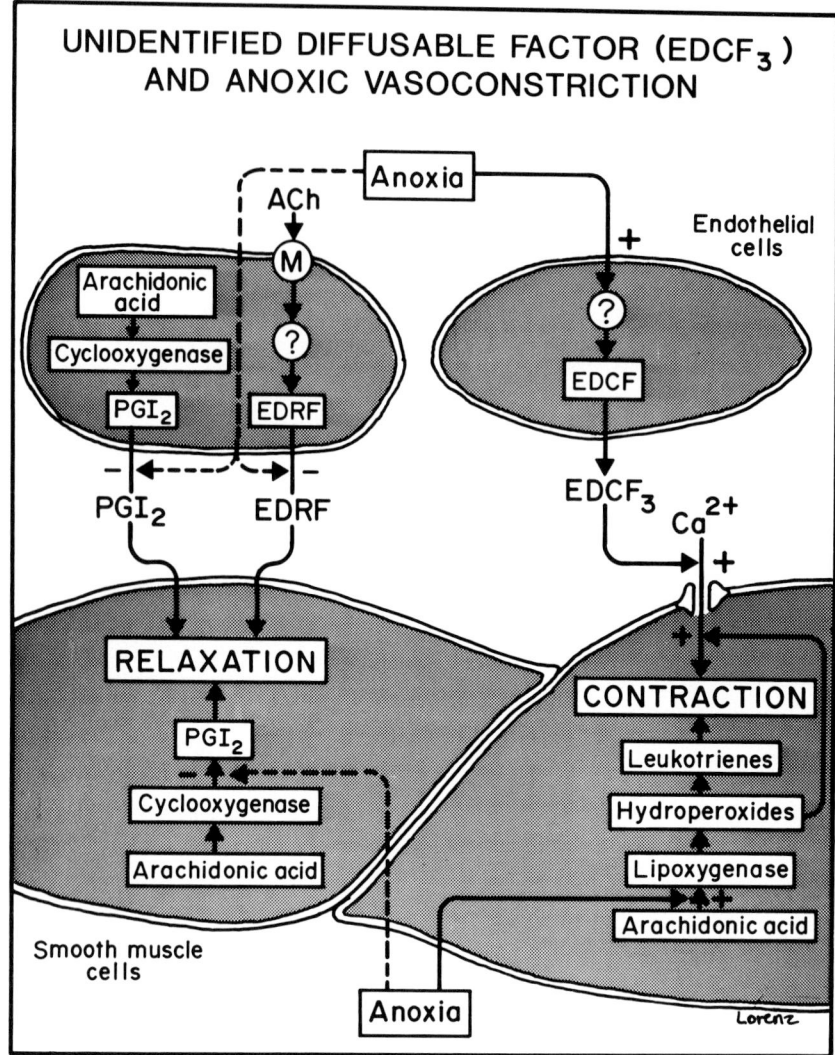

FIGURE 3.   Hypothetical mechanisms involved in anoxic/hypoxic vasoconstriction. In most canine blood vessels the response consists of an endothelium-mediated component and a direct action on vascular smooth muscle. Endothelium-mediated vasoconstriction is the consequence of (1) depressed release of vasodilator mediators (e.g., prostacyclin [PGI_2] and still unidentified endothelium-derived relaxing factor(s) [EDRF]) and (2) the release of a still unidentified diffusable constrictor factor ($EDCF_3$) which either facilitates (permits) the direct action of anoxia or by itself triggers vasoconstriction, presumably by enhancing Ca^{2+}-influx into vascular smooth muscle cells. The direct action of anoxia on smooth muscle may be due to inhibition of cyclooxygenase (and the production of vasodilator prostanoids, like PGI_2) and a consequent shift of arachidonic acid metabolism to the lipoxygenase pathway, which produces hydroperoxide intermediates (which may exert a positive feedback on Ca^{2+} entry) and vasoconstrictor leukotrienes.

C. Unidentified Diffusable Vasoconstrictor Factor ($EDCF_3$) Released from Native Endothelial Cells during Anoxia and Hypoxia

The important role played by the endothelium in anoxic facilitation of contractile responses in several canine blood-vessel preparations was explained by the following two equally possible mechanisms: (1) anoxia inhibits the tonic synthesis/release of relaxing mediators (e.g., PGI_2, EDRF) from endothelial cells; and (2) anoxia stimulates the release of a diffusable vasoconstrictor mediator from endothelial cells which facilitates or triggers smooth muscle contraction[6,23,24] (Figure 3). Production of prostacyclin in coronary arteries is inhibited at low oxygen tension (less than 5 mmHg).[54,55] However, moderate reduction of oxygen tension (to 35 to 75 mmHg, which does not inhibit, but rather facilitates cyclooxygenase activity[26,54,55]) also evoked endothelium-dependent contractions in canine coronary and femoral arteries.[23] In addition, endothelium-dependent anoxic facilitation of contractions in canine femoral,[23,24] coronary,[22,23] and basilar arteries[25] was not affected by different inhibitors of cyclooxygenase. Anoxia prevents endothelium-dependent relaxations evoked by acetylcholine in the rabbit aorta[2] and canine femoral artery,[24] and by vasopressin and thrombin in the canine basilar artery.[25] It was postulated that the synthesis/release of EDRF requires the presence of oxygen in several (but not all, e.g., canine coronary artery[23]) blood vessels, suggesting that the anoxic facilitation may be the consequence of depressed basal release of EDRF. However, inhibitors of production (e.g., quinacrine) or action on smooth muscle of EDRF (e.g., methylene blue), and inactivators of EDRF in transit (e.g., phenidone), did not affect endothelium-dependent contractions to anoxia in canine coronary arteries.[23] These observations strongly supported the hypothesis that endothelium-dependent anoxic facilitation is not due to reduced release of a relaxing mediator, but rather due to the production of a facilitatory modulator and/or a vasoconstrictor factor by native endothelial cells. The hypothesis that the endothelium produces and releases a diffusable vasoconstrictor factor during anoxia was directly substantiated by transfer experiments, where a circumferential coronary artery strip without endothelium was layered (intimal surface against intimal surface) with a longitudinal strip of the same artery with endothelium.[23] Anoxia facilitated the contractile response of the circumferential strip without endothelium only when it was layered with a longitudinal strip with endothelium, demonstrating that anoxic facilitation must be mediated (at least in part) by a diffusable vasoconstrictor substance.[23] Inhibitors of phospholipase A_2 (quinacrine), cyclooxygenase (indomethacin), and lipoxygenase (NDGA, phenidone) did not affect endothelium-dependent anoxic contractions in layered canine coronary arteries.[23] It was therefore postulated that the diffusable endothelium-derived vasoconstrictor factor released during anoxia is not a vasoactive cyclooxygenase or lipoxygenase metabolite of arachidonic acid. Similarly, the possibility was ruled out that the vasoconstrictor factor is norepinephrine, acetylcholine, serotonin, histamine, adenine nucleotide, or a free oxygen radical, by using appropriate pharmacological antagonists.[23,25,56] Although trypsin and α-chymotrypsin treatment of layered coronary or femoral artery preparations did not block the endothelium-dependent contractions to anoxia,[56] the possibility that the vasoconstrictor factor is similar to the polypeptide-like constrictor factor produced in cultured endothelial cells[50] cannot be ruled out with certainty. Indeed, hypoxia stimulates the production of the vasoconstrictor polypeptide factor by cultured bovine aortic endothelial cells[58] (Figure 2). Thus, the exact chemical nature of the diffusable vasoconstrictor mediator released from anoxic/hypoxic native endothelial cells remains to be determined.

IV. POSSIBLE MECHANISMS BY WHICH ENDOTHELIUM-DERIVED VASOCONSTRICTOR FACTORS FACILITATE CONTRACTION OF VASCULAR SMOOTH MUSCLE

A. Metabolites of Arachidonic Acid

Indomethacin sensitive endothelium-dependent contractions in canine basilar arteries in response to high K^+[12] or stretch[18] can be prevented by the Ca^{2+}-channel antagonist diltiazem. It was therefore postulated that either (1) entry of Ca^{2+} into endothelial cells is an essential step in the production of vasoconstrictor *cyclooxygenase* metabolites, and/or (2) endothelium-derived cyclooxygenase products induce contraction by facilitating the influx of extracellular Ca^{2+} into vascular smooth muscle. *Lipoxygenase* products of arachidonic acid metabolism (suggested to be involved in endothelium-mediated contraction to arachidonic acid in canine pulmonary veins[40] see also Section III.A.1) are potent vasoconstrictors[46-48] and the hydroperoxide intermediate products act as calcium ionophores.[44,45] It is thus conceivable to assume that endothelium-dependent contractions mediated by lipoxygenase products may also be the consequence of facilitated Ca^{2+}-entry into vascular smooth muscle cells.[43]

B. Vasoconstrictor Polypeptide Factor

The vasoconstriction induced by conditioned culture media of bovine aortic endothelial cells in bioassay porcine coronary artery rings was of slow onset.[50] Most of the bioassay vessels did not relax completely to baseline after extensive washing, illustrating a substantial tonic component of the contractile response.[50] These findings suggest that the polypeptide-like vasoconstrictor factor produced by cultured endothelial cells may facilitate vascular smooth muscle contraction via cellular mechanisms different from those triggered by the endothelium-derived vasoconstrictor factor released from native endothelium in response to anoxia/hypoxia, since anoxic contractions developed more rapidly and were immediately reversed upon readdition of oxygen to the organ bath.[23,24] The polypeptide vasoconstrictor apparently requires the influx of extracellular Ca^{2+} into vascular smooth muscle cells of the bioassay coronary artery rings, since (1) they were unresponsive to the conditioned medium when incubated in Ca^{2+}-free Krebs solution containing 1 mM EGTA, and (2) the contractile response was blocked by the Ca^{2+}-channel antagonist verapamil[50] (Figure 2).

C. Anoxic Vasoconstriction

Studies on different isolated canine veins and arteries suggested that anoxia can cause contraction not only through the endothelium but also by a direct action on the vascular smooth muscle. The latter mechanism is seen only when the smooth muscle is partially activated by a constrictor agonist or by indomethacin, but not in unstimulated (quiescent) blood vessel preparations.[6,22,25] The possibility of tonic release of a modulator factor which facilitates (or permits) this direct action of anoxia on vascular smooth muscle cells was proposed on the basis of the following observations: (1) in quiescent femoral artery rings with endothelium (in the absence of basal myogenic tone) anoxia did not evoke contractions[24] and (2) in most canine blood vessels, removal of the endothelium did not abolish but only depressed the anoxic facilitation.[22,24] In contrast to the endothelium-dependent component,[23] the direct smooth muscle action of anoxia could be abolished by inhibition of lipoxygenase.[22,43] Thus, in some vascular smooth muscle prepartions, inhibition of cyclooxygenase (by anoxia or by indomethacin) probably directs endogenous arachidonic acid metabolism into the lipoxygenase pathway, and the formed metabolites activate the contractile process, presumably by acceleration of calcium entry[22,25,60,61] (Figure 3). Anoxic facilitation of the contractile response to acetylcholine in canine saphenous veins is inhibited by the calcium-channel antagonist verapamil.[59] Calcium-entry blockers also depress the augmentation by anoxia of contractions in canine coronary arteries[21,22] and in canine basilar arteries.[25] Removal of calcium from the Krebs Ringer solution for 40 min also abolished the anoxia-induced contractions in canine coronary arteries, which could be restored by readdition of calcium.[22]

A recent study in isolated perfused rat lungs demonstrated that the 1,4-dihydropyridine derivative compound Bay K 8644, which promotes Ca^{2+} influx through voltage-dependent Ca^{2+} channels,[62-64] significantly augmented hypoxic pulmonary vasoconstriction,[65] which is mediated (at least in part) by endothelium-derived vasoconstrictor factor(s).[6,27] In addition, hypoxic pulmonary vasoconstriction is inhibited by various Ca^{2+}-channel antagonists.[66-68] These data strongly support the hypothesis, that the mechanism of anoxic vasoconstriction involves activation of voltage-dependent Ca^{2+} channels in the vascular smooth muscle (Figure 3). It remains to be determined whether the Ca^{2+}-channel antagonists act on the endothelial cells as well, and prevent anoxia-induced influx of Ca^{2+}, which may be an initial step in the production of the still unidentified endothelium-derived vasoconstrictor factor(s).

Anoxic facilitation cannot be observed in rings of canine femoral veins with endothelium.[6,23] However, experiments with mixed layered preparations (using coronary artery strips without endothelium and femoral vein strips with endothelium) demonstrated that anoxia stimulated the production of a diffusable vasoconstrictor factor also in endothelial cells of this vein.[23] This observation implies that although similar vasoconstrictor factor(s) may be released from anoxic arterial and venous endothelium, differences exist in the responsiveness of arterial and venous smooth muscle to the vasoconstrictor mediator(s).

Thus, in canine arteries and veins, anoxia/hypoxia facilitates contractions by acting on endothelial cells and the vascular smooth muscle. In endothelial cells of some (but not all) blood vessels, it presumably depresses the basal production of prostacyclin and EDRF, and at the same time triggers the production/release of still unidentified endothelium-derived constrictor factor(s) ($EDCF_3$). It is still uncertain whether this factor initiates contractions or facilitates (permits) the direct action of anoxia/hypoxia on vascular smooth muscle. The prevention of anoxic contraction by Ca^{2+}-channel antagonists and removal of extracellular Ca^{2+} suggests that either (1) anoxia stimulates the influx of Ca^{2+} into endothelial cells, which is essential for the production/release of EDCF(s); and/or (2) EDCF activates contractions by enhancing Ca^{2+} influx into vascular smooth muscle cells. In the already activated smooth muscle (either by the endothelium-dependent mechanism or by exogenous vasoconstrictors), anoxia presumably facilitates the formation of lipoxygenase products which may exert a positive feedback on the entry of calcium (Figure 3).

V. CONCLUDING REMARKS

Ample experimental evidence obtained in various blood vessels from different species in the past 5 years now clearly demonstrates that the endothelium can modulate the tone of underlying vascular smooth muscle not only by the synthesis and release of relaxing mediators but also by the production of factors which facilitate or trigger smooth muscle contraction. It also became obvious that endothelium-dependent facilitation of vascular smooth muscle contraction is a heterogenous phenomenon with regard to (1) the species, (2) the anatomical origin of the blood vessel in the same species, (3) the agonists and experimental conditions which evoke the response, and (4) the chemical nature of endothelium-derived vasoconstrictor factors which mediate the response. Although transfer experiments clearly indicated that endothelial cells (in culture of *in situ*) can produce and release diffusable vasoconstrictor mediators, their nature is still uncertain. The term ''endothelium-derived constrictor factor (EDCF)'' represents at least three different classes of vasoactive substances: (1) metabolites of arachidonic acid ($EDCF_1$), (2) a polypeptide-like factor (or factors) produced by cultured endothelial cells ($EDCF_2$), and (3) a still unidentified diffusable factor (or factors) released from anoxic/hypoxic native endothelial cells ($EDCF_3$). With the exception of the polypeptide-like factor, it is still unknown whether (1) the endothelium facilitates the contractions by the continuous release of facilitatory modulators (which augment or permit the direct smooth muscle action of various agonists and anoxia), or (2) certain agonists, stretch, and anoxia stimulate the production of endothelial factors which can (by themselves) trigger smooth muscle contraction. The prevention of endothelium-dependent facilitation of contractions in various blood vessels by Ca^{2+}-channel antagonists and by removal of extracellular Ca^{2+} suggest that either (1) influx of Ca^{2+} into endothelial cells is an essential initial step in the production or release of the vasoconstrictor factors, and/or (2) the factors facilitate or activate contractions by enhancing Ca^{2+} influx into vascular smooth muscle cells.

It is too early to evaluate the physiological and pathological significance of endothelium-mediated contraction of blood vessels. First it must be demonstrated that this indirect mechanism for contracting vascular smooth muscle is involved in the actions of various stimuli (such as increase in transmural pressure, elevated extracellular K^+, liberation of endogenous arachidonic acid, anoxia, etc.) on resistance vessels that modulate blood flow in various vascular beds. If such a mechanism is involved, then one can speculate that it may play a role in the regulation of vascular smooth muscle tone and hence vascular resistance under normal and pathological conditions.

Already there has been some speculation based on this concept. Endothelium-mediated contraction in response to stretch and rapid increase in transmural pressure was postulated to be involved in the so-called "mygogenic response", which may contribute to autoregulation of blood flow in certain vascular beds.[18,20] The loss of the ability of endothelial cells in a given vascular bed to generate EDCF(s) in response to stretch (e.g., as a consequence of chronic exposure to high blood pressure) would explain the lack of autoregulation observed in perfused hearts isolated from SHR.[69]

Excessive production of vasoconstrictor mediators in combination with depressed release of EDRFs may contribute to the development and/or maintenance of elevated peripheral vascular resistance in certain forms of hypertension.[16,17] The initiation of contraction of coronary and cerebral arteries by anoxia/hypoxia would favor the occurrence or the maintenance of vasospasm in proximal arterial segments of these vascular beds under hypoxic/ischemic conditions.

The possibility should also be considered that certain drugs may produce vasoconstriction via endothelial cells. For instance, in canine coronary arterial rings[70] (and presumably in the pulmonary arteries of perfused rat lungs[65]) the contractile effect of the Ca^{2+}-channel agonist drug Bay K 8644 is dependent in part on the presence of endothelium.

Irrespective of the physiological, pathological, and pharmacological significance of endothelium-mediated vasoconstriction, characterization of the cellular mechanisms of endothelial production and action on vascular smooth muscle of endothelium-derived vasoconstrictor factors and their chemical identification remain the major goals for future studies in this new area of cardiovascular research.

REFERENCES

1. **Furchgott, R. F.,** Role of the endothelium in responses of vascular smooth muscle, *Circ. Res.,* 53, 557, 1983.
2. **Furchgott, R. F. and Zawadzki, J. V.,** The obligatory role of endothelial cells in the relaxation of arterial smooth muscle by acetylcholine, *Nature (London),* 288, 373, 1980.
3. **Peach, M. J., Loeb, A. L., Singer, H. A., and Saye, J. A.,** Endothelium-derived vascular relaxing factor, *Hypertension,* 7 (Suppl. 1), 194, 1985.
4. **Vanhoutte, P. M., Rubanyi, G. M., Miller, V. M., and Houston, D. S.,** Modulation of vascular smooth muscle contraction by the endothelium, *Ann. Rev. Physiol.,* 48, 307, 1986.
5. **De Mey, J. G. and Vanhoutte, P. M.,** Role of the intima in cholinergic and purinergic relaxation of isolated canine femoral arteries, *J. Physiol.,* 316, 347, 1981.
6. **De Mey, J. G. and Vanhoutte, P. M.,** Heterogeneous behavior of the canine arterial and venous wall: importance of the endothelium, *Circ. Res.,* 51, 439, 1982.
7. **Bevan, J. A. and Duckles, S. P.,** Evidence for alpha-adrenergic receptors on intimal endothelium, *Blood Vessels,* 12, 307, 1975.
8. **Kalsner, S.,** Differential activation of the inner and outer muscle cells layers of the rabbit ear artery, *Eur. J. Pharmacol.,* 20, 122, 1972.
9. **Pascual, R. and Bevan, J. A.,** Asymmetry of consequences of drug disposition mechanisms in the wall of the rabbit aorta, *Circ. Res.,* 46, 22, 1980.
10. **Garland, C. J. and Keatinge, W. R.,** Constrictor actions of acetylcholine, 5-hydroxytryptamine and histamine on bovine coronary artery inner and outer muscle, *J. Physiol. (London),* 327, 363, 1982.
11. **Cohen, R. A., Shepherd, J. T., and Vanhoutte, P. M.,** Endothelium and asymmetrical responses of the coronary arterial wall, *Am. J. Physiol.,* 247, H403, 1984.
12. **Katusic, Z. S., Shepherd, J. T., and Vanhoutte, P. M.,** Potassium induces endothelium-dependent rhythmic activity in the canine basilar artery, *The Pharmacologist,* 27 (Abstr.), 223, 1985.

13. **Katusic, Z. S.,** personal communication, 1986.
14. **Konishi, M. and Su, C.,** Role of endothelium in dilator responses of spontaneously hypertensive rat arteries, *Hypertension,* 5, 881, 1983.
15. **Winquist, R. J., Bunting, P. B., Baskin, E. P., and Wallace, A. A.,** Decreased endothelium-dependent relaxation in New Zealand genetic hypertensive rats, *J. Hypertension,* 2, 541, 1984.
16. **Luscher, T. F. and Vanhoutte, P. M.,** Endothelium-dependent contractions to acetylcholine in the aorta of the spontaneously hypertensive rat, *Hypertension,* 8, 344, 1986.
17. **Luscher, T. F. and Vanhoutte, P. M.,** Endothelium-dependent responses to platelets and serotonin in spontaneously hypertensive rats, *Hypertension,* 8(Suppl. II), 55, 1986.
18. **Katusic, Z. S., Shepherd, J. T., and Vanhoutte, P. M.,** Endothelium-dependent contraction to stretch in canine basilar arteries, *Fed. Proc.,* 289 (Abstr.), 784, 1986.
19. **Stroudgaard, S. and Pauson, O.,** Cerebral autoregulation, *Stroke,* 15, 413, 1984.
20. **Rubanyi, G. M.,** Release of vasoactive mediators from endothelial cells by changes in transmural pressure — potential role in autoregulation (abstract), 10th World Congress of Cardiology, Washington, D. C., 1986.
21. **Van Nueten, J. M. and Vanhoutte, P. M.,** Effect of the Ca^{2+} antagonist lidoflazine on normoxic and anoxic contractions of canine coronary arterial smooth muscle, *Eur. J. Pharmacol.,* 64, 173, 1980.
22. **Rimele, T. J. and Vanhoutte, P. M.,** Effects of inhibitors of arachidonic acid metabolism and calcium entry on hypoxic contractions of the isolated canine coronary artery, in *Calcium Antagonists: Mechanism of Action on Cardiac Muscle and Vascular Smooth Muscle,* Sperelakis, N. and Canfield, J. B., Eds., Martinus Nijhoff, The Hague, 1984, 303.
23. **Rubanyi, G. M. and Vanhoutte, P. M.,** Hypoxia releases a vasoconstrictor substance from the canine vascular endothelium, *J. Physiol.,* 364, 45, 1985.
24. **De Mey, J. G. and Vanhoutte, P. M.,** Anoxia and endothelium dependent reactivity of the canine femoral artery, *J. Physiol.,* 335, 65, 1983.
25. **Katusic, Z. S. and Vanhoutte, P. M.,** Anoxic contractions in isolated canine cerebral arteries. Contribution of endothelium-derived factors, metabolites of arachidonic acid and calcium entry, *J. Cardiovasc. Pharmacol.,* in press.
26. **Rubanyi, G. and Paul, R. J.,** Two distinct effects of oxygen on vascular tone in isolated porcine coronary arteries, *Circ. Res.,* 56, 1, 1985.
27. **Holden, W. E. and McCall, E.,** Hypoxic vasoconstriction of porcine pulmonary aortic strips in vitro requires an intact endothelium, *Am. Rev. Respir. Dis.,* 127 (Abstr.), 301, 1983.
28. **Moncada, S., Herman, A. G., Higgs, E. A., and Vane, J. R.,** Differential formation of prostacyclin (PGX or PGI_2) by layers of the arterial wall. An explanation for the antithrombotic properties of vascular endothelium, *Thromb. Res.,* 11, 323, 1977.
29. **De Mey, J. G., Claeys, M., and Vanhoutte, P. M.,** Endothelium-dependent inhibitory effects of acetylcholine, adenosine triphosphate, thrombin and arachidonic acid in the canine femoral artery, *J. Pharmacol. Exp. Ther.,* 222, 166, 1982.
30. **MacIntyre, D. E., Pearson, J. D., and Gordon, J. L.,** Localization and stimulation of prostacyclin production in vascular cells, *Nature (London),* 271, 549, 1978.
31. **Kuhn, H., Ponicke, K., Halle, W., Schewe, T., and Forster, W.,** Evidence for the presence of lipoxygenase pathway in cultured endothelial cells, *Biomed. Biochim. Acta,* 42, K1, 1983.
32. **Johnson, A., Revtyak, G., and Campbell, W.,** Arachidonic acid metabolites and endothelial injury: studies with cultures of human endothelial cells, *Fed. Proc.,* 44, 19, 1985.
33. **Abraham, N. G., Pinto, A., Mullane, K. M., Levere, R. D., and Spokas, E.,** Presence of cytochrome P-450-dependent monooxygenase in intimal cells of the hog aorta, *Hypertension,* 7, 899, 1985.
34. **Rubanyi, G. M. and Vanhoutte, P. M.,** Ouabain inhibits endothelium-dependent relaxations to arachidonic acid in canine coronary arteries, *J. Pharmacol. Exp. Ther.,* 235, 81, 1985.
35. **Dusting, G. J., Moncada, S., and Vane, J. R.,** Prostacyclin (PGI_2) is a weak contractor of coronary arteries of the pig, *Eur. J. Pharmacol.,* 45, 301, 1977.
36. **Gordon, J. L. and Martin, W.,** Stimulation of endothelial prostacyclin production plays no role in endothelium-dependent relaxation of the pig aorta, *Br. J. Pharmacol.,* 80, 179, 1983.
37. **Levy, J. V.,** Contractile responses to prostacyclin (PGI_2) of isolated human saphenous and rat venous tissue, *Prostaglandin,* 16, 93, 1978.
38. **Pomerantz, K., Sintetos, A., and Ramwell, P.,** The effect of prostacyclin on the human umbilical artery, *Prostaglandins,* 15, 1035, 1978.
39. **Salzman, P. M., Salmon, J. A., and Moncada, S.,** Prostacyclin and thromboxane A_2 synthesis by rabbit pulmonary artery, *J. Pharmacol. Exp. Ther.,* 215, 240, 1980.
40. **Miller, V. M. and Vanhoutte, P. M.,** Endothelium-dependent contractions to arachidonic acid are mediated by products of cyclooxygenase, *Am.J. Physiol.,* 248, H432, 1985.
41. **Singer, H. A. and Peach, M. J.,** Endothelium-dependent relaxation of rabbit aorta. I. Relaxation stimulated by arachidonic acid, *J. Pharmacol. Exp. Ther.,* 226, 790, 1983.

42. **Skidgel, R. A. and Printz, M. P.,** PGI_2 production by rat blood vessels: diminished prostacyclin formation in veins compared to arteries, *Prostaglandins,* 16, 1, 1978.

43. **Rimele, T. J. and Vanhoutte, P. M.,** Effects of inhibitors of arachidonic acid metabolism and calcium entry on responses to acetylcholine, potassium and norepinephrine in the isolated canine saphenous vein, *J. Pharmacol. Exp. Ther.,* 225, 720, 1983.

44. **Asano, M. and Hidaka, H.,** Contractile response of isolated rabbit aortic strips to unsaturated fatty acid peroxides, *J. Pharmacol. Exp. Ther.,* 208, 347, 1979.

45. **Koide, T., Neichi, T., Takato, M., Matsushita, H., Sugioka, K., Nakano, M., and Hata, S. I.,** Possible mechanisms of 15-hydroperoxyarachidonic acid-induced contraction of the canine basilar artery in vitro, *J. Pharmacol. Exp. Ther.,* 221, 481, 1982.

46. **Hanna, C. J., Bach, M. K., Pare, P. D., and Schellenberg, R. R.,** Slow-reacting substances (leukotrienes) contract human airway and pulmonary vascular smooth muscle *in vitro, Nature (London),* 290, 343, 1981.

47. **Woodman, D. L. and Dusting, G. J.,** Coronary vasoconstriction induced by leukotrienes in the anesthetized dog, *Eur. J. Pharmacol.,* 86, 125, 1983.

48. **Forstermann, U. and Neufang, B.,** C-6-sulfidopeptide leukotrienes are unlikely to be involved in the endothelium dependent relaxation of rabbit aorta by acetylcholine, *Prostaglandins,* 27, 181, 1984.

49. **Kontos, H. A., Nei, E. P., Povlistock, J. T., Dietrich, W. D., Magiene, C. T., and Ellis, E. F.,** Cerebral arteriolar damage by arachidonic acid and prostaglandin G_2, *Science,* 209, 1242, 1980.

50. **Hickey, K. A., Rubanyi, G. M., Paul, R., and Highsmith, R. F.,** Characterization of a coronary vasoconstrictor produced by cultured endothelial cells, *Am. J. Physiol.,* 248, C550, 1985.

51. **O'Brien, R. F. and McMurtry, I. F.,** Endothelial cell supernates contract bovine pulmonary artery rings *Am. Rev. Respir. Dis.,* 129 (Abstr.), A337, 1984.

52. **Ryan, U. S.,** Processing of angiotensin and other peptides by the lungs, in *Handbook of Physiology: The Respiratory System I,* 351.

53. **Dzau, V. J.,** Vascular wall renin-angiotensin pathway in control of the circulation, *Am. J. Med.,* 77, 31, 1984.

54. **Kalsner, S.,** The effect of hypoxia on prostagland in output and on tone in isolated coronary arteries, *Can. J. Physiol. Pharmacol.,* 55, 882, 1977.

55. **Roberts, A. M., Messina, E. J., and Kaley, G.,** Prostacyclin (PGI_2) mediates hypoxic relaxation of bovine coronary arterial strips, *Prostaglandins,* 21, 555, 1981.

56. **Rubanyi, G. M.,** unpublished data, 1986.

57. **Rubanyi, G. M.,** unpublished data, 1986.

58. **Highsmith, R. F.,** personal communication, 1986.

59. **Vanhoutte, P. M.,** Effects of anoxia and glucose depeletion on isolated veins of the dog, *Am. J. Physiol.,* 230, 1261, 1976.

60. **Higgs, G. A., Mugridige, K. G., and Moncada, S.,** Arachidonic acid metabolism and calcium flux, in *Calcium Entry Blockers and Tissue Protection,* Godfraind, T. et al., Eds., Raven Press, New York, 1985, 51.

61. **Vanhoutte, P. M., Rimele, T. J., and Flavahan, N. A.,** Lipoxygenase and calcium entry in vascular smooth muscle, *J. Cardiovasc. Pharmacol.,* in press.

62. **Schramm, M., Thomas, G., Towart, R., and Franckowiak, G.,** Activation of calcium channels by novel 1,4-dihydropyridines, a new mechanism for positive inotropics or smooth muscle stimulants, *Arzneim. Forsch.,* 33, 1268, 1983.

63. **Su, C. M., Swamy, V. C., and Triggle, D. J.,** Calcium channel activation in vascular smooth muscle by Bay K 8644, *Can. J. Physiol. Pharmacol.,* 62, 1401, 1984.

64. **Yamamoto, H., Hwang, O., and Van Breemen, C.,** Bay K 8644 differentiates between potential and receptor operated Ca^{2+} channels, *Eur. J. Pharmacol.,* 102, 555, 1984.

65. **McMurtry, I. V.,** Bay K 8644 potentiates and A23187 inhibits hypoxic vasoconstriction in rat lungs, *Am. J. Physiol.,* 249, H741, 1985.

66. **McMurtry, I. F., Davidson, A. B., Reeves, J. T., and Grover, R. F.,** Inhibition of hypoxic pulmonary vasoconstriction by calcium antagonists in isolated rat lungs, *Circ. Res.,* 38, 99, 1976.

67. **Tucker, A., McMurtry, I. F., Grover, R. F., and Reeves, J. T.,** Attenuation of hypoxic pulmonary vasoconstriction by verapamil in intact dogs, *Proc. Soc. Exp. Biol. Med.,* 151, 611, 1976.

68. **Simmoneau, G., Escourrou, P., Duroux, P., and Lockhart, A.,** Inhibition of hypoxic pulmonary vasoconstriction by nifedipine, *N. Engl. J. Med.,* 304, 1582, 1981.

69. **Edoute, Y., Luscher, T. F., and Rubanyi, G. M.,** Autoregulation and vascular reserve in the cornary circulation of the spontaneously hypertensive rat, *J. Hypertension,* 4 (Suppl. 5), S290, 1987.

70. **Rubanyi, G. M.,** unpublished data, 1986.

Chapter 33

INTERCELLULAR JUNCTIONS AND JUNCTIONAL TRANSFER IN THE BLOOD VESSEL WALL

David M. Larson

TABLE OF CONTENTS

I. INTRODUCTION

The vascular endothelium provides a dynamic interface between two compartments: plasma and interstitial fluid. The other cellular and acellular components of the blood vessel wall interact with the endothelium to modulate, regulate, and maintain this interface. Central to the interrelationships of the cellular components of the vessel wall are intercellular junctions. Of the various types of defined junctions in vertebrate tissues,[1] tight and gap junctions are of particular importance in the normal function of the vessel wall.

Tight junctions are stuctures located at the interfaces between epithelial cells and between endothelial cells. The major biological functions of tight junctions are (1) to form a barrier to paracellular permeability,[1-3] (2) to maintain cellular apical-basal polarity,[3-5] and (3) to aid in intercellular adhesion.[1,6] Gap junctions, on the other hand, are nearly ubiquitous among animal cell types,[1,7] and provide adjacent cells with a direct pathway for exchange of ions and small molecules.[1,7,8]

This review is concerned with the morphology and function of junctions between endothelial cells, smooth muscle cells, and pericytes, and heterocellular junctional interactions among these cell types. However, the morphology and physiology of interendothelial tight junctions has been recently reviewed,[3,5,9-12] and will not be described in depth here.

II. INTERENDOTHELIAL JUNCTIONS

Interendothelial gap and tight junctions show a variable distribution throughout the vascular tree.[9,10,13-19] In general, gap junctions are large and frequent on the arterial side of the circulation, smaller and more rare on the venous side, and absent, in most cases, from capillaries and pericytic venules. Tight junctions are generally extensive in arteries, arterioles, and capillaries and veins, and discontinuous in venules.

A. Endothelial Tight Junctions

Both endothelial and epithelial tight junctions in thin-section electron micrographs appear as punctate contacts or fusions of the outer leaflets of the membranes of two adjacent cells[1,3,13-15,20,21] (Figure 1A). However, freeze-fracture electron microscopy reveals some differences between the two cell types. Epithelial tight junctions appear as belt-like, interlacing networks of strands on the P-face (protoplasmic) with complementary grooves on the E-face (exoplasmic).[1,3] In contrast, endothelial tight junctions appear as a series of E-face grooves, with (often) discontinuous rows of particles, and complementary P-face ridges, relatively devoid of particles[13-15,19,21] (Figure 1B). These ultrastructural differences in epithelial and endothelial tight junctions may reflect differences in tight junctional structure and composition, a subject about which there is some controversy.[3]

The variable distribution of tight junctions throughout the vascular tree[9,10,13-17] seems to be generally correlated with the variable permeability characteristics of endothelia.[3,5,11,22] For instance, brain microvascular endothelium, the site of the blood-brain barrier, is a tight endothelium,[5,11,23-25] and has extensive tight junctional networks.[23,24,26] On the other hand, the endothelium of muscle capillaries[27] and arteries[28] is leaky. These leaks have been attributed to discontinuities in the tight junctional networks.[12,22,28] In general, venular endothelium is very leaky, with very discontinuous tight junctions.[29]

B. Endothelial Gap Junctions

Intercellular gap junctions[30] are commonly distributed in animal tissues.[7,31] Although a great deal of information on gap junction structure and composition is known,[1,7,32-34] their biological roles are still only poorly understood.

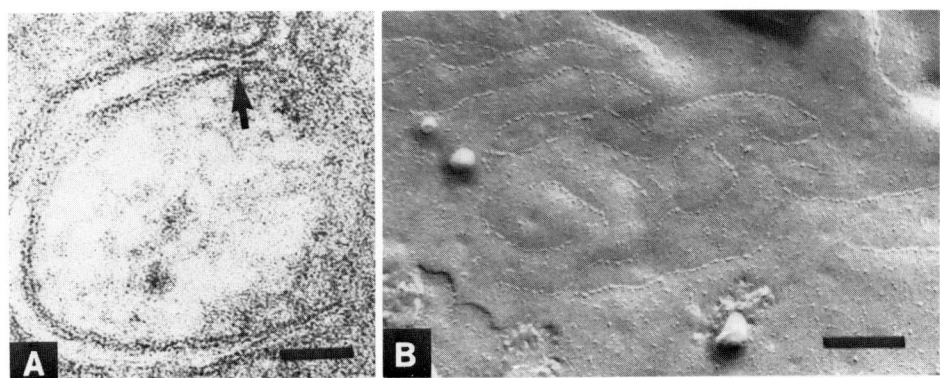

FIGURE 1.   Interendothelial tight junctions. (A) Thin-section electron micrograph of a tight junction (arrow) between two freshly isolated bovine umbilical vein endothelial cells (bar = 50 nm; photo courtesy of J. D. Sheridan). (B) Freeze-fracture micrograph of tight junctions in a bovine brain microvascular endothelial culture; note E-face grooves with particles (bar = 200 nm).

1. Structural Studies

The ultrastructural appearance of gap junctions (Figure 2A) in optimal thin-section electron micrographs is that of a septilaminar structure (15 to 19 nm thick) composed of the plasma membranes of two cells separated by a 2- to 3-nm "gap".[1,30] Freeze-fracture replicas show polygonal arrays of P-face particles (8 to 10 nm in diameter) with complementary E-face pits,[1,7,16,17,23,25] usually in macular plaques (Figure 2B and D). Detailed structural analyses have resulted in models for the gap junctional morphological unit, the connexon.[32,35] It is a paired, symmetric hexamer of protein subunits (about 27 kdaltons in liver, 47 kdaltons in heart)[36-38a] with a hydrophilic core channel (1.5 to 3 nm in diameter)[35] that links the cytoplasm of adjacent cells. Antibodies and cDNA probes have been produced to gap junction proteins[33,36,37,39] and message[39a,b] and show variable species and tissue cross-reactivity.

Gap junction distribution is variable throughout vascular endothelia.[9,10,13-15,18,19] In general,[9] gap junctions are large and frequent in arteries and arterioles,[13,14,18,19,26,28,40] and generally smaller and more rare in veins and muscular venules.[13,14,19] The general observation that gap junctions are absent in post capillary venules and capillaries[16,23,28,48,49] has several exceptions.[41-44]

There is also variability in the association of gap and tight junctions in different endothelia[13,14] (Figure 2B through D). Arterial gap junctions are predominantly intercalated into tight junctional networks while venous gap junctions are isolated.[9,13,14] In some instances, endothelial gap junctions are seen in freeze-fracture replicas as anastomosing ribbons of P-face particles[14,21] (Figure 2C). The significance of these and other[13-15,19] variant junctional morphologies is unclear.

Intercellular transfer of ions and small molecules has consistantly been correlated with the presence of gap junctions.[1,7,8,31,34,45-47] However, it is important to note that the ultrastructural detection of gap junctions does not necessarily mean that the cells involved were in direct communication since the patency of gap junctional channels is known to be under complex control.[8,31,32,35]

2. Functional Studies

Three different types of assays for junctional transfer have been widely used.[48] Electrical coupling, detected using intracellular microelectrodes, is the most sensitive assay for junctional transfer,[34,45,48] and can provide information on junctional ionic conductance. Dye transfer assays involve intracellular injection of fluorescent tracer molecules,[21,48-51] and monitoring of the spread of dye to adjacent cells. In addition to demonstrating transfer,

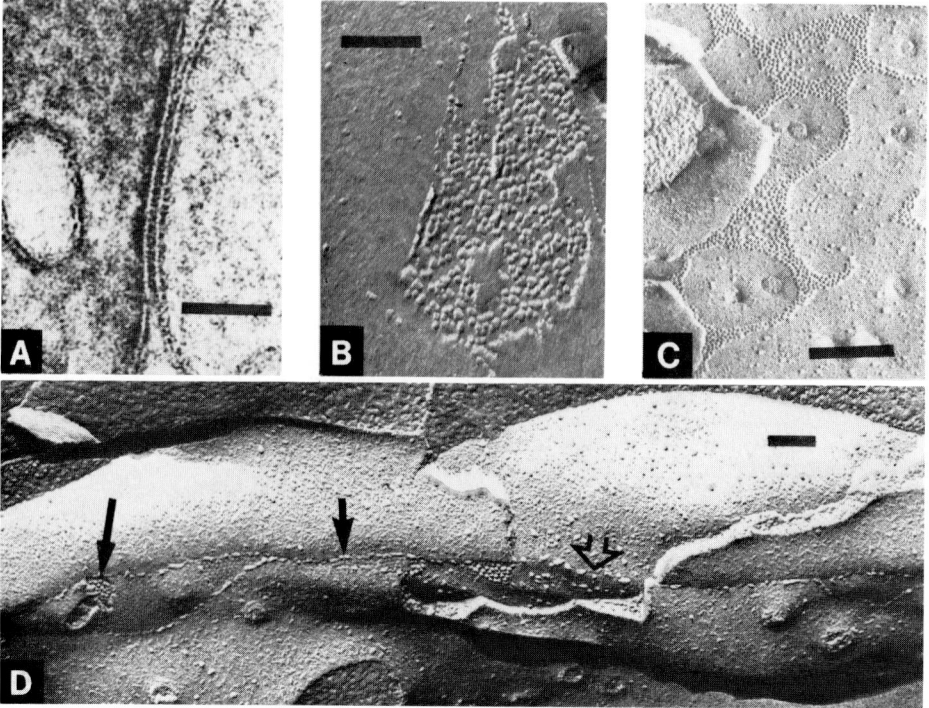

FIGURE 2.   Interendothelial gap junctions. (A) Thin-section micrograph of gap junction between two bovine aortic endothelial cells in vivo (bar = 50 nm). (B) Freeze-fracture micrograph of gap junction between cultured bovine brain microvascular endothelial cells (bar = 100 nm). (C) Freeze-fracture micrograph of P-face gap junction particles in anastomosing strands, freshly isolated bovine umbilical vein endothelium (bar = 200 nm). (D) Freeze-fracture micrograph of gap junctions (P-face, long arrow; E-face, short arrow) intercalated into tight junction network; note fine groove in P-face tight junction ridge (open arrow), cultured bovine aortic endothelium (bar = 100 nm). (Photos A, B, and D reproduced from *J. Cell. Biol.*, 1982, 92, 183 by copyright permission of The Rockefeller University Press.)

these assays have provided useful information on the molecular size and charge limits for junctional permeation. In mammalian cells, the studies of Flagg-Newton et al.[51] suggested an upper limit of less than 1000 daltons, which corresponds well with channel size estimates from structural studies.[35]

A variety of other assays for junctional transfer have been used in various systems. One of the most useful of these, although it has generally only been applied to cultured cells, is the autoradiographic detection of cellular metabolite transfer, "metabolic cooperation".[52] Originally, this assay could only be used for certain cells deficient in some useful enzyme (e.g., HGPRT);[52] a modification by Pitts and Simms[53] using "donor" cells preloaded with [3H]-uridine, can be used with any combination of cells.

Junctional transfer in endothelium *in situ* has been demonstrated by Sheridan,[54,55] who reported dye transfer between endothelial cells in the microvasculature of the rat omentum (Figure 3). Interestingly, dye transfer occurred between endothelial cells in pericytic venules and capillaries, vessels in which interendothelial gap junctions have not been ultrastructurally demonstrable.[13] We have recently confirmed these observations using freshly isolated bovine brain capillaries.[55a] There are several potential explanations for this discrepancy,[55,56] including dye transfer by endothelial tight junctions, the difficulty of distinguishing very small gap junctions in freeze-fracture replicas, and the possibility of transfer via unaggregated connexons. Williams and DeHaan[57] have demonstrated what might be a similar phenomenon in reaggregated cardiac muscle cells where electrotonic coupling was detected in the absence of ultrastructurally demonstrable gap junctions.

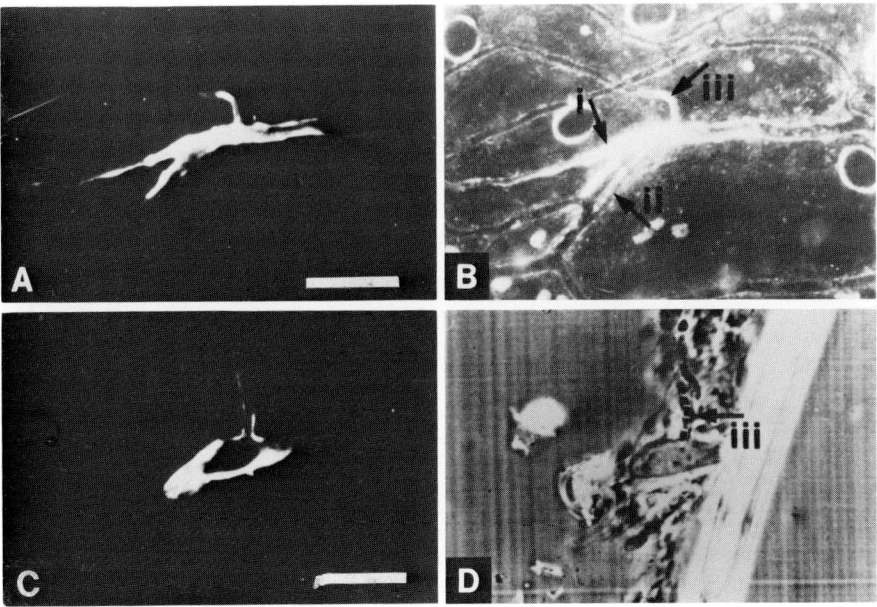

FIGURE 3.   Dye transfer in rat omental microvasculature. (A) and (B) Darkfield fluorescence and white light images of injected venular endothelial cell, with transfer to cells in branches (i, ii, iii). The preparation was fixed, embedded in plastic and sectioned (bar = 100 μm). (C) and (D) Section of above, showing transfer from venular to capillary endothelium (iii); (bar = 50 μm). (From Sheridan, J. D. and Larson, D. M., *The Functional Integration of Cells in Animal Tissues,* Pitts, J. D. and Finbow, M. E., Eds., Cambridge University Press, New York, 1982, 263. With permission.)

The potential roles for junctional communication in endothelium[9,21,49,55,58] parallel those mentioned for other cell types. These include intercellular exchange of nutrients,[47] regulation of growth,[45,46] and differentiation[59] (e.g., maintenance of the monolayer topology in endothelium),[60] cooperative responsiveness to exogenous or endogenous stimuli,[47,61,62] and maintenance of tissue homeostasis.[52] The difficulty of assessing any of these possibilities in vivo has lead to reliance on culture models.

3. Studies on Cultured Endothelium

Gap junctions have been demonstrated in vascular endothelial cultures derived from large[21,64-66] and small[50,56,67,68] vessels. In general, gap junctions in cultured endothelium are similar to those in vivo,[21] with some reductions in size and frequency (possibly due to the use of proteolytic enzymes in culture techniques).[65] One series of studies has provided parallel quantitative ultrastructural[21] and transfer (electrical, dye, and metabolite)[49,58] data on primary cultures of large vessel (bovine aortic and umbilica vein) endothelium. Studies on electrotonic coupling in these cultures[21,58] (Figure 4A) provided estimates of the passive membrane characteristics of the cells, including junctional (aortic, 3.7 × 10^6 Ω) and nonjunctional (6.1 × 10^8 Ω) resistances and specific resistances (12.4 Ωcm²; 1.1 × 10^4 Ωcm²).[58] Dye transfer was used to investigate possible culture-induced changes in junctional transfer.[21,49] Dye (Lucifer yellow) injection in sheets of endothelial cells, freshly isolated from the vessel walls, was compared with injection in established primary monolayer cultures (Figure 4B). Transfer was more extensive in the freshly isolated sheets, indicating that these cells had a greater capacity for junctional transfer, presumably reflecting in part a greater number of patent channels.

Recently, Larson et al.[50,56] have reported on junctional transfer in cultures of bovine brain microvascular endothelial cells.[69] The brain endothelial cells were found to express tight

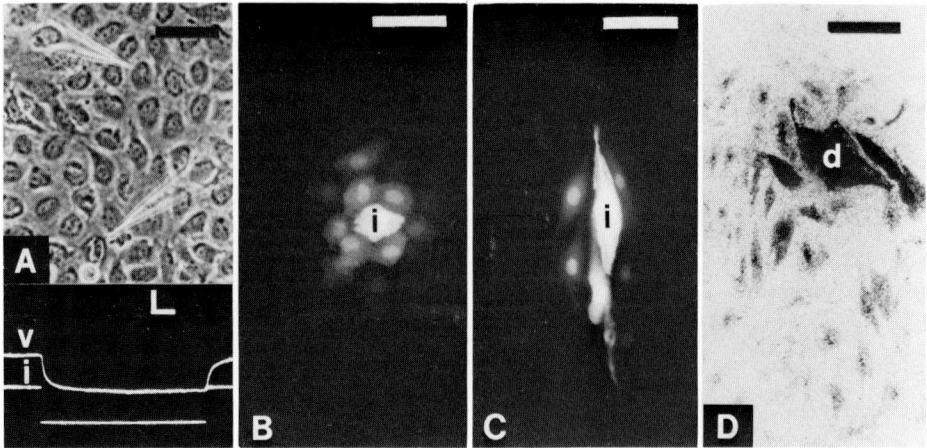

FIGURE 4.   Junctional transfer between cultured endothelial cells. (A) Electrotonic coupling. Current (trace i) injected into one cell via microelectrode produces a membrane potential change (trace v) in another cell; in lower panel, current pulse = 6×10^{-9} A, calibration bracket for voltage trace, horizontal = 40 msec, vertical = 10 mV. (B) Dye transfer. Lucifer yellow injected into bovine aortic endothelial cell (i) transfers to adjacent cells. (C) Dye transfer in cultured bovine brain microvascular endothelial cells. (D) Nucleotide transfer autoradiography. Preloaded (^{3}H-uridine) donor bovine brain microvascular endothelial cell (d) showing transfer to adjacent endothelial cells. (All bars = 40 μm.)

and gap junctions (Figures 1B and 2B) and to engage in interendothelial junctional transfer of Lucifer yellow (Figure 4C) and uridine nucleotides (Figure 4D). The particular significance of this study is that these cultured cells, based on the best available characterization, were derived from capillaries and postcapillary (pericytic) venules,[56,69] vessels in which interendothelial gap junctions have not been ultrastructurally demonstrable.[13,24] Although it is clear that there are culture-induced changes in endothelial junctions, these studies indicate that capillary and pericytic venular endothelium can express functional gap junctions under certain conditions.[54,55]

Junctional transfer studies on cultured endothelium may also shed some light on the role of junctional communication in growth control and the maintenance of the endothelial monolayer topography in vivo. Some studies have suggested that loss of gap junctions is an early event in the regenerative response of endothelium to intimal denudation,[60,70,71] paralleling similar findings in regenerating liver epithelium.[72] However, other research has demonstrated increases in gap junctions in regenerating endothelium at later time points.[73] Preliminary studies in this author's laboratory have suggested that the capability for junctional transfer is only slightly diminished in wounded aortic endothelial cultures,[74] since dye transfer occurred at nearly the same frequency in cells at the wound edge and in cells in control cultures. Endothelial cells seem to maintain patent junctional contact in this stressed condition even while actively migrating to fill the wound. Since cultured endothelial cells hundreds of micrometers from the wound edge are "activated" and show an increased cellular motility,[75,76] it seems likely that a signal is transmitted via gap junctions from cells at the wound edge back to physically unperturbed cells in the monolayer, "informing" them of the disaster.

Hypertension has also been shown to affect interendothelial junctions.[6,77] Studies using both experimentally induced[77] and spontaneous hypertensive[6] rats have suggested that both tight and gap junctions are altered in these animals. These alterations in junctional morphology may represent functional adaptations to increased transmural pressure. Certainly this seems reasonable for the tight junctions.

III. PERIVASCULAR CELL JUNCTIONS

Perivascular cells with close associations with endothelium include smooth muscle (in arteries, arterioles, muscular venules, and veins) and pericytes (in capillaries and postcapillary venules).[9,78] The distribution, morphology, and physiology of gap junctions between smooth muscle cells have received some attention, but junctions between pericytes have not been well defined.

A. Smooth Muscle Junctions

Burnstock's[79] model for autonomic neuromuscular function in blood vessels involves signal conduction through gap junctions between smooth muscle cells. It has been clearly shown that smooth muscle cells are electrically coupled,[80-82] and many studies have demonstrated gap junctions in vascular smooth muscle.[9,16,83-87] Recently, Segal and Duling[88,89] have demonstrated propagation of vasodilation from arterioles to feed arteries with a time course consistant with electrotonic, not diffusional, signal transmission.[88] These data support the earlier suggestion of Hilton[90] concerning conduction of vasodilatory signals through vascular smooth muscle. However, it must be noted that some studies have reported an absence of detectable gap junctions in smooth muscle known to be electrically coupled.[91] This is a conceptually difficult problem which will require further experimental verification.

Junctional transfer in vascular smooth muscle is known to be altered under certain circumstances. For instance, vascular smooth muscle in hypertensive rats appears to have a reduced core resistance in several different models;[92,93] this effect has been postulated to be due to an increased number of intercellular contacts (exact morphology undetermined) in these cells.[94] Grünwald and co-workers[95] reported qualitative increases in the frequency and size of gap junctions between smooth muscle cells cultured from the aortas of hypertensive versus control rats. Changes in the junctions between vessel wall cells in hypertension may therefore be a generalized phenomenon (see Section II). In addition, subendothelial (myointimal) hyperplastic smooth muscle cells express abundant gap junctions,[84,86] and increases in gap junctions have also been noted in hypertrophic nonvascular smooth muscle.[96] These examples of increased frequency and size of gap junctions and increased coupling of smooth muscle cells may be a response to stress.

Aside from the example mentioned above,[95] very few studies on gap junctions or junctional transfer have been reported for cultured vascular smooth muscle cells. The capability for nucleotide transfer,[21,49,97] electrical coupling,[98] and dye transfer[98] (Figure 5C) have been shown for certain arterial and venous smooth muscle cells. Draznin and co-workers have recently published a preliminary report[99] on dye and nucleotide transfer in cultured bovine aortic smooth muscle cells that includes the fascinating observation that insulin treatment blocked dye transfer. Whether this is due to a mitogenic effect of the insulin is not clear, but this report has obvious implications for the pathogenesis of macrovascular disease in diabetes.

B. Pericyte Junctions

Pericytes are anatomically defined cells found in capillaries and postcapillary venules in most tissues. An excellent recent review by Sims[78] covers what is known about the distribution, structure, and functions of pericytes. There are no known definite reports of intercellular junctions between pericytes in vivo. However, Sheridan's studies on microvessels in the rat omentum[54,55] have demonstrated dye transfer between pericytes *in situ*.

Cultured retinal and brain vascular pericytes have recently become readily available.[69,100] Cultured bovine brain pericytes[69] have been shown to express gap junctions and to transfer radiolabeled nucleotides (Figure 5B) and fluorescent dye (Figure 5C).[50,56] Similarly, dye transfer has been observed between cultured retinal pericytes.[112] The intimate association of

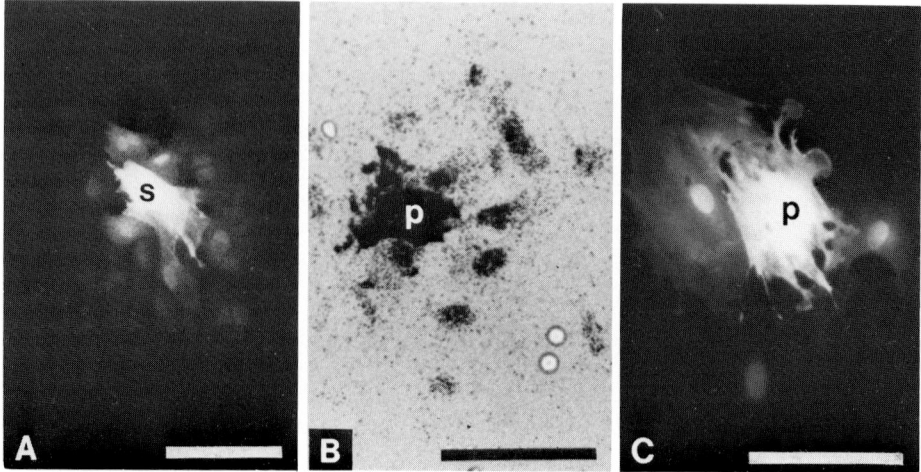

FIGURE 5.   Junctional transfer in cultured perivascular cells. (A) Dye transfer between rat mesenteric artery smooth muscle cells (injected cell = s). (B) Nucleotide transfer between bovine brain pericytes (donor cell = p). (C) Dye transfer between bovine brain pericytes (injected cell = p). (All bars = 100 μm.)

pericytes with the microvasculature and relative paucity of information on their functions suggest that the distribution of gap junctions and the capacity for intercellular transfer in pericytes is worthy of further investigation; culture systems may be very useful in future studies.

C. Heterocellular Junctions

A large number of descriptions of myoendothelial[10,16,18,87,101,102] and endothelial-pericyte bridges[17,20,26,103-105] have been published. These bridges (often called "junctions", and composed of processes from one cell type making apparent contact with the other) have been suggested to be adhesive[106] or communication structures.

The distribution of myo- and periendothelial bridges in the microvasculature was described in Rhodin[16,17] in his comprehensive series on microvascular ultrastructure. He commented that myoendothelial bridges were especially prevalent in precapillary sphincters and suggested a role in transmission of signals from the blood to the smooth muscle cells via endothelium; others have proposed similar hypotheses.[18,21,49,55,87,97,101,102,107] A logical transduction mechanism would, of course, be myoendothelial gap junctions, and some recent evidence suggests that this hypothesis may be correct. Several authors have published micrographs showing gap junctions between endothelium and smooth muscle[87,101,102] and between endothelium and pericytes[103,105] in various vessels. It appears that myo- and periendothelial gap junctions do exist in at least some vessels, and may be quite common.[101] Experimental data from Sheridan's[54,55] and this author's[55a] laboratories from dye transfer studies on *ex vivo* preparations of rat omentum and bovine brain microvessels provide additional evidence, since dye transfer was noted between endothelial cells and pericytes and (possibly) smooth muscle cells.

In culture, vascular smooth muscle and endothelium have been shown to engage in heterocellular nucleotide transfer[21,49,55,97] (Figure 6A), and both nucleotide transfer (Figure 6B) and dye transfer have been demonstrated between cultured microvascular endothelium and pericytes.[50,56] While the demonstration of heterocellular junctional transfer in vitro does not prove that this activity occurs in vivo, it does indicate that the cells have the capability to produce permeable intercellular contacts.

If gap junctional transfer between endothelium and perivascular cells is a generalized

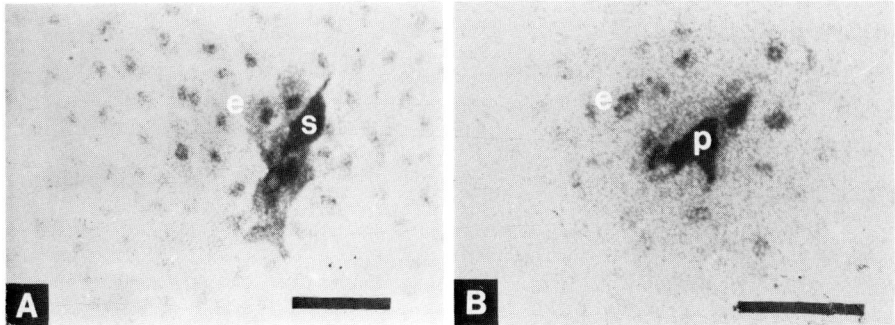

FIGURE 6. Heterocellular nucleotide transfer autoradiography. (A) Transfer from donor bovine aortic smooth muscle cell(s) to bovine aortic endothelial cells (e). (B) Transfer from donor bovine brain pericyte (p) to bovine brain microvascular endothelial cells (e). (All bars = 100 μm.)

phenomenon, then what are the biological roles of this transfer? Modulation of vessel tone in response to blood-borne or endothelial-generated signals remains a possibility,[16,17,21,55] although endothelial-dependent vasodilation is mediated by an extracellular signal, endothelium-derived relaxing factor.[108,109] Endothelial cells are known to have receptors and other cellular machinery for processing vasoactive substances;[110,111] further work is needed to determine whether they transfer information to underlying smooth muscle cells or pericytes. Another possible role for these junctions is suggested by studies indicating altered junctions and junctional transfer in endothelial and smooth muscle cells in hypertension,[6,77,92-95] myointimal hyperplasia,[84,86] and endothelial denudation.[60,70,71,73,74] Junctional mediation of coordinate responsiveness in adapations to vascular disease is a reasonable possibility.

IV. CONCLUSIONS

Although the physiological significance of interendothelial tight junctions is clear, we only have the slightest hints about the physiologic and pathophysiologic role(s) of gap junctions and junctional transfer in the vessel wall. It is clear, however, that cells in the vessel wall form a highly interdependent, integrated network. The challenge is to increase our understanding of the roles that junctions play in the functional integration of vessel wall cells. Certain questions and approaches may be fruitful avenues to explore in future research.

First, do capillary and pericytic venular endothelial cells express gap junctions *in vivo;* if not, what is the basis for dye transfer[54,55] in these cells? Do endothelial tight junctions mediate transfer? These questions might be addressed using antibodies to junctional proteins.[33,36,37,39]

Second, what are the roles of junctions in the pathogenesis of vascular disease; how do physically uninvolved endothelial cells "know" about a wound; what is the nature of any signals involved?

Third, what is the physiological role of myo- and periendothelial gap junctions; can the transfer of identified, physiologically relevant signals be demonstrated in vivo or in vitro?

Finding answers to these and other similar questions will not be easy. However, with new, refined technical approaches and the growing body of information on junctional and endothelial biology, a great deal may be accomplished.

ACKNOWLEDGMENTS

The work presented in this review was supported in part by the following N.I.H. grants:

HL06314, HL21166, HL33442, HL23567, HL268951, and BRSG SO7 RR05883. My thanks to Albert Wallstrom and Susan Carlson for their assistance, and to those individuals who supplied information, reprints, and preprints.

REFERENCES

1. **McNutt, N. S. and Weinstein, R. S.,** Membrane ultrastructure at mammalian intercellular junctions, *Prog. Biophys. Mol. Biol.,* 26, 45, 1973.
2. **Powell, D.,** Barrier function of epithelia, *Am. J. Physiol.,* 241, G275, 1981.
3. **Schneeberger, E. E. and Lynch, R. D.,** Tight junctions: their structure, composition and function, *Circ. Res.,* 55, 723, 1984.
4. **Sabatini, D. D., Griepp, E. B., Rodriguez-Boulan, E. J., Dolan, W. J., Robbins, E. S., Papadopoulos, S., Ivanov, I. E., and Rindler, M. E.,** Biogenesis of epithelial cell polarity, *Mod. Cell Biol.,* 2, 419, 1983.
5. **Pardridge, W. M.,** Brain metabolism: a perspective from the blood-brain barrier, *Physiol. Rev.,* 63, 1481, 1983.
6. **McGuire, P. G. and Twietmeyer, T. A.,** Aortic endothelial junctions in development hypertension, *Hypertension,* 7, 483, 1985.
7. **Larsen, W. J.,** Biological implications of gap junction structure, distribution and composition. A review, *Tissue Cell,* 15, 645, 1983.
8. **Spray, D. C. and Bennett, M. V. L.,** Physiology and pharmacology of gap junctions, *Annu. Rev. Physiol.,* 47, 281, 1985.
9. **Simionescu, N. and Simionescu, M.,** The cardiovascular system, in *Histology. Cell and Tissue Biology,* 5th ed., Weiss, L., Ed., Elsevier, New York, 1983, 371.
10. **Bundgaard, M. and Frøkjaer-Jensen, J.,** Functional aspects of the ultrastructure of terminal blood vessels: a quantitative study on consecutive segments of the frog mesenteric microvasculature, *Microvasc. Res.,* 23, 1, 1982.
11. **Fenstermacher, J. D. and Rapoport, S. I.,** The blood-brain barrier, in *Handbook of Physiology, The Microcirculation,* Renkin, E. M. and Michel, C. C., Eds., American Physiological Society, Washington, D.C., 1984, 969.
12. **Bundgaard, M.,** The three-dimensional organization of tight junctions in a capillary endothelium revealed by serial-section electron microscopy, *J. Ultrastruct. Res.,* 88, 1, 1984.
13. **Simionescu, M., Simionescu, N., and Palade, G. E.,** Segmental differentiations of cell junctions in the vascular endothelium. The microvasculature, *J. Cell. Biol.,* 67, 863, 1975.
14. **Simionescu, M., Simionescu, N., and Palade, G. E.,** Segmental differentiations of cell junctions in the vascular endothelium. Arteries and veins, *J. Cell. Biol.,* 68, 705, 1976.
15. **Schneeberger, E. E.,** Segmental differentiations of endothelial intercellular junctions in intra-acinar arteries and veins of the rat lung, *Circ. Res.,* 49, 1102, 1981.
16. **Rhodin, J. A. G.,** The ultrastructure of mammalian arterioles and precapillary sphincters, *J. Ultrastruct. Res.,* 18, 181, 1967.
17. **Rhodin, J. A. G.,** Ultrastructure of mammalian venous capillaries, venules and small collecting veins, *J. Ultrastruct. Res.,* 25, 452, 1968.
18. **Hüttner, I., Boutet, M., and More, R. H.,** Gap junctions in arterial endothelium, *J. Cell Biol.,* 57, 247, 1973.
19. **Schneeberger, E. E. and Karnovsky, M. J.,** Substructure of intercellular junctions in freeze-fractured alveolar capillary membranes of mouse lung, *Circ. Res.,* 38, 404, 1976.
20. **Bruns, R. R. and Palade, G. E.,** Studies on blood capillaries. I. General organization of blood capillaries in muscle, *J. Cell Biol.,* 37, 244, 1968.
21. **Larson, D. M. and Sheridan, J. D.,** Intercellular junctions and transfer of small molecules in primary vascular endothelial cultures, *J. Cell Biol.,* 92, 183, 1982.
22. **Wissig, S. L.,** Identification of the small pore in muscle capillaries, *Acta Physiol. Scand.,* 463, 33, 1979.
23. **Tani, E., Yamagata, S., and Ito, Y.,** Freeze-fracture of capillary endothelium in rat brain, *Cell Tissue Res.,* 176, 157, 1977.
24. **Nagy, Z., Peters, H., and Hüttner, I.,** Fracture faces of cell junctions in cerebral endothelium during normal and hyperosmotic conditions, *Lab. Invest.,* 50, 313, 1984.
25. **Crone, C. and Olesen, S. P.,** Electrical resistance of brain microvascular endothelium, *Brain Res.,* 241, 49, 1982.

26. **Dermietzel, R.,** Junctions in the central nervous system of the cat. IV. Interendothelial junctions of cerebral blood vessels from selected areas of the brain, *Cell Tissue Res.,* 164, 45, 1975.

27. **Karnovsky, M. J.,** The ultrastructural basis of capillary permeability studied with peroxidase as a tracer, *J. Cell Biol.,* 35, 213, 1967.

28. **Hüttner, I., Boutet, M., and More, R. H.,** Studies on protein passage through arterial endothelium. I. Structural correlates of permeability in rat arterial endothelium, *Lab. Invest.,* 28, 672, 1973.

29. **Simionescu, N., Simionescu, M., and Palade, G. E.,** Open junctions in the endothelium of the postcapillary venules of the diaphragm, *J. Cell Biol.,* 79, 27, 1978.

30. **Revel, J. P. and Karnovsky, M. J.,** Hexagonal array of subunits in intercellular junctions of mouse heart and liver, *J. Cell Biol.,* 33, C7, 1967.

31. **Peracchia, C.,** Structural correlates of gap junction permeation, *Int. Rev. Cytol.,* 66, 81, 1980.

32. **Zampighi, G. A. and Simon, S. A.,** The structure of gap junctions as revealed by electron microscopy, in *Gap Junctions,* Bennett, M. V. L. and Spray, D. C., Eds., Cold Spring Harbor Laboratory, Cold Spring Harbor, N. Y., 1985, 13.

33. **Paul, D. L.,** Antibody against liver gap junction 27-kD protein is tissue specific and cross-reacts with a 54-kD protein, in *Gap Junctions,* Bennett, M. V. L. and Spray, D. C., Eds., Cold Spring Harbor Laboratory, Cold Spring Harbor, N. Y., 1985, 107.

34. **Loewenstein, W. R.,** Junctional intercellular communication. The cell-to-cell membrane channel, *Physiol. Rev.,* 61, 829, 1981.

35. **Makowski, L.,** Structural domains in gap junctions: implications for the control of intercellular communication, in *Gap Junctions,* Bennett, M. V. L. and Spray, D. C., Eds., Cold Spring Harbor Laboratory, Cold Spring Harbor, N. Y., 1985, 5.

36. **Hertzberg, E. L. and Skibbens, R. V.,** A protein homologous to the 27,000 dalton liver gap junction protein is present in a wide variety of species and tissues, *Cell,* 39, 61, 1984.

37. **Hertzberg, E. L. and Spray, D. C.,** Studies of gap junctions: biochemical analysis and use of antibody probes, in *Gap Junctions,* Bennett, M. V. L. and Spray, D. C., Eds., Cold Spring Harbor Laboratory, Cold Spring Harbor, N. Y., 1985, 57.

38. **Revel, J. P., Nicholson, B. J., and Yancey, S. B.,** Chemistry of gap junctions, *Annu. Rev. Physiol.,* 47, 263, 1985.

38a. **Manjunath, C. K. et al.,** The cardiac gap junction protein (M_r 47,000) has a tissue specific cytoplasmic domain of M_r 17,000 at its carboxy-terminus, *Biochem. Biophys. Res. Commun.,* 142, 228, 1987.

39. **Dermietzel, R., Leibstein, A., Frixen, U., Janssen-Timmen, U., Traub, O., and Willecke, K.,** Gap junctions in several tissues share antigenic determinants with liver gap junctions, *EMBO J.,* 3, 2261, 1984.

39a. **Paul, D. L.,** Molecular cloning of cDNA for rat liver junctional protein, *J. Cell Biol.,* 103, 123, 1986.

39b. **Beyer, E. C., Paul, D. L., and Goodenough, D. A.,** Connexin43, a protein from rat heart homologous to a gap junction protein from liver, *J. Cell Biol.,* in press.

40. **Mink, D., Schiller, A., Kriz, W., and Taugner, R.,** Interendothelial junctions in kidney vessels, *Cell Tissue Res.,* 236, 567, 1984.

41. **Freddo, T. F. and Raviola, G.,** Freeze-fracture analysis of the interendothelial junctions in the blood vessels of the iris in *Macaca mulatta, Invest. Ophthalmol. Vis. Sci.,* 23, 154, 1982.

42. **Heinrich, D., Metz, J., Raviola, E., and Forssmann, W. G.,** Ultrastructure of perfusion-fixed fetal capillaries in the human placenta, *Cell Tissue Res.,* 172, 157, 1976.

43. **Firth, J. A., Bauman, K. F., and Sibley, C. P.,** The intercellular junctions of guinea-pig placental capillaries: a possible structural basis for endothelial solute permeability, *J. Ultrastruct. Res.,* 85, 45, 1983.

44. **Raviola, G. and Raviola, E.,** Paracellular route of aqueous outflow in the trabecular network and canal of Schlemm. A freeze-fracture study of the endothelial junctions in the sclerocorneal angle of the macaque monkey eye, *Invest. Ophthalmol. Vis. Sci.,* 21, 52, 1981.

45. **Loewenstein, W. R.,** Junctional intercellular communication and the control of growth, *Biochim. Biophys. Acta,* 560, 1, 1979.

46. **Sheridan, J. D.,** Cell coupling and cell communication during embryogenesis, in *Cell Surface in Animal Embryogenesis and Development,* Poste, G. and Nicholson, G. L., Eds., North-Holland, Amsterdam, 1976 409.

47. **Pitts, J. D.,** Direct interaction between animal cells, in *Cell Interactions,* Silvestri, L. G., Ed., North-Holland, Amsterdam, 1972, 227.

48. **Socolar, S. J. and Loewenstein, W. R.,** Methods for studying transmission through permeable cell-to-cell junctions, *Methods Membr. Biol.,* 10, 123, 1979.

49. **Larson, D. M. and Sheridan, J. D.,** Junctional transfer in cultured vascular endothelium. II. Dye and nucleotide transfer, *J. Membr. Biol.,* 83, 157, 1985.

50. **Larson, D. M., Carson, M. P., and Haudenschild, C. C.,** Junctional transfer in cultured bovine brain microvascular endothelial cells and pericytes, *J. Cell Biol.,* 101, 179a, 1985.

51. **Flagg-Newton, J., Simpson, I., and Loewenstein, W. R.,** Permeability of the cell-to-cell membrane channels in mammalian cell junction, *Science,* 205, 404, 1979.

52. **Subak-Sharpe, J. H., Burk, R. R., and Pitts, J. D.,** Metabolic co-operation by cell to cell transfer between genetically different mammalian cells in tissue culture, *Heredity,* 21, 342.

53. **Pitts, J. D. and Simms, J. W.,** Permeability of junctions between animal cells. Intercellular transfer of nucleotides but not of macromolecules, *Exp. Cell Res.,* 194, 153, 1977.

54. **Sheridan, J. D.,** Dye transfer in small vessels from the rat omentum: homologous and heterologous junctions, *J. Cell Biol.,* 87, 61a, 1980.

55. **Sheridan, J. D. and Larson, D. M.,** Junctional communication in the peripheral vasculature, in *The Functional Integration of Cells in Animal Tissues,* Pitts, J. D. and Finbow, M. E., Eds., Cambridge University Press, Cambridge, 1982, 263.

55a. **Larson, D. M., Carson, M. P., and Haudenschild, C. C.,** Gap junctions in endothelial cells and pericytes in *Proceedings of the 4th World Congress for Microcirculation,* Elsevier, Amsterdam, in press.

56. **Larson, D. M., Carson, M. P., and Haudenschild, C. C.,** Junctional transfer of small molecules in cultured bovine brain microvascular endothelial cells and pericytes, *Microvasc. Res.,* 34, 184, 1987.

57. **Williams, E. H. and DeHaan, R. L.,** Electrical coupling among heart cells in the absence of ultrastructurally defined gap junctions, *J. Membr. Biol.,* 60, 237, 1981.

58. **Larson, D. M., Kam, E. Y., and Sheridan, J. D.,** Junctional transfer in cultured vascular endothelium. I. Electrical coupling, *J. Membr. Biol.,* 74, 103, 1983.

59. **Caveney, S.,** The role of gap junctions in development, *Annu. Rev. Physiol.,* 47, 319, 1985.

60. **Spagnoli, L. G., Pietra, G. G., Villaschi, S., and Johns, L. W.,** Morphometric analysis of gap junctions in regenerating arterial endothelium, *Lab. Invest.,* 46, 139, 1982.

61. **Lawrence, T. S., Beers, W. H., and Gilula, N. B.,** Transmission of hormonal stimulation by cell-to-cell communication, *Nature (London),* 272, 501, 1978.

62. **Murray, S. A. and Fletcher, W. H.,** Hormone-induced intercellular signal transfer dissociates cyclic AMP-dependent protein kinase, *J. Cell Biol.,* 98, 1710, 1984.

63. **Sheridan, J. D., Finbow, M. E. and Pitts, J. D.,** Metabolic interactions between animal cells through permeable intercellular junctions, *Exp. Cell Res.,* 123, 111, 1979.

64. **Shaklai, M., Loskutoff, D., and Tavassoli, M.,** Membrane characteristics of cultured endothelial cells: identification of gap junction, *Isr. J. Med. Sci.,* 14, 306, 1978.

65. **Ryan, U. S., Mortara, M., and Whitaker, C.,** Methods for microcarrier culture of bovine pulmonary artery endothelial cells avoiding the use of enzymes, *Tissue Cell,* 12, 619, 1980.

66. **Shivers, R. R. and Bowman, P. D.,** A freeze-fracture paradigm of the mechanism for delivery and insertion of gap junction particles into the plasma membrane, *J. Submicrosc. Cytol.,* 17, 199, 1985.

67. **Pauli, B. U., Anderson, S. N., Memoli, V. A., and Kuettner, K.,** The isolation and characterization *in vitro* of normal epithelial cells, endothelial cells and fibroblasts from rat urinary bladder, *Tissue Cell,* 12, 419, 1980.

68. **Diglio, C. A., Grammas, P., Giacomelli, F., and Wiener, J.,** Primary culture of rat cerebral microvascular endothelial cells. Isolation, growth, and characterization, *Lab. Invest.,* 46, 554, 1982.

69. **Carson, M. P. and Haudenschild, C. C.,** Microvascular endothelium and pericytes: high yield, low passage cultures, *In Vitro Cell. Dev. Biol.,* 22, 344, 1986.

70. **Schwartz, S. M., Stemerman, M. B., and Benditt, E.P.,** The aortic intima. II. Repair of the aortic lining after mechanical denudation, *Am. J. Pathol.,* 81, 15, 1975.

71. **Schwartz, S. M., Haudenschild, C. C., and Eddy, E. M.,** Endothelial regeneration. I. Quantitative analysis of initial stages of endothelial regeneration in rat aortic intima, *Lab. Invest.,* 38, 568, 1978.

72. **Yee, A. G. and Revel, J. P.,** Loss and reappearance of gap junctions in regenerating liver, *J. Cell Biol.,* 78, 554, 1978.

73. **Hüttner, I., Walker, C., and Gabbiani, G.,** Aortic endothelial cell during regeneration. Remodeling of cell junctions, stress fibers, and stress fiber-membrane attachment domains, *Lab. Invest.,* 53, 287, 1985.

74. **Larson, D. M. and Haudenschild, C. C.,** Junctional transfer in wounded cultures of bovine aortic endothelial cells, *J. Cell Biol.,* 103, 192a, 1986.

75. **Haudenschild, C. C. and Harris-Hooker, S.,** Injury affects endothelial integrity beyond physical location of the damage, *Circulation,* 66(2), 205, 1982.

76. **Haudenschild, C. C. and Harris-Hooker, S.,** Endothelial cell motility, in *The Biology of Endothelial Cells,* Jaffe, E. A., Ed., Martinus Nijhoff, The Hague, 1984, 74.

77. **Hüttner, I., Costabella, P. M., De Chastonay, C., and Gabbiani, G.,** Volume, surface, and junctions of rat aortic endothelium during experimental hypertension, a morphometric and freeze fracture study, *Lab. Invest.,* 46, 489, 1982.

78. **Sims, D. E.,** The pericyte-a review, *Tissue Cell,* 18, 153, 1986.

79. **Burnstock, G.,** Ultrastructure of autonomic nerves and neuroeffector junctions; analysis of drug action, in *Methods Pharmacol.,* 3, 113, 1975.

80. **Kajiwara, M., Kitamura, K., and Kuriyama, H.,** Neuromuscular transmission and smooth muscle membrane properties in the guinea-pig ear artery, *J. Physiol. (London),* 315, 283, 1981.

81. **Neild, T. O.,** The relation between the structure and innervation of small arteries and arterioles and the smooth muscle membrane potential changes expected at different levels of sympathetic nerve activity, *Proc. R. Soc. London Ser. B.,* 220, 237, 1983.

82. **Tomita, T.,** Electrical properties of mammalian smooth muscle, in *Smooth Muscle,* Bulbring, E., Brading, A. F., Jones, A. W., and Tomita, T., Eds., Edward Arnold, London, 1970, 197.

83. **Iwayama, T.,** Nexuses between areas of the surface membrane of the same arterial smooth muscle cell, *J. Cell Biol.,* 49, 521, 1971.

84. **Cuevas, P. and Gutierrez Diaz, J. A.,** Smooth muscle cells in "venous patches" grafted into the rat common carotid artery. A structural study, *Experientia,* 38, 122, 1982.

85. **Litwin, J. A.,** Cell membrane features of rabbit arterial smooth muscle. A freeze-fracture study, *Cell Tissue Res.,* 212, 341, 1980.

86. **Cuevas, P., Gutierrez Diaz, J. A., and Reimers, D.,** Intercellular junctions between smooth muscle cells in myointimal hyperplasia, *Acta Anat.,* 114, 22, 1982.

87. **Metz, J. and Weihe, E.,** Intercellular junctions in the full term human placenta. II. Cytotrophoblast cells, intravillous stroma cells and blood vessels, *Anat. Embryol.,* 158, 167, 1980.

88. **Segal, S. S. and Duling, B. R.,** Arteriolar dilation propagates in an electronic fashion, *Fed. Proc.,* 45, 1158, 1986.

89. **Segal, S. S. and Duling, B. R.,** Flow control among microvessels coordinated by intercellular conduction, *Science,* 234, 868, 1986.

90. **Hilton, S. M.,** Peripheral arterial conducting mechanism underlying dilation of the femoral artery and concerned in functional dilation in skeletal muscle, *J. Physiol. (London),* 149, 93, 1959.

91. **Daniel, E. E., Saniel, V. P., Duchon, G., Garfield, R. E., Nichols, M., Malhotra, S. K., and Oki, M.,** Is the nexus necessary for cell-to-cell coupling of smooth muscle?, *J. Memb. Biol.,* 28, 207, 1976.

92. **Holloway, E. T. and Bohr, D. F.,** Reactivity of vascular smooth muscle in hypertensive rats, *Circ. Res.,* 33, 678, 1973.

93. **Pegram, B. L. and Ljung, B.,** Neuroeffector functions of isolated portal vein from spontaneously hypertensive and Wistar-Kyoto rats: Dependence on external calcium concentration, *Blood Vessels,* 18, 89, 1981.

94. **Thayer, E. S. and Bandick, N. R.,** Intercellular junctions between femoral arterial smooth muscle cells of renal hypertensive rats, *Paroi Arterielle,* 7, 135, 1981.

95. **Grünwald, J., Robenek, H., Mey, J., and Hauss, W. H.,** In vivo and in vitro cellular changes in experimental hypertension: electronmicroscopic and morphometric studies of aortic smooth muscle cells, *Exp. Mol. Pathol.,* 36, 164, 1982.

96. **Gabella, G.,** Hypertrophic smooth muscle. III. Increase in number and size of gap junctions, *Cell Tissue Res.,* 201, 263, 1979.

97. **Davies, P. F., Ganza, P., and Diehl, P. S.,** Reversible microcarrier-mediated junctional communication between endothelial and smooth muscle cell monolayers: an in vitro model of vascular cell interactions, *Lab. Invest.,* 85, 710, 1985.

98. **Blennerhassett, M. G., Kannan, M. S., and Garfield, R. E.,** Functional characterization of cell-to-cell coupling in cultured rat aortic smooth muscle, *Am. J. Physiol.,* 252, C555, 1987.

99. **Draznin, M. B., El-Fouly, M. H., Erickson, L. M., and Trosko, J. E.,** Insulin blocks cell-cell communication in cultured bovine aortic smooth muscle cells: application of a rapid in vitro assay to study hormonal acceleration of atherogenesis, *Diabetes,* 35 (Suppl. 1), 213a, 1986.

100. **Gitlin, J. D. and D'Amore, P. A.,** Culture of retinal capillary cells using selective growth media, *Microvasc. Res.,* 26, 74, 1983.

101. **Spagnoli, L. G., Villaschi, S., Neri, L., and Palmieri, G.,** Gap junctions in myo-endothelial bridges of rabbit carotid arteries, *Experientia,* 38, 124, 1982.

102. **Taugner, R., Kirchheim, H., and Forssmann, W. G.,** Myoendothelial contacts in glomerular arterioles and in renal interlobular arteries of rat, mouse and *Tupaia belangeri, Cell Tissue Res.,* 235, 319, 1984.

103. **Spitznas, M. and Reale, E.,** Fracture faces of fenestrations and junctions of endothelial cells in human choroidal vessels, *Invest. Ophthalmol.,* 14, 98, 1975.

104. **Tilton, R. G., Kilo, C., and Williamson, J. R.,** Pericyte-endothelial relationships in cardiac and skeletal muscle capillaries, *Microvasc. Res.,* 18, 325, 1979.

105. **Cuevas, P., Gutierrez-Diaz, J. A., Reimers, D., Dujovny, M., Diaz, F. G., and Ausman, J. I.,** Pericyte endothelial gap junctions in human cerebral capillaries, *Anat. Embryol.,* 170, 155, 1984.

106. **Courtoy, P. J. and Boyles, J.,** Fibronectin in the microvasculature: localization in the pericyte-endothelial interstitium, *J. Ultrastruct. Res.,* 83, 258, 1983.

107. **Richardson, J. B. and Beaulnes, A.,** The cellular site of action of angiotensin, *J. Cell Biol.,* 51, 419, 1971.

108. **Furchgott, R. F.,** Role of endothelium in responses of vascular smooth muscle, *Circ. Res.,* 53, 557, 1983.

109. **Vanhoutte, P. M. and Rimele, T. J.,** Role of endothelium in the control of vascular smooth muscle function, *J. Physiol. (Paris),* 78, 681, 1983.

110. **Shepro, D. and D'Amore, P. A.,** Physiology and biochemistry of the vascular wall endothelium, in *Handbook of Physiology, The Microcirculation,* Renkin, E. M. and Michel, C. C., Eds., American Physiological Society, Washington, D. C., 1984, 103.

111. **Ryan, U. S. and Ryan, J. W.,** Cell biology of pulmonary endothelium, *Circulation,* 70 (Suppl. 3), 46, 1984.

112. **Larson, D. M. et al.,** unpublished observations.

Single Capillary and Whole Animal Methods

Chapter 34

ENDOTHELIAL PROPERTIES ASSESSED FROM SINGLE-CAPILLARY METHODS

Christian Crone, Magnus Bundgaard, and Søren-Peter Olesen

TABLE OF CONTENTS

I. BACKGROUND

Single-capillary methods have developed in two distinct waves, each connected with a specific experimental technique. In the 1920s Eugene Landis developed the microocclusion technique permitting determination of fluid filtration in single capillaries.[1] This technique gave the first information on the hydraulic conductivity of the capillary membrane (L_p). The other aspect of capillary function, diffusion of solutes between blood and tissues, yielded to single-vessel methods only in the late 1970s, when electrophysiological techniques were introduced.[2-4] These methods gave values for the permeability coefficient (P_d).

Apart from the fact that it has proven possible to study the function of extremely small blood vessels, the question may be asked what has been gained from this kind of research in contrast to what can be known from less demanding experiments on the entire capillary bed — so-called "whole organ" experiments. In this chapter we shall try to evaluate the situation as we see it. It should be emphasized that the physiological emphasis in this kind of microvascular research has been on mechanisms and structures that allow communication between blood and cells. Since all hydrophilic solutes (both small molecules and macromolecules) have to pass by paracellular routes, research in this area has mainly focused on the properties of the paracellular pathways which played a decisive role in the formulation of the main paradigm in transcapillary transport — the "pore theory".[5] In all likelihood the pores are located *between* the endothelial cells. This view agrees with the calculated fractional pore area which is about 1/10,000 of the total surface area.[6]

II. STRUCTURE AND FUNCTION IN PERMEABILITY

A simple calculation shows that known dimensions of paracellular pathways (or interendothelial clefts) comply with what is required to explain exchange velocities. If P_d is permeability (i.e., diffusion rate through 1 cm^2 under unit of concentration difference), the following equation applies:

$$P_d = D \times \frac{L \times w}{\Delta x}$$

where D is the diffusion coefficient, L is the interendothelial cleft length per square centimeter, w is the average width of the cleft, and Δx is the depth (Figure 1). With diffusion coefficients of small solutes about 10^{-5} cm^2/sec and experimentally determined permeabilities of mesenteric capillaries to such solutes close to 5×10^{-4} cm/sec,[3] it is seen that the factor $(L \times w)/\Delta x$ has a value of 50 cm^{-1}.

Let us evaluate this figure. The length of the interendothelial cleft has been determined experimentally by stereological methods to be ~2000 cm/cm^2.[7] The average cleft width is about 10^{-6} cm ($-$ 100 Å), and a reasonable value for the diffusion distance through the cleft is 4×10^{-5} cm (0.4 μm). The combination of these figures gives a value for $(L \times w)/\Delta x$ of 50 cm^{-1}. This calculation shows that very simple reasoning leads to reasonable estimates of permeability in terms of structure. The permeability of the capillaries in muscle tissue is about ten times lower than in mesenteric capillaries,[8,9] and it is necessary to postulate that only 10% of the cleft length is effectively open. In brain, capillary permeability is further reduced to about 1/100 of that in muscle,[10] implying that in this organ only a tiny fraction of the interendothelial cleft is accessible to diffusion of hydrophilic solutes (about 0.1%). Figure 1 illustrates the simple model behind the reasoning. Table 1 summarizes permeability figures obtained with single capillary methods.

A number of experimental results focusing on other descriptors of capillary permeability require that the simple "pore" model of a cleft with plane parallel walls be modified somewhat. When it is said that the average cleft width is about 100 Å, the truth is that a

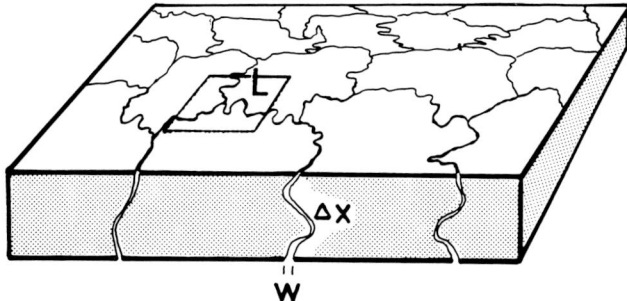

$$P = D \cdot \frac{L \cdot W}{\Delta X}$$

FIGURE 1.   Basis for morphological analysis of capillary permeability.

Table 1
ELECTRICAL RESISTANCES AND ION
PERMEABILITIES OF VASCULAR
ENDOTHELIA

Tissue	Electrical resistance (Ω cm^2)	Small ion permeability (10^{-5} cm sec^{-1})	Ref.
Brain	2000	0.05	45
Skin	24—70	2—5	46
Muscle	23—33	5	9
Mesentery	1—3	70	4
Connective tissue			
Arterioles	18	5.5	24
Venules	3.5	28	24
Aorta	10.0	—	25

major part of the cleft has a width of 200 Å, while about 10% of the cleft has a constriction with a width of 50 to 60 Å. Such a "constriction" explains a reflection coefficient to small solutes in capillaries in mesentery and muscle of 0.1[21] and takes care of the hydraulic conductance.[6]

Can we take this simple model of the interendothelial cleft literally? Recent morphological studies of the details of the cleft morphology in heart capillaries[12] indicate how to interpret cleft morphology, and how to understand capillary pore morphology. On sections of capillaries, the interendothelial cleft appears closed and one almost always observes one, two, or three fusion points (cell contacts) between neighboring cell membranes. However, the distance of the fusion point from a reference level, say, the basement membrane, varies. What this signifies was explored in studies[12] where capillary endothelium was cut in series and the three-dimensional organization reconstructed. In heart capillaries this gave the picture shown in Figure 2. In other words, the cleft is not a regular space with an open part and a "constriction"; rather it has interruptions in the fusion lines here and there, permitting solutes to find their way through the "junction". Bundgaard's study shows how electron-microscopial pictures in two dimensions may give misleading information about the three-dimensional structure. Although it is not possible to give exact figures for the distribution

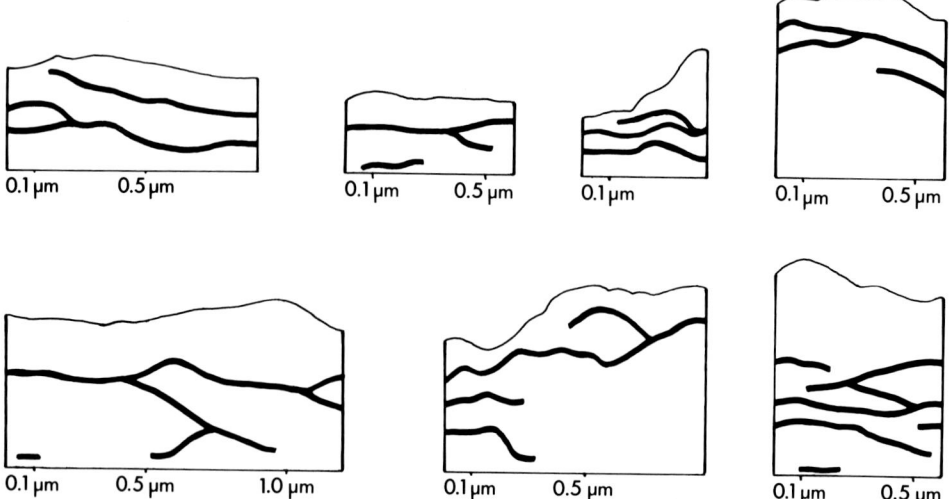

FIGURE 2.   Seven reconstructed junctional segments and venular endothelia. The heavy lines are the contact lines. The thin horizontal and the upper irregular lines represent the abluminal and luminal openings of the cleft. It is seen that the contact lines in most cases have a finite length, leaving an opening for diffusion. Due to the fact that not more than 10 to 28 consecutive segments were contained in each reconstruction, the ending of a contact line is not always identified. (From Bundgaard, M., *J. Ultrastruct. Res.*, 88, 1, 1984. With permission.)

of pore sizes, the morphological findings agree surprisingly well with narrow interruptions having dimensions in the neighborhood of the required 50 to 100 Å.[11]

Microvascular permeability changes in a variety of conditions. Inflammation is known to augment permeability also to macromolecules.[13] It has recently been shown that the permeability response to a variety of putative mediators of the inflammatory response is extremely fast,[14] and when the application period is short, it is reversible within minutes. This observation places limits to interpretations of mechanisms of changes of capillary permeability.

A major field of interest in current research is to understand modulation of capillary permeability as a reflection of alterations in the microtopography of interendothelial clefts.[15] An obvious guess suggests that changes in cytoskeletal components of the endothelial cell take place in response to ligand-receptor interaction at the endothelial cell surface. To this may be added a possible supplementary mechanism involving cell adhesion molecules anchoring two adjoining endothelial cells together (E-CAMs).

These possibilities are discussed in more detail below following a short summary of single capillary techniques that have allowed assessment of rapid permeability changes.

III. ELECTROPHYSIOLOGICAL APPROACH TO CAPILLARY PERMEABILITY

In experiments on whole organs a bolus of known composition is injected into the artery of an organ. The venous blood is collected fractionally and analyzed.[16,17] Although this so-called ''indicator diffusion'' method has yielded much information about capillary permeability in a number of organs,[6] it has repeatedly been objected that due to heterogeneities in the microvascular bed and due to unproven assumptions, the method as such may not give strict quantitative information — the published permeability coefficients might even be off by unacceptable factors.

The only response to this stalemate was the development of methods allowing determination of permeability in single vessels. Two roads were followed:

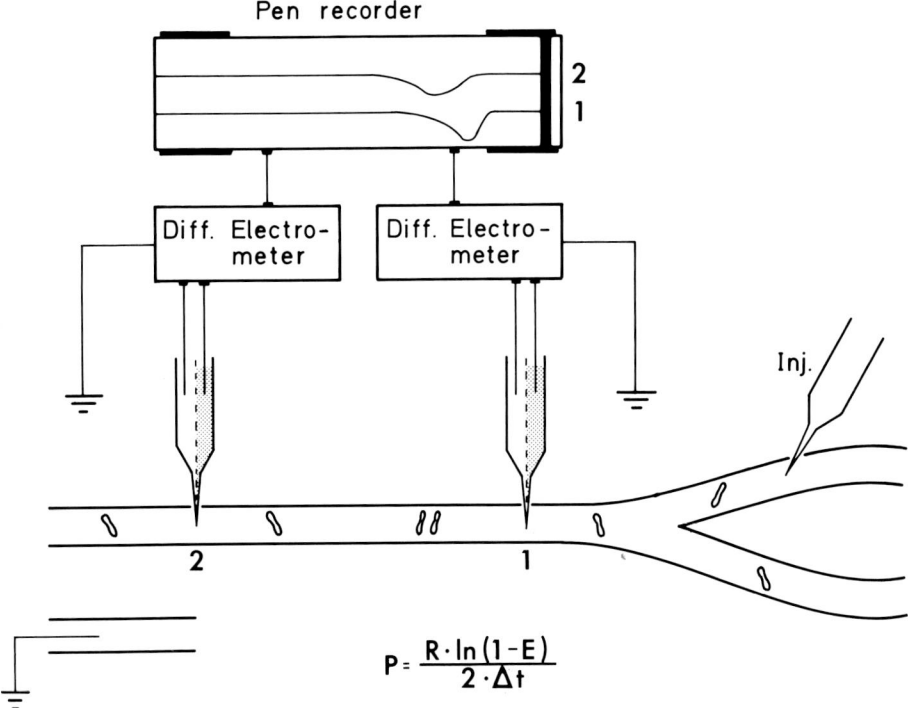

FIGURE 3.   Microinjection of a test solution into the microvascular bed. Two ion-sensitive microelectrodes are placed in a nearby capillary. (From Crone, C. and Friedman, J. J., *Acta Physiol. Scand.*, 96, 13A, 1976. With permission.)

1.   Determination of ion permeability with ion-sensitive microelectrodes.[3]
2.   Determination of the electrical resistance of the capillary wall.[4]

If a bolus of a high-potassium solution is injected into an arteriole feeding a capillary bed, the excess potassium diffuses through the capillary wall according to the permeability and driving forces. Figure 3 shows the technique. The formalism applied in indicator diffusion experiments on whole organ was directly applicable to single vessels.[3] The first experiments gave values for K^+ permeability that appeared too high by a factor of ten. This might indicate that the "whole organ" assumptions were unapplicable, but the true explanation was different. The first experiments were performed on mesenteric microvessels. Later experiments on single microvessles in muscle[8,9] showed that mesenteric capillaries are, in fact, much more permeabile than muscle capillaries. The permeabilities to K^+ in muscle agreed very well with earlier values obtained on whole organs.[18]

Apart from demonstrating the feasibility of diffusion studies on single vessels and giving a reasonable explanation for the early discordance in permeability values (in terms of a real difference between mesenteric and muscle capillaries), the electrophysiological methods opened the way to studies of modulation.

An improvement of the methodology was developed which made it more powerful. Instead of injecting fluids with excess concentration of specific ions, electrical resistance was determined by injection of current into a microvessel.

Current will leave the capillary through the wall according to the resistance that ions meet. In neurophysiology this method of determining the electrical resistance of a cell membrane is well known as "cable analysis". Without going into details, what is determined is the electrical-potential profile within the microvessel in response to current injections as illustrated in Figure 4.

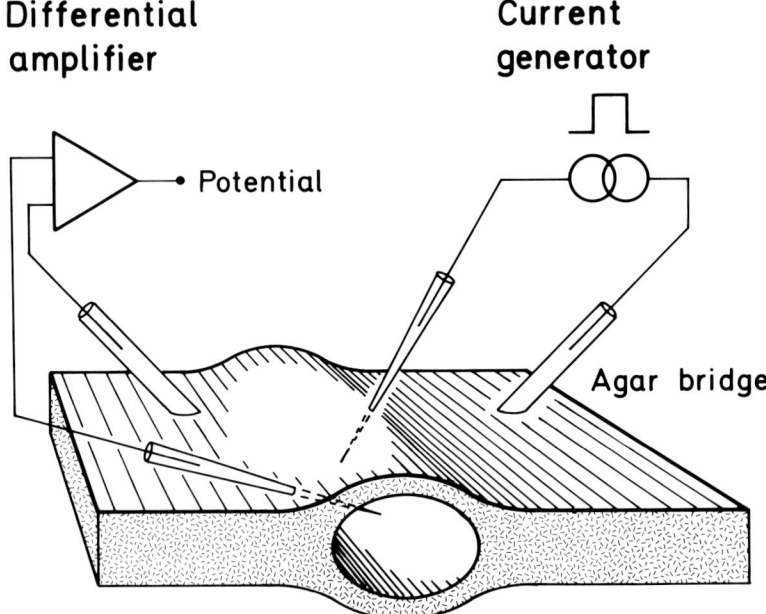

FIGURE 4.   Square-wave current pulses are injected into a capillary via glass microelectrode. Electrical potential is measured with another glass microelectrode connected to a high-impedance electrometer. (Reproduced from *J. Gen. Physiol.*, 1981, 77, 394, by copyright permission of the Rockefeller University Press.)

Table 1 summarizes determinations of electrical resistance in three different capillary beds (mesentery, muscle, and brain). It is clear that even within the group of "continuous" capillaries, resistance varies significantly (by a factor of 1000). Since the electrical current is carried by ions, the determinations reflect ion permeability in the microvascular membrane.

How can we explain the fact that continuous capillaries differ so much in permeability? Although the interendothelial cleft length must be very similar in different capillary beds, it is nevertheless possible to explain differences in permeability by postulating differences in the degree that endothelial cells are bound together. Essentially, what we postulate is that the interendothelial cleft is zippered up more or less fully in different organs. This explains the two important observations that (1) diffusional permeability and hydraulic conductance vary in parallel, and (2) the reflection coefficient stays constant.[21] A rigorous proof of this hypothesis will be difficult to deliver because of the paramount difficulties in morphometric analysis of such a problem.

The question now is how to explain rapid changes in permeability.

IV. MODULATION OF MICROVASCULAR PERMEABILITY

The reason for using the terminology "microvascular permeability" rather than "capillary permeability" is found in the studies of Majno et al.,[19] who showed that the leakage provoked by serotonin was located in the venular portion of the microvasculature. This conclusion has not been seriously questioned. Bundit and Wissig[20] repeated earlier studies indicating a gradient of permeability along the consecutive microvascular segments, but could not support this earlier notion. The claim that macromolecular permeability in unexposed venules is higher than in "upstream" portions of the microcirculation seems no longer tenable; if vessels were fixed *in situ* before exposure, a macromolecular tracer (Evans blue dye) did not permeate. The conclusion is important because it touches one of the basic dogmas in microcirculation.

Electrical resistance

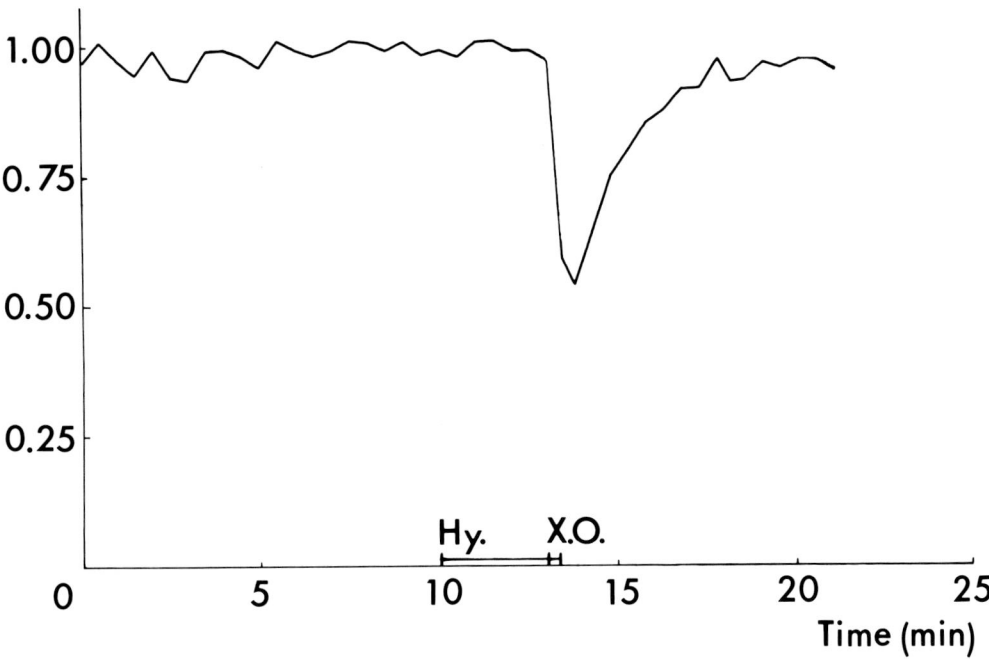

FIGURE 5.   Continuous measurement of electrical resistance of venular endothelium. Application of xanthine oxidase (X.O.) with hypoxanthine (Hy.) present leads to an immediate lowering of the electrical resistance and augmentation of ion permeability.

This notwithstanding, permeability changes largely take place in the postcapillary venules. Studies of the permeation of macromolecular fluorescent tracers in vivo in response to various pharmacological stimuli have amply supported this notion.[22,23] In this connection it should be noticed that the permeability of arterioles and large venules is surprisingly close to that of capillaries. Olesen[24] took advantage of the single-vessel methods to determine electrical resistance of arterioles and venules in the hamster cheek pouch. He found arteriolar resistances to be 19 Ω cm^2 and venular resistances to be 3.3 Ω cm^2. Converted to equivalent permeabilities of Na$^+$, these values correspond to P$_{Na^+}$ of 4×10^{-5} cm/sec^{-1} and 23×10^{-5} cm/sec^{-1}, respectively — values that are close to values for capillary permeability to this ion.[6] It is interesting that the electrical resistance of endothelium in large vessels is close to that of small vessels. O'Donnell and Vargas[25] report 10 Ω cm^2 in aortic endothelium, and thus it is likely that the endothelium lining the entire vascular system has almost similar passive properties.

V. SINGLE-VESSEL STUDIES OF MODULATION

The standard methods to assess changes in the permeability of small blood vessels are (1) measurement of lymph flow, (2) electronmicroscopical studies of macromolecular tracers, and (3) in vivo microscopy of exposed vessels filled with fluorescent tracers. These methods all have a relatively low time resolution — on the order of minutes. The electrophysiological methods, on the contrary, have an extremely high time resolution and it has now been shown that many "classical" modulators of capillary permeability (serotonin, bradykinin, histamine) augment permeability within seconds[26] (Figure 5). This speaks in favor of receptor-mediated events. The idea that endothelial cells possess specific receptors is relatively new;

in retrospect it is surprising that this idea only lately became evident. The fact that low-molecular-weight substances of a specific nature change permeability, plus the observation that the effects can be counteracted by other specific agents, suggest receptor-mediated events.[27,28] This concept has the important consequence that knowledge from other cells showing receptor-mediated phenomena can be applied to the endothelial cell.

Crone[15,29] suggested that Ca^{2+} may be an important second messenger in endothelial cells, and the evidence that this may be so is now forthcoming.[30] A calcium transient in response to histamine has been reported.[31] Bradykinin also elicits a calcium transient.[32] The application of the calcium ionophore A 23187 leads to significant augmentation of the electrical conductance of the capillary wall.[14] Michel and Phillips[33] observed an increase in hydraulic conductance under the same stimulus, but instead proposed magnesium to be the important ion.[34]

Recently, Olesen[35] has investigated whether the ionophore acts via calcium or via magnesium. Using a ionophore with higher Ca^{2+} specificity (ETH 1001), he concludes that calcium is the important messenger. The situation may be somewhat complex, because cytosolic calcium may increase either due to a sudden influx of calcium or due to release from the endoplasmic reticulum. Similar problems have been studied in the activation of thrombocytes by ADP[36] where a dual mode of recruitment of cytosolic calcium is suggested. In all likelihood, the endothelial cell shares mechanisms for calcium mobilization with many other cells in which cytosolic calcium is the messenger that activates various cellular mechanisms (skeletal muscle cells, smooth muscle cells, neurosecretory cells, salivary glands cells, etc.). There can hardly be any doubt that the endothelial cytoskeleton is involved in permeability changes, although the details of the mechanism are not yet known. A simple picture envisages traction by actomyosin on the interendothelial cleft creating subtle changes in permeability. Some of the substances that Olesen and Crone[14] found to increase microvascular permeability only provoked rather small changes in electrical resistance, in keeping with the idea of "traction" on the junctional region. In other instances, an impressive, irreversible permeability increase was elicited (with neuraminidase, trypsin, snake venoms) suggesting different mechanisms of "opening".

The events that follow ligand-receptor interaction on the endothelial surface are only vaguely known. Olesen[35] systematically investigated a number of substances known to act in intracellular signaling in various cell types, such as cyclic AMP, cyclic GMP, phorbol myristate acetate (a protein kinase C stimulator), but found them ineffective in altering capillary permeability. The significance of the inositol phosphate system is unknown, although one might guess that since calcium is important and since inositol phosphates release calcium from endoplasmic reticulum,[37] inositol phosphates may act as messengers also in endothelial cells.

Bundgaard[38] looked for the presence of an endoplasmic reticulum in endothelial cells using a serial-section technique with three-dimensional reconstruction,[38] and identified this structure in capillaries from rat diaphragm. The likelihood is that it acts in conjunction with the plasmalemmal invaginations.[15,29] The proposal that the invaginations serve purposes different from those postulated by Bruns and Palade[39] opens a new alley of investigations with emphasis on the significance of this system in control of endothelial cell functions and capillary permeability.

VI. EDRF AND MODULATION OF CAPILLARY PERMEABILITY

The discovery of an endothelial-derived relaxing factor (EDRF) by Furchgott and Zawadski,[40] released when acetylcholine acts on endothelial cells, has triggered many studies that have led to the realization that a long list of different substances release EDRF and produce vasodilatation. The surprising fact that has come out of such studies and those on

Table 2
SUBSTANCES THAT RELEASE EDRF, AUGMENT MICROVASCULAR PERMEABILITY, AND INDUCE FORMATION OF PROSTAGLANDINS

EDRF[42,43]	Permeability increase[14]	Prostaglandin formation[41,44]
Histamine	Histamine	Histamine
Bradykinin	Bradykinin	Bradykinin
ATP, ADP	ATP, ADP, AMP	ATP, ADP
Oxygen radicals	Oxygen radicals	?
Arachidonic acid	Arachidonic acid	Arachidonic acid
Hypoxia	Hypoxia	?
Cyanide	Cyanide	?
Melittin	Melittin	Melittin
Ionophore A 23187	Ionophore A 23817	Ionophore A 23817
Acetylcholine	Protamine sulfate	Insulin
Thrombin	Venoms	Angiotensin II
Substance P	Phospholipase A2	Phospholipase A2
	LTC_4	LTC_4, LTD_4

permeability-increasing substances[14] is a great overlap between factors eliciting the two different effects (Table 2). This concordance may reflect a simple "economy" in cell regulation, but it may have a deeper significance inasmuch as vasodilatation and permeability increase may occur together. This, of course, is the case in overt inflammation, but vasodilatation and (slight) augmentation of capillary permeability (facilitating exchange of substances) as a physiological mechanism is a possibility that should be entertained experimentally. A further perspective emerges when a list of substances that induce formation of prostaglandins[41] is compared with the two preceding lists (Table 2). Again an interesting correspondence is seen.

What these "overlaps" and concordances mean and where they will lead is not clear at the moment, but some "unifying" hypothesis may lie ahead of us.

REFERENCES

1. **Landis, E. M.,** Micro-injection studies of capillary permeability. II. The relation between capillary pressure and the rate at which fluid passes through the walls of single capillaries, *Am. J. Physiol.,* 82, 217, 1927.
2. **Crone, C. and Friedman, J. J.,** A method for determining potassium permeability of a single capillary, *Acta Physiol. Scand.,* 96, 13A, 1976.
3. **Crone, C., Frøkjær-Jensen, J., Friedman, J. J., and Christensen, O.,** The permeability of single capillaries to potassium ions, *J. Gen. Physiol.,* 71, 195, 1978.
4. **Crone, C. and Christensen, O.,** Electrical resistance of a capillary endothelium, *J. Gen. Physiol.,* 77, 349, 1981.
5. **Pappenheimer, J. R.,** Passage of molecules through capillary walls, *Physiol. Rev.,* 33, 387, 1953.
6. **Crone, C. and Levitt, D. G.,** Capillary permeability to small solutes, in *Handbook of Physiology. The Cardiovascular System. IV. Microcirculation,* Renkin, E. M. and Michel, C. C. Eds., American Physiological Society, Bethesda, Md., 1984, chap. 10.
7. **Bundgaard, M. and Frøkjær-Jensen, J.,** Functional aspects of the ultrastructure of terminal blood vessels: a quantitative study on consecutive segments of the frog mesenteric microvasculature, *Microvasc. Res.,* 23, 1, 1982.
8. **Frøkjær-Jensen, J.,** Permeability of single muscle capillaries to potassium ions, *Microvasc. Res.,* 24, 168, 1982.
9. **Olesen, S.-P. and Crone, C.,** Electrical resistance of muscle capillary endothelium, *Biophys. J.,* 42, 31, 1983.

10. **Crone, C.,** Lack of selectivity to small ions in paracellular pathways in cerebral and muscle capillaries of the frog, *J. Physiol. (London)*, 353, 317, 1984.

11. **Haraldsson, B. and Rippe, B.,** Restricted diffusion of Cr-EDTA and cyanocobalamine across the exchange vessels in rat hindquarters, *Acta Physiol. Scand.*, 127, 359, 1986.

12. **Bundgaard, M.,** The three-dimensional organization of tight junctions in a capillary endothelium revealed by serial-section electron microscopy, *J. Ultrastruct. Res.*, 88, 1, 1984.

13. **Arfors, K.-E., Rutili, G., and Svensjö, E.,** Microvascular transport of macromolecules in normal and inflammatory conditions, *Acta Physiol. Scand. Suppl.*, 463, 93, 1979.

14. **Olesen, S.-P. and Crone, C.,** Substances that rapidly augment ionic conductance of endothelium in cerebral venules, *Acta Physiol. Scand.*, 127, 233, 1986.

15. **Crone, C.,** Modulation of solute permeability in microvascular endothelium, *Fed. Proc.*, 45, 77, 1986.

16. **Crone, C.,** The permeability of capillaries in various organs as determined by use of the "indicator diffusion" method, *Acta Physiol. Scand.*, 58, 292, 1963.

17. **Chinard, F. P., Vosburgh, G. J., and Enns, Th.,** Transcapillary exchange of water and of other substances in certain organs of the dog, *Am. J. Physiol.*, 1983, 221, 1955.

18. **Tancredi, R. G., Yipintsoi, T., and Bassingthwaighte, J. B.,** Capillary and cell wall permeability to potassium in isolated dog hearts, *Am. J. Physiol.*, 229, 537, 1975.

19. **Majno, G., Palade, G. E., and Schoefl, G. I.,** Studies on inflammation. II. The site of action of histamine and serotonin along the vascular tree: a topographical study, *J. Biophys. Biochem. Cytol.*, 11, 607, 1961.

20. **Bundit, V. and Wissig, S. L.,** Surgical exposure induces formation of an arteriovenous permeability gradient for macromolecules in the microcirculation of muscle, *Microvasc. Res.*, 21, 235, 1986.

21. **Curry, F. E. and Frøkjær-Jensen, J.,** Water flow across the walls of single muscle capillaries in the frog, Rana Pipiens, *J. Physiol. (London)*, 350, 293, 1984.

22. **Svensjö, E., Arfors, K. E., Raymond, R. M., and Grega, G. J.,** Morphological and physiological correlation of bradykinin-induced macromolecular efflux, *Am. J. Physiol.*, 236, H600, 1979.

23. **Fox, J., Galey, F., and Wayland, H.,** Action of histamine on the mesenteric microvasculature, *Microvasc. Res.*, 19, 108, 1980.

24. **Olesen, S.-P.,** Electrical resistance of arterioles and venules in the hamster cheek pouch, *Acta Physiol. Scand.*, 123, 12, 1985.

25. **O'Donnell, M. P. and Vargas, F. F.,** Electrical conductivity and its use in estimating an equivalent pore size for arterial endothelium, *Am. J. Physiol.*, 250, H16, 1986.

26. **Olesen, S.-P.,** A calcium-dependent reversible permeability increase in microvessels in frog brain, induced by serotonin, *J. Physiol. (London)*, 361, 103, 1985.

27. **Svensjö, E. and Grega, G. J.,** Evidence for endothelial cell-mediated regulation of macromolecular permeability by postcapillary venules, *Fed. Proc.*, 45, 89, 1986.

28. **Clerck, F. De and Reneman, R. S.,** Serotonin and microvascular permeability, *Prog. Appl. Microcirc.*, 10, 32, 1986.

29. **Crone, C.,** From 'Porositates Carnis' to cellular microcirculation, *Int. J. Microcirc. Clin. Exp.*, 6, 101, 1986.

30. **Haraldsson, B., Zackrisson, U., and Rippe, B.,** Calcium dependence of histamine-induced increases in capillary permeability in isolated perfused rat hindquarters, *Acta Physiol. Scand.*, 128, 247, 1986.

31. **Rotrosen, D. and Gallin, J. I.,** Histamine increases cytosolic free Ca^{++} fluxes and intracellular pools, *Fed. Proc.*, 45 (Abstr.), 1852, 1986.

32. **Morgan-Boyd, R. and Hassid, A.,** Bradykinin modulation of cytosolic free Ca in cultured endothelial cells, *Fed. Proc.*, 45 (Abstr.), 790, 1986.

33. **Michel, C. C. and Phillips, M. E.,** Effects of the divalent cation ionophore, A 23187, on the permeability of single frog mesenteric capillaries, *J. Physiol. (London)*, 311, 111P, 1984.

34. **Michel, C. C.,** Vascular permeability — the consequences of Malpighi's hypothesis, *Int. J. Microcirc. Clin. Exp.*, 4, 265, 1985.

35. **Olesen, S.-P.,** Regulation of ion permeability in frog brain venules. Significance of calcium, cyclic nucleotides and protein kinase C, *J. Physiol. (London)*, 1987, in press.

36. **Hallam, T. J. and Rink, T. J.,** Responses to adenosine diphosphate in human platelets loaded with the fluorescent calcium indicator quin2, *J. Physiol.*, 368, 131, 1985.

37. **Gill, D. L., Ueda, T., Chueh, S.-H., and Noel, M. W.,** Ca^{2+} release from endoplasmic reticulum is mediated by a guanine nucleotide regulatory mechanism, *Nature (London)*, 320, 461, 1986.

38. **Bundgaard, M.,** Invaginations of the endothelial cell membrane — a possible clue to their functional significance, *Int. J. Microcirc. Clin. Exp.*, 5, 1209, 1986.

39. **Bruns, R. R. and Palade, G. E.,** Studies on blood capillaries. II. Transport of ferritin molecules across the wall of muscle capillaries, *J. Cell Biol.*, 37, 277, 1968.

40. **Furchgott, R. F. and Zawadski, J. V.,** The obligatory role of endothelial cells in the relaxation of arterial smooth muscle by acetylcholine, *Nature (London)*, 288, 373, 1980.

41. **Gerritsen, M. E. and Cheli, C. D.,** Arachidonic acid and prostaglandin endoperoxidase metabolism in isolated rabbit and coronary microvessels and isolated and cultivated coronary microvessel endothelial cells, *J. Clin. Invest.,* 72, 1658, 1983.
42. **Furchgott, R. F.,** Role of endothelium in responses of vascular smooth muscle, *Circ. Res.,* 53, 557, 1983.
43. **Vanhoutte, P. M., Rubanyi, G. M., Miller, V. M., and Houston, D. S.,** Modulation of vascular smooth muscle contraction by the endothelium, *Annu. Rev. Physiol.,* 48, 307, 1986.
44. **Johnson, A. R., Revtyak, G., and Campbell, W. B.,** Arachidonic acid metabolites and endothelial injury: studies with cultures of human endothelial cells, *Fed. Proc.,* 44, 19, 1985.
45. **Crone, C. and Olesen, S.-P.,** Electrical resistance of brain microvascular endothelium, *Brain Res.,* 241, 49, 1982.
46. **Olesen, S.-P., Saint-Aubain, de M. L., and Bundgaard, M.,** Permeabilities of single arterioles and venules in the frog skin: a functional and morphological study, *Microvasc. Res.,* 28, 1, 1984.

Chapter 35

ENDOTHELIAL CELL REACTIONS TO INFLAMMATORY MEDIATORS ASSESSED IN VIVO BY FLUID AND SOLUTE FLUX ANALYSIS

George J. Grega, Carl G. A. Persson, and Erik Svensjö

TABLE OF CONTENTS

I. INTRODUCTION

Vital activity of endothelial cells had already been suggested[1,2] before the turn of the century. However, the primary concept developed during much of this century suggested that endothelial cells were functionally inert and that the microvascular endothelium only served as a simple semipermeable membrane. In modern times, the stimulus to consider endothelial cells as important functional units of the vasculature was first provided by the classical study of Majno and Palade.[3] These investigators demonstrated that inflammatory mediators caused endothelial cell separation in the postcapillary venules[4-6] which permitted carbon particles to be trapped in venular gaps by the basement membrane. The formation of the venular large-junctional gaps soon became the accepted morphological basis of inflammatory mediator-stimulated increases in vascular permeability. The response suggested that vascular permeability was a physiologically regulated mechainsm rather than a non-specific response to vessel injury, and provided a functional rationale for distinguishing between venules and capillaries. Studies of vital activities of endothelial cells have come to the fore of vascular research during the past decade.[7-10]

It is evident that endothelial cell function may be assessed by studying the effects produced by changes in endothelial cell activity in addition to direct study of the cell itself. The aim of this paper is to review the evidence from analysis of fluid and solute fluxes demonstrating that macromolecular permeability is subject to direct physiological regulation, and to correlate the findings from these studies with those of the morphology of the vascular leakage sites and cellular mechanisms which affect the contractile state of the endothelial cell.

II. THE MAJNO-PALADE FINDINGS

Histamine-type inflammatory mediators were found to induce the formation of venular junctional gaps which were large enough to frequently trap platelets. It was eventually concluded that the formation of the junctional gaps was attributable to endothelial cell contraction.[3-6] This conclusion was supported by findings of functionally important changes in venular endothelial cell shape including a thickening of the cell and a bulging of the nucleus toward the lumen. In contrast to the smooth nucleus of the elongated endothelial cell at rest, the nucleus of the contracted endothelial cell had a wrinkled appearance suggesting contraction. Endothelial cell contraction implied the existence of receptors, contractile proteins, and receptor-activated modulation of the porosity of the microvascular endothelium. Although these studies provided the first evidence of a large pore system and its precise localization along the vascular tree, the findings stimulated little interest and lay largely dormant for more than a decade.

III. STARLING HYPOTHESIS

According to the Starling hypothesis,[11] fluid transfer is determined by the balance of hydrostatic and osmotic forces across the microvascular endothelium. The transmural hydrostatic pressure gradient promotes fluid filtration, whereas the transmural colloid osmotic pressure gradient opposes fluid filtration. Net fluid filtration occurs when the transmural hydrostatic pressure gradient exceeds the transmural colloid osmotic pressure gradient, and may be brought about by an increase in the former, a decrease in the latter, or by a combination of both. An increase in vascular permeability to macromolecules, if sufficiently large, virtually eliminates the transmural colloid osmotic gradient — the force opposing fluid filtration. Net fluid filtration occurs at almost any microvascular pressure (Pmv) under these conditions, and the filtration response is greatly amplified if Pmv increases. Edema develops when the rate of fluid filtration exceeds the rate of lymphatic drainage. The studies of Starling demonstrated that transvascular fluid and solute exchange were passive events regulated by imbalances in physical forces. The microvascular endothelium appeared to be functionally analogous to that of an artificial semipermeable membrane perforated with static pores having fixed hydraulic conductivities. Indeed, fluid and solute transport under basal conditions could be satisfactorily explained by filtration/diffusion across a static heteroporous microvascular endothelium. However, a static heteroporous model of the microvascular endothelium cannot explain the extravasation of plasma with a high protein concentration in the absence of vessel injury.[12] The emphasis on studies demonstrating that fluid and solute fluxes were passive events determined by imbalances in physical forces inadvertently diverted attention from the endothelial cell. Surprisingly, few considered the possibility that although fluid and solute fluxes were passive events, the porosity of the microvascular endothelium might be subject to active physiological regulation.

In contrast to classical pore theory, the Majno-Palade studies demonstrated that the microvascular endothelium also has a system of variable large pores. The opening of the large pores (venular junctional gaps) disturbs the equilibrium of the Starling forces, virtually eliminating the transmural colloid osmotic pressure gradient by permitting plasma proteins to flood into the interstitium.

IV. INFLAMMATORY MEDIATOR-STIMULATED VASCULAR LEAKAGE

A. Intravital Light Microscopy

Intravital light microscopy has proved invaluable in studies of the mechanism by which inflammatory mediators produce increases in the flux of macromolecules[13] across the microvascular endothelium. Following the intravenous (i.v.) injection of a suitable macromolecular tag such as FITC-dextran (70,000 or 150,000 daltons), intravital microscopy permits the precise localization of sites of tracer extravasation along the vascular tree, and study of the time course of the formation and disappearance of vascular leakage sites. The check pouch of the hamster is ideally suited for studies of vascular permeability by intravital light microscopy, and this preparation was used in many of the studies discussed in this review. In addition, the simultaneous measurement of the number of the vascular leakage sites and plasma to suffusate tracer fluxes permits determination of the relationship between the increase in tracer efflux and the formation of the vascular leakage sites, and the duration of the increase in vascular permeability.

B. Morphology of Venular Leakage Sites

Following exposure to various inflammatory mediators, macromolecular tags extravasate from postcapillary venules 8 to approximately 30 μm in diameter.[14] Histamine, leukotrienes, platelet activating factor, immune complexes, bradykinin, complement fragments, other plasma-derived mediators, and free radicals are among the many inflammatory mediators known to induce visible leakage of macromolecules from postcapillary venules.[15-20] These mediators fail to induce leakage of macromolecules from other microvessels (arterioles or capillaries), even if the mediator is directly applied in high concentrations. Thus, inflammatory mediators of widely diverse chemical composition produce a characteristic pattern of vascular leakage from postcapillary venules. The formation of the venular leakage sites is dose dependent, subject to inhibition by receptor antagonists which are specific for that particular mediator, and independent of the hemodynamic actions of the inflammatory mediators. Histamine-type inflammatory mediators produce arteriolar vasodilation. Arteriolar vasodilation increases Pmv which tends to promote net fluid filtration. However, noninflammatory vasodilators such as papaverine or isoproterenol, agents with hemodynamic profiles similar to that of histamine, fail to induce the formation of vascular leakage sites.[21-23] These findings suggest that mediator-stimulated venular macromolecular leakage results from the stimulation of a physiological mechanism rather than from a hemodynamic effect or injury to the vascular endothelium.

Electron-microscopic study of the mediator-stimulated venular leakage sites reveals gaps between endothelial cells 0.2 to 1.1 μm in width.[4-6,24-26] Precipitates of the tracer macromolecules are found in the vascular lumen, the venular junctional gaps, and in the interstitium. In control animals, venular junctional gaps are not observed, and precipitates of the macromolecular tag are found only in the vascular lumen.

C. Endothelial Cell Receptors and Contractile Proteins

A wide variety of endothelial cell receptors have been identified, including receptors for inflammatory mediators and antipermeability agents.[7,9,27-30] The highest density of histamine receptors, for example, are found in postcapillary venules.[29] Capillaries are essentially devoid of histamine receptors. The histamine receptors are strategically localized near venular endothelial cell junctions in proximity to the contractile proteins. Histamine-stimulated increases in vascular permeability to macromolecules are primarily due to the stimulation of H_1 receptors.[31-33] The increase in vascular permeability may be prevented by treatment with selective H_1 receptor antagonists. Specific H_2 receptor antagonists reduce histamine-stimulated increases in vascular permeability in some tissues, perhaps due to crossover blockade of H_1 receptors or a diminution of flow.

Various cytoplasmic filaments are found in endothelial cells, including microtubules, intermediate filaments, microfilaments (actin), and myosin-like filaments.[8,34-38] The distribution of the contractile proteins in venular endothelial cells differs from that in arterial endothelial cells. In venular endothelial cells, the contractile proteins are localized along cell junctions, regions with high densities of mediator receptors.[38] It is interesting to note that mediator-stimulated, receptor-activated junctional-gap formation has been demonstrated in cultured endothelial cells.[39] The strength of the endothelial cell contraction is sufficient to cause retraction of clots,[40] which is further evidence that the contractile system is functional. Mediators which produce increases in venular permeability decrease the formation of F-actin cables in endothelial cells which results in the loss of junctional integrity[41] and the formation of junctional gaps. Cytochalasin B, a substance which inhibits F-actin polymerization, inhibits mediator-stimulated increases in vascular permeability.[42] Calcium is also intimately linked to the formation of the venular leakage sites.[42,43] Some calcium channel entry blockers inhibit the formation of mediator-stimulated venular leakage sites.[44] In contrast, the calcium ionophore A23187 induces the formation of venular leakage sites and plasma extravasation,[45] thus mimicking the effects produced by the inflammatory mediators. These findings are consistent with those demonstrating a role for cytosolic calcium in the reversible modulation of junctional dimensions in monolayers of epithelial cells. The junctions are opened by increases in calcium in the epithelial cells.[46] Melittin, which forms channels in lipid bilayers and permits calcium entry,[47] also produces increases in vascular permeability.[43] These observations support the hypothesis that the dimensions of the junctional gaps in the postcapillary venules are subject to physiological modulation via receptor-operated endothelial cell contraction/relaxation mechanisms.

D. Relationship between the Formation of the Venular Leakage Sites and the Increase in Macromolecular Efflux

Most inflammatory mediators produce a prompt increase in the formation of leakage sites[48-52] as evidenced by the extravasation of labeled macromolecular tracers from postcapillary venules in as little as 20 sec. The peak vascular leakage response usually occurs within 5 min. The venular leakage sites initially appear as intensely bright, discrete spots. Thereafter, the venular leakage sites slowly disperse, fade, and are eventually cleared by the suffusion solution.[18,20,21] The formation of the venular leakage sites is associated with increases in the concentration of the macromolecular tag in the effluent suffusion solution. The increase in the tracer concentration of the effluent suffusate and the density of the venular macromolecular leakage sites are dose dependent, and the correlation coefficient between the increase in efferent suffusate tracer concentration and the formation of the venular leakage sites is highly significant.[18,52]

E. Duration of the Increase in Venular Permeability

The duration of mediator-stimulated increases in venular permeability is not readily determined from intravital light-microscopic observations alone. The venular leakage sites may persist for prolonged periods of time before fading and eventually disappearing altogether. Inferring the duration of the increased venular permeability only by noting the time course of the formation and disappearance of the venular leakage sites may result in large overestimations of the duration of the increased venular permeability. However, intravital light microscopy combined with direct measurements of plasma to suffusate tracer fluxes permits determination of the duration of the increase in venular permeability by varying the time of the i.v. injection of the macromolecular tracer relative to the start of the mediator suffusion.[20,21] If the tracer is injected i.v. during the time which reveals the formation of the venular leakage sites, then the tracer concentration in the effluent suffusate is markedly increased. However, if the tracer is injected at a time which fails to reveal the formation of venular leakage sites (usually after 10 to 20 min), then the tracer concentration in the effluent suffusate falls to increase.[20,21] Thus, mediator-stimulated increases in macromolecular efflux are dependent on the transient formation of the venular leakage sites. The persistence of the venular leakage sites after the closing of the venular junctional gaps simply reflects the exceedingly slow washout of the extravasated tracer. The period of time that the venular leakage sites persist before fading and disappearing varies inversely with the suffusate flow rate.[49] If the suffusate flow rate is high, then the venular leakage sites appear relatively short lived. In contrast, the venular leakage sites appear to persist for prolonged periods of time before fading and disappearing when suffusate flow rate is zero.

The duration of the increase in vascular permeability in various pathophysiological states may be either transient or sustained.[12] However, it is not always clear whether evidence that vascular leakage sites may be revealed over prolonged periods of time reflects a sustained increase in vascular permeability in the same vessel or simply the spreading of the inflammatory response to adjacent vessels. If the latter, then the duration of the increase in vascular permeability may be transient, but high vascular permeability may be maintained in the inflamed area by the recruitment of other vessels which were not previously affected. Although all known inflammatory mediators produce only transient increases in vascular permeability, a sequential release of mediators could result in prolonged increases in vascular permeability, either by inducing vascular leakage in adjacent vessels which were not previously affected or by keeping the vascular permeability increased in a given vessel. This conclusion is supported by the finding that after becoming refractory to one mediator, the microvascular endothelium responds to a different mediator. Activation of extravasated plasma proteins by proteases and other factors which are abundant in the interstitial fluid of inflamed tissues may result in the production of potent inflammatory mediators. A vicious cycle may be established by such a positive feedback system resulting in sustained increases in vascular permeability.

These observations also fail to support the contention that inflammatory mediators stimulate protein transport via vesicular transport or another active transport mechanism after the venular junctional gaps close. This conclusion is supported by physiological, pharmacological, and morphological findings. Free vesicles are abundant in endothelial cells prepared for electron-microscopic study by conventional sectioning. Ultra-thin serial sectioning of the endothelial cells reveals few if any free vesicles.[43] The vesicles revealed by ultra-thin serial sectioning were found to be formed by invaginations of the endothelial plasmalemmal surface, and were directly connected to the vessel lumen. The absence of free vesicles strongly argues against the theory of vesicular protein transport. The function of the plasmalemmal invaginations is not known, but is probably unrelated to the transport of macromolecules across the vascular endothelium.

Continuous exposure of the microvascular endothelium to histamine-type inflammatory mediators results in tachyphylaxis to that particular mediator.[18,23,49-52] However, if the exposure time is short (~5 min) and if the period of time between exposures is at least 20 to 30 min, then the microvasculature responds to repeated applications of the mediator usually without appreciable diminution of effect. In contrast, the microvascular endothelium of immunized animals is refractory to repeated exposure to antigen.

F. Potentiation of Mediator-Stimulated Venular Leakage

The prostaglandins either exert no effect or produce only feeble increases in the formation of venular macromolecular leakage sites. However, the prostaglandin Es, at least, dramatically potentiate the formation of venular leakage sites produced by other inflammatory mediators.[53,54] The potentiation is independent of the vasodilator action of the prostaglandins. The vasodilators papaverine and acetylcholine fail to potentiate the formation of mediator-stimulated venular leakage sites, whereas the equipotent vasodilators isoproterenol and terbutaline inhibit mediator-stimulated formation of venular leakage sites. The potentiation must be ascribed to a nonhemodynamic action which enhances the venular endothelial response to the inflammatory mediator. It is interesting to note that of the many endogenous inflammatory mediators, the prostaglandins are the only mediators known to potentiate the formation of venular macromolecular leakage sites. However, in underperfused tissues, vasodilation will enhance mediator-stimulated extravasation of macromolecules.[55] The enhanced extravasation results from an increased perfused surface area and Pmv. The prostaglandins, in addition to producing vasodilation, potentiate mediator-stimulated formation of venular leakage sites, which is a nonhemodynamic effect.

V. INFLAMMATORY MEDIATOR-STIMULATED INCREASES IN FLUID FILTRATION, PROTEIN EFFLUX, AND EDEMA FORMATION

The canine forelimb perfused either naturally or at constant flow was employed as the test organ in many of the studies discussed in this section. Canine and feline hindlimbs, hindquarters, or specific hindlimb skeletal muscles were also used in some of the studies.

A. Hemodynamic vs. Nonhemodynamic Effects

It would be predicted that the formation of the venular leakage sites would have a dramatic impact on transvascular fluid and protein fluxes resulting in massive edema formation. This prediction is supported by experimental findings.[56-63] Local i.a. infusions of inflammatory mediators such as histamine or bradykinin into the canine forelimb produce vasodilation and marked increases in fluid filtration, protein clearance, lymph total protein concentration, and edema formation. The lymph to plasma ratio for total plasma proteins (L/P protein ratio) increases from control values of approximately 0.35 to 0.85, which is evidence of an increase in vascular permeability. The vasodilation and increase in protein clearance and edema formation are dose dependent. The edema produced by local infusions of high concentrations of the mediators is visibly evident as limb volume may be more than doubled. Lymph flow rate is increased initially, but may return to control levels as the massive edema may physically compress the lymphatic vessels thus blocking lymphatic drainage. Small vein pressure, which represents a minimum for Pmv, is increased by local i.a. infusions of histamine-type mediators subsequent to arteriolar vasodilation. Thus, the increase in fluid filtration is attributable, in part, to an increase in the transmural hydrostatic pressure gradient.

B. Noninflammatory Vasoactive Agents

Local i.a. infusions of noninflammatory vasodilators such as papaverine or acetylcholine produce increases in blood flow and small vein pressure comparable to that produced by histamine-type mediators.[56-63] However, the increase in fluid filtration, protein clearance, and edema formation is only a small fraction of that produced by the inflammatory mediators, and lymph protein concentration and, inferentially, interstitial fluid protein concentration fail to increase. The noninflammatory vasodilators fail to increase the L/P protein ratio; thus, the filtrate must be protein poor. Vasoconstrictors such as norepinephrine and serotonin also fail to produce increases in the L/P protein ratio despite producing venoconstriction which increases Pmv and net fluid filtration under constant flow conditions.[64,65]

C. Nonhemodynamic Effects of the Inflammatory Mediators

If limb blood flow is held constant, noninflammatory vasodilators fail to promote increases in Pmv, fluid filtration, protein clearance, or edema formation. In contrast, inflammatory mediators promote increases in fluid filtration, protein clearance, the L/P ratio, and edema formation under constant flow. The edema is comparable to that produced under natural flow conditions, although the filtration rate is considerably reduced when flow and Pmv are prevented from increasing. Hence, more time is required to produce a comparable degree of edema under constant flow conditions. Since the inflammatory mediators produced increase the L/P protein ratio in the absence of an increase in blood flow and Pmv, the increase in protein efflux and the L/P protein ratio must result from a direct action[69-71] on the venular endothelium, which virtually eliminates the transmural colloid osmotic pressure gradient. Thus, mediator-stimulated edema formation is primarily attributable to a direct action on the microvascular endothelium which results in the formation of a hydraulically conductive macromolecular transport pathway. However, the rate of filtration and edema formation varies directly with Pmv,[72,73] which amplifies the response produced by the direct action of the inflammatory mediators.

D. Potentiation of Mediator-Stimulated Fluid Filtration and Protein Efflux

Prostaglandin E_1, for example, greatly amplifies the increase in fluid filtration, protein clearance, and edema formation produced by various inflammatory mediators.[74,75] The potentiation must be ascribed to a nonhemodynamic action of the inflammatory mediators, as it may be demonstrated under conditions which prevent blood flow and Pmv from increasing. Moreover, other potent vasodilators such as acetylcholine or papaverine fail to potentiate mediator-stimulated increases in fluid filtration, protein clearance, and edema formation in forelimbs perfused at constant flow.

E. Duration of Mediator-Stimulated Increases in Protein Transport

Local i.a. infusions of inflammatory mediators produce sustained increases in the L/P protein ratio, which many have interpreted as evidence for a sustained increase in protein efflux. Since microscopic data suggest that inflammatory mediators produce marked but transient increases in venular permeability, some have suggested that the mediators may stimulate an active protein transport mechanism to account for the apparently sustained increase in protein efflux. However, a sustained increase in the L/P ratio is not necessarily evidence of a sustained increase in protein transport. Slow lymphatic drainage of the engorged interstitium by the lymphatics could readily explain the sustained increase in the L/P ratio, and this conclusion is supported by the following observations.

The findings from studies in which an independent macromolecular tracer was injected i.v. at varying intervals after the start of a prolonged histamine infusion demonstrates that prolonged histamine infusions produce transient increases in macromolecular permeability lasting less than 30 min.[18] Second, prolonged histamine infusions produce only transient increases in the capillary filtration coefficient (CFC) lasting 10 to 20 min, and this is evidence of a transient increase in vascular permeability.[69-70] Finally, both short-term and prolonged local infusions of inflammatory mediators produce similar changes in limb weight, lymph flow, and lymph protein concentration, suggesting that the increase in vascular permeability is transient.[23] Thus, the persistent elevation of the L/P protein ratio reflects the slow lymphatic removal of the extravasated proteins rather than a sustained increase in vascular permeability.

F. Evidence that the Increase in Macromolecular Efflux is due to the Stimulation of an Active Physiological Mechanism

Histamine increases fluid filtration and the CFC in the maximally dilated rat hindquarter preparation,[76] as evidence of an increase in vascular permeability. If the temperature of the perfusate is reduced to 5°C, then histamine fails to promote increases in fluid filtration and CFC.[76] In contrast, the increase in protein clearance produced by increases in Pmv are not affected by a reduction in perfusate temperature to 5°C.[77-79] Thus, the histamine-stimulated macromolecular transport pathway differs from the transport pathway utilized under basal conditions. These findings provide evidence that the microvascular endothelium regulates transvascular protein movement via a system of static and variable large pores, and that the variable large pores are subject to direct regulation.

VI. VASCULAR PERMEABILITY INCREASES IN PATHOPHYSIOLOGICAL STATES

Various patterns of vascular leakage are reported to occur in pathophysiological states associated with increases in vascular permeability.[12,80] Severe trauma or thermal injury may damage a portion of the microvascular endothelium resulting in the extravasation of macromolecules from arterioles, capillaries, and venules. Electron-microscopic findings provide evidence of frank endothelial cell damage or destruction under these conditions. Less severe injury results in vascular leakage from interendothelial cell gaps, usually in postcapillary venules, and endothelial cell damage is virtually nonexistent in these vessels. The formation of the venular gaps is attributable to the action of various inflammatory mediators on the venular endothelial cells. Interendothelial cell gap formation in the capillaries, when it occurs, cannot be ascribed to the action of any known inflammatory mediator. It has been suggested that increases in capillary permeability may result from an osmotic swelling of the cell which produces a functionally important change in cell shape resulting in the detachment of adjacent cells.

Bronchial airway disease is associated with increases in macromolecular permeability.[81-82] Inflammatory provocation and plasma excudation may not produce marked bronchial edema, as the inflamed epithelial lining may be freely pervious to macromolecules. Superfusion of the guinea pig tracheal mucosa with a variety of inflammatory mediators or following the induction of a local Ige-mediated allergic reaction results in the recovery of a significant amount of macromolecules from the tracheal lumen. The route for this passage is short. In many species, there is a richly developed capillary-venular barrier situated just beneath the epithelium. Following exposure to an inflammatory stimulus, extruded plasma proteins and its activated products may be readily recovered by washing the mucosal surface. Although two barriers are involved, experiments of this type nevertheless provide a measure of changes in the permeability of the endothelial-epithelial barrier. In man, nasal provocations followed by nasal washings may be used to gain important information about the regulation of microvascular permeability in humans in health and disease. In allergic airway disease, plasma excudation may be an important factor in the pathogenesis of the disease. The physical and physiological aspects of an increased microvascular and epithelial bronchial permeability suggest involvement of this mechanism in several facets of asthma pathology.

Ischemia may also result in marked increases in macromolecular permeability. In canine forelimb or hindlimb, several hours of ischemia may be tolerated[84] before increases in macromolecular permeability are observed.[85] In tissues such as the hamster cheek pouch, ischemia for as little as 15 min results in marked increases in macromolecular permeability subsequent to the formation of venular leakage sites.[86]

VII. ENDOTHELIAL CELL STABILIZERS

A wide variety of inflammatory mediators appear to participate in pathophysiological states associated with increases in macromolecular permeability and, therefore, blockade of the receptor of one or even several mediators may not exert a significant inhibitory effect on pathophysiological-stimulated increases in vascular permeability. This conclusion is supported by the failure of inflammatory mediator receptor antagonists to prove clinically effective in the management of various inflammatory states.

In the mid 1970s, agents were discovered which inhibited mediator-stimulated formation of venular leakage sites[19,23,87] and increases in fluid filtration, protein clearance, the L/P protein ratio, and edema formation.[88-92] The inhibition was independent of changes in hemodynamics, and was attributable to an action on the microvascular endothelium which effectively counteracted the direct action of the inflammatory mediator. In contrast to the inhibition produced by specific mediator receptor antagonists, agents such as the beta-adrenergic stimulants and the glucocorticoids were found to inhibit increases in macromolecular permeability produced by inflammatory mediators of widely differing chemical classes and composition. These unique mediator inhibitors may be termed "endothelial cell stabilizers", and function as physiological rather than pharmacological receptor antagonists of mediator-stimulated increases in venular permeability. That is, the endothelial cell stabilizers trigger an endothelial cell response which is opposite to that produced by the inflammatory mediators. The inhibition is receptor operated, but involves the activation of a receptor different from that stimulated by the inflammatory mediator.

A. Pharmacological Inhibition of Mediator-Stimulated Venular Leakage Sites

The suffused hamster cheek pouch was employed in many of the intravital light-microscopic studies discussed in this section. The canine forelimb perfused either naturally or at constant flow was used as the test organ in most of the studies which focused on changes in filtration rates, protein clearance, and L/P ratio, and edema formation. In some studies, canine or feline hindlimbs or hindquarters or specific hindlimb skeletal muscles were employed as the test organ.

1. Venular Leakage and Macromolecular Efflux

Norepinephrine, other drugs which stimulate beta-adrenergic stimulants, vasopressin, glucocorticoids, and xanthines are among the substances which may function as physiological antagonists of mediator-stimulated increases in venular permeability.[33,50,52,93,94] Isoproterenol or terbutaline, for example, produce a dose-dependent inhibition of mediator-stimulated formation of venular leakage sites. In fact, treatment with 10^{-5} M concentrations of the β-adrenoceptor stimulant isoproterenol completely prevents the formation of venular leakage sites produced by inflammatory mediators[21] and immune complexes.[20] In mild airway inflammation, treatment with terbutaline inhibited vascular leakage already at 10^{-7} M, which corresponds to 10^{-8} M of isoproterenol. Ischemia-induced increases in macromolecular permeability are also susceptible to inhibition by terbutaline.[86] The inhibition of the venular leakage is triggered by stimulation of β-adrenoceptors,[23] but is independent of the hemodynamic actions of the beta-adrenoceptor stimulant.[23,88] The β-adrenoceptor stimulants are potent vascular smooth muscle relaxants whereas vasopressin produces vasoconstriction. Therefore, the inhibition of venular leakage is independent of vasoconstriction or vasodilation per se, and may be attributed to a direct action on the microvascular endothelium. DDAVP, an analog of vasopressin with ADH activity but devoid of vasoconstrictor activity, also inhibits mediator-stimulated formation of venular leakage sites.[93]

The inhibition of vascular leakage produced by the β-adrenoceptor stimulant terbutaline is stereospecific, and the antipermeability effect resides in the l-isomer.[50,97] The d-isomer of terbutaline is inactive, and fails to inhibit mediator-stimulated venular leakage in concentrations 100 times greater than the effective concentration of l-terbutaline. The inhibition of mediator-stimulated formation of venular leakage sites produced by the β-adrenoceptor stimulants may be reversed by treatment with β₂-adrenoceptor blocking agents.[94] If the venular leakage is attributable to endothelial cell contraction, it is reasonable to assume that the endothelial cell stabilizers such as the β-adrenoceptor stimulants produce a receptor-mediated physiological antagonism by relaxing the endothelial cell contractile elements and thereby negating the action of the inflammatory mediators. Beta-adrenoceptor stimulation increases cAMP. It is interesting to note that xanthines and stimulation of the ADH receptor also increase cAMP.[95] Bottaro et al.[95] demonstrated that agents which increase cAMP increase the formation of actin cables in the endothelial cell. The actin cable bundles maintain the relatively tight junctional apposition of adjacent endothelial cells. In contrast, inflammatory mediators decrease the density of the actin cables resulting in loss of the integrity of the junctional barrier. Receptor-mediated modulation of permeability is not unique to vascular endothelium. The discovery of catecholamine-induced modulation of venular permeability was soon followed by the reporting of similar findings in renal tubular epithelial cells. Catecholamines were shown to directly modulate the epithelial transport of sodium subsequent to a direct action on renal tubule completely independent of changes in systemic or renal hemodynamics.[96]

2. Fluid and Solute Flux Analysis

Endothelial cell stabilizers such as isoproterenol, vasopressin, terbutaline, norepinephrine, and methylprednisolone which inhibit the formation of venular leakage sites also inhibit mediator-stimulated increases in fluid filtration, protein clearance, the L/P protein ratio, and edema formation.[97-100] The inhibition of mediator-stimulated increases in vascular permeability is attributable to a receptor-mediated, nonhemodynamic action of the endothelial cell stabilizers. The inhibition may be demonstrated under natural flow or constant flow conditions with either vasodilator or vasoconstrictor stabilizers. Vasoconstrictor and vasodilator agents (angiotensin II, acetylcholine, papaverine) which fail to inhibit mediator-stimulated formation of venular leakage sites also fail to inhibit mediator-stimulated increases in fluid filtration, protein clearance, and edema formation. For example, the vasoconstrictors norepinephrine and vasopressin, and the vasodilators isoproterenol and terbutaline, are potent endothelial cell stabilizers. Norepinephrine stabilizes the venular endothelial cell even following alpha-adrenoceptor blockade which inhibits the vasoconstrictor action of this catecholamine. The endothelial cell stabilization produced by norepinephrine and the β-adrenoceptor stimulants is a β-adrenoceptor effect which is negated by beta-adrenoceptor blockade. Angiotensin II or dopamine, in doses which produce an equal or greater degree of vasoconstriction than vasopressin, fail to inhibit mediator-stimulated increases in vascular permeability.[97,98,100] The antiinflammatory glucocorticoid methylprednisolone fails to produce measurable hemodynamic effects, yet inhibits mediator-stimulated increases in vascular permeability. Isoproterenol and vasopressin, however, are far more potent endothelial cell stabilizers than methylprednisolone in the canine forelimb. Methylprednisolone inhibits mediator-stimulated increases in fluid filtration, protein clearance, and edema formation produced by low but not high doses of the mediators. The dramatic inhibition of mediator-stimulated increases in vascular permeability emphasizes the primary importance of the increase in vascular permeability to the formation of inflammatory edema.

B. Endogenous Mechanisms Inhibiting Mediator-Stimulated Venular Leakage
1. Local vs. Intravenous Infusions of Mediators

It has long been known that i.v. infusions of mediators[101] fail to produce increases in vascular permeability even though the estimated plasma concentration of the mediator may exceed that achieved by local infusions which produce marked increases in vascular permeability. Intravenous infusions of vasodilator inflammatory mediators produce profound hypotension whereas local infusions of inflammatory mediators in doses which produce increases in vascular permeability exert modest effects on blood pressure. Hypotension is a stimulus for an increased sympathoadrenalmedullary discharge, and many of the inflammatory mediators cause a direct release of adrenal catecholamines. These observations suggest the possibility that circulating catecholamines may antagonize the direct effect of the mediators on the microvascular endothelium, thereby accounting for the differential effects on vascular permeability produced by intravenous and local infusions of the inflammatory mediators. Following β-adrenoceptor blockade, intravenous infusions of inflammatory mediators produce increases in the L/P protein ratio[102] in forelimbs perfused at constant flow. Thus, endogenous catecholamines released by intravenous infusions of large doses of histamine inhibit the direct action of this mediator on the microvascular endothelium, and account, in part, for the differential effect of intravenous and local infusions of histamine on vascular permeability.

2. Sympathoadrenalmedullary-Mediated Inhibition of Vascular Permeability

The inhibitory effect of increased sympathoadrenalmedullary activity on mediator-stimulated increases in vascular permeability is also revealed in experiments in which canine forelimbs are perfused at constant flow with autologous blood from either normotensive animals or animals previously hemorrhaged.[90-92] Hemorrhage is a potent stimulus for increased sympathoadrenalmedullary activity. The local i.a. infusion of the inflammatory mediator into the forelimbs of normotensive control animal perfused at constant flow marked vasodilation and increases in fluid filtration, the L/P protein ratio, and edema formation. In contrast, the local i.a. infusion of the inflammatory mediator into the forelimbs of the previously hemorrhaged animals perfused at constant flow produces marked vasodilation, but fails to increase fluid filtration, the L/P protein ratio, or edema formation. These observations suggest that something released into the blood of the hemorrhaged animals effectively negates the direct action of the inflammatory mediator on the microvascular endothelium. Several substances found in blood from hemorrhaged animals — catecholamines, vasopressin, and serotonin (in the dog) — function as endothelial cell stabilizers. It appears unlikely that adrenal cortical hormones contribute inportantly to the inhibition of mediator-stimulated increases in vascular permeability, as the exogenous administration of high pharmacological doses of glucocorticoids (40 mg/kg) fail to modify the increases in vascular permeability produced by the doses of inflammatory mediators that were used in these studies. Further experimentation will be necessary to determine whether the adrenal cortical hormones function as important endogenous endothelial cell stabilizers.

3. Kininases and Catecholamines

Bradykinin is rapidly metabolized by kininase II. This is evidenced by a rapid waning of the vasodilator response during prolonged infusions of this mediator. The failure of intravenously infused bradykinin to produce increases in vascular permeability in limbs perfused at constant flow comparable to that produced by local infusions of this agent is primarily related to the inactivation by kininase II. The breakdown of bradykinin by kininase II may be prevented by treatment with the kininase II inhibitor captopril. Following treatment with captopril, i.v. infusions of even low concentrations of bradykinin now produce profound systemic hypotension subsequent to arteriolar vasodilation. Although bradykinin produces arteriolar vasodilation, it fails to produce increases in vascular permeability under these conditions. Following treatment with both captopril and propranolol, i.v. infusions of low concentrations of bradykinin now produce marked increases in vascular permeability despite producing severe hypotension.[103] These observations suggest that following treatment with captopril, intravenously infused bradykinin triggers an increased sympathoadrenalmedullary discharge which antagonizes the direct effect of bradykinin on the microvascular endothelium. It is well documented that bradykinin is a potent releaser of adrenal catecholamines. Therefore, if the metabolism by kininase II is prevented, i.v. infusions of bradykinin then produce both profound hypotension and the release of adrenal catecholamines. Blockade of the beta-adrenoceptors by treatment with propranolol prevents catecholamine-induced inhibition of the direct action of bradykinin on the microvascular endothelium. Under these conditions, the intravenous infusion of bradykinin now produces marked increases in fluid filtration, the L/P protein ratio, and edema formation in canine forelimbs perfused at constant flow.

4. Safety Factor Against Edema Formation

The observations that (1) the failure of i.v. infusions of certain inflammatory mediators to produce increases in macromolecular permeability comparable to that produced by local infusions is related, in part, to an endogenous release of catecholamines (which may be reversed by β-adrenoceptor blockade), and (2) that perfusion of limbs with blood from hemorrhaged animals (a potent stimulus for increased sympathoadrenalmedullary activity) inhibits the increase in macromolecular permeability produced by local infusions of inflammatory mediators suggest that the sympathoadrenalmedullary hormones function as endogenous endothelial cell stabilizers. Thus, the sympathoadrenalmedullary hormones may function as an important safety factor protecting against serious plasma volume loss in pathophysiological states associated with a generalized increase in inflammatory mediators as in anaphylaxis. Many of the endogenous endothelial cell stabilizers are also vasoconstrictors which decrease blood flow, Pmv, and perfused surface area which minimizes the hydrostatic component of edema formation. Second, the endothelial cell stabilizing properties of these agents prevent large decreases in the transmural colloid osmotic pressure gradient, thus minimizing the permeability component of mediator-edema formation. The fact that many inflammatory mediators induce a direct release of adrenalmedullary catecholamines may be by design rather than by happenstance. Localized inflammatory responses would not trigger the release of adrenalmedullary catecholamines thus permitting local inflammatory responses to develop. In fact, many of the inflammatory mediators which are potent releasers of adrenalmedullary catecholamines inhibit neurotransmitter release from adrenergic nerve endings via presynaptic actions. The local inhibition of adrenergic neurotransmitter release from adrenergic neurons would permit local inflammatory responses to continue unabated. The role, if any, of the adrenal cortical hormones in the endogenously-mediated inhibition of mediator stimulated increases in macromolecular permeability cannot be determined from these studies.

VIII. CORRELATION OF MORPHOLOGICAL, PHYSIOLOGICAL, AND PHARMACOLOGICAL EVIDENCE FOR THE REGULATION OF VENULAR PERMEABILITY

Venular large junctional gap formation is a characteristic feature of pathophysiological states associated with increases in vascular permeability. Inflammatory mediators of diverse chemical classes and composition trigger receptor-activated formation of venular macromolecular leakage sites. Electron-microscopic analysis of the venular leakage sites reveals the formation of large junctional gaps in the postcapillary venules. The formation of the venular macromolecular leakage sites and the receptor-mediated increases in fluid filtration, protein clearance, the L/P protein ratio, and edema formation are independent of the hemodynamic actions of the inflammatory mediators, and may be ascribed to a direct effect on the venular endothelium. Inflammatory mediators which induce the formation of venular macromolecular leakage sites also produce increases in fluid filtration, protein clearance, and the L/P protein ratio, thus promoting massive edema formation. Noninflammatory vasoactive agents fail to induce the formation of venular macromolecular leakage sites or to produce increases in the L/P protein ratio. Prostaglandins which potentiate mediator-stimulated formation of venular leakage sites also dramatically potentiate mediator-stimulated increases in fluid filtration, protein clearance, and edema formation. Endothelial cell stabilizers which suppress mediator-stimulated formation of venular leakage sites also inhibit mediator-stimulated increases in fluid filtration, protein clearance, the L/P ratio, and edema formation. Nonendothelial cell stabilizers which fail to suppress mediator-stimulated formation of venular leakage sites also fail to prevent mediator-stimulated increases in fluid filtration, protein clearance, the L/P protein ratio, or edema formation. Inflammatory mediators produce marked but transient increases in macromolecular permeability as assessed either by analysis of the venular leakage sites or by changes in the plasma-to-lymph transport of macromolecules, but sufficient to result in massive edema formation.

Mediator-stimulated increased macromolecular transport occurs via a hydraulically conductive pathway, as protein clearance is dramatically facilitated by increases in Pmv. Studies of endothelial cell receptors, contractile proteins, and stimulus-response coupling mechanisms strongly suggest that macromolecular permeability is subject to direct receptor-activated modulation subsequent to venular endothelial cell contraction/relaxation.[103] Under basal conditions, macromolecules are primarily transported via a static large pore system. In inflammatory states, macromolecules are primarily transported via a receptor-activated variable large-pore system. The correlation among physiological, pharmacological, and morphological findings suggest that the venular large-junctional gaps are the equivalent of the receptor-activated, hydraulically conductive variable macromolecular transport pathway revealed by studies of fluid and protein transport across the microvascular endothelium.

REFERENCES

1. **Heidenhain, R.,** Versuche und Fragen Zur Lehre von der Lymphbildung, *Pfluegers Arch. Gesamte Physiol. Menschen Tiere,* 49, 209, 1891.
2. **Krogh, A.,** *The Anatomy and Physiology of Capillaries,* Yale University Press, New Haven, Conn., 1922.
3. **Majno, G. and Palade, G. E.,** Studies on inflammation. I. The effect of histamine and serotonin on vascular permeability: an electron microscopic study, *J. Biophys. Biochem. Cytol.,* 11, 571, 1961.
4. **Majno, G., Palade, G. E., and Schoefl, G. I.,** Studies on inflammation. II. The site of action of histamine and serotonin along the vascular tree: A topographic study, *J. Biophys. Biochem. Cytol.,* 11, 607, 1961.
5. **Majno, G., Gilmore, V., and Leventhal, M.,** On the mechanism of vascular leakage caused by histamine-type mediators, *Circ. Res.,* 21, 833, 1967.
6. **Majno, G., Shea, S. M., and Leventhal, M.,** Endothelial contraction induced by histamine-type mediators, an electron microscopic study, *J. Cell Biol.,* 42, 647, 1969.
7. **Furchgott, R. F.,** Role of the endothelium in responses of vascular smooth muscle, *Circ. Res.,* 53, 557, 1983.
8. **Shepro, D. and D'Amore, P. A.,** Physiology and biochemistry of the vascular wall endothelium, in *Handbook of Physiology, Section 2, The Cardiovascular System, Vol. IV, Microcirculation, Pt. 1* Renkin, E. M. and Michel, C. D., Eds., American Physiological Society, Washington, D.C., 1984, 103.

9. **Furchgott, R. F.**, The role of the endothelium in the responses of vascular smooth muscle to drugs, *Annu. Rev. Pharmacol. Toxicol.*, 24, 175, 1984.

10. **Ryan, U. S.**, Structural basis for metabolic activity, *Annu. Rev. Physiol.*, 44, 223, 1982.

11. **Starling, E.**, On the absorption of fluids from the connective tissue spaces, *J. Physiol. (London)*, 19, 312, 1896.

12. **Hurley, J. V.**, Inflammation, in *Edema*, Staub, N. C. and Taylor, A. E., Eds., Raven Press, New York, 1984, 463.

13. **Svensjö, E., Arfors, K.-E., Arturson, G., and Rutili, G.**, The hamster cheek pouch preparation as a model for studies of macromolecular permeability of the microvasculature, *Upsala J. Med. Sci.*, 83, 71, 1978.

14. **Svensjö, E. and Arfors, K.-E.** Dimensions of postcapillary venules sensitive to bradykinin and histamine-induced leakage of macromolecules, *Upsala J. Med. Sci.*, 84, 47, 1979.

15. **Björk, J. and Smedegard, G.**, Acute microvascular effects of PAF-acether, as studied by intravital microscopy, *Eur. J. Pharmacol.*, 96, 87, 1983.

16. **Dahlén, S.-E., Björk, J., Hedqvist, P., Arfors, K.-E., Hammarström, S., Lindgren, J. A., and Samuelsson, B.**, Leukotrienes promote plasma leakage and leukocyte adhesion in postcapillary venules: in vivo effects with relevance to the acute inflammatory response, *Proc. Natl. Acad. Sci. U.S.A.*, 78, 3887, 1981.

17. **Del Maestro, R. F., Björk, J., and Arfors, K.-E.**, Increase in microvascular permeability induced by enzymatically generated free radicals. II. Role of superoxide anion radicals, hydrogen peroxide, and hydroxyl radical, *Microvas. Res.*, 22, 255, 1981.

18. **Svensjö, E., Adamski, S. W., Su, K., and Grega, G. J.**, Quantitative physiological and morphological aspects of microvascular permeability changes induced by histamine and inhibited by terbutaline, *Acta Physiol. Scand.*, 116, 265, 1982.

19. **Svensjö, E., Persson, C. G. A., and Rutili, G.**, Inhibition of bradykinin induced macromolecular leakage from postcapillary venules by a β_2-adrenoceptor stimulant, terbutaline, *Acta Physiol. Scand.*, 101, 504, 1977.

20. **Adamski, S. W., Langone, J. J., Grega, G. J.**, Modulation of macromolecular permeability by immune complexes and a β-adrenoceptor stimulant, *Am. J. Physiol.*, in press.

21. **Horan, K. L., Adamski, S. W., Ayele, W., Langone, J. J., and Grega, G. J.**, Evidence that prolonged histamine suffusions produce transient increases in vascular permeability subsequent to the formation of venular macromolecular leakage sites: proof of the Majno-Palade hypothesis, *Am. J. Pathol.*, 123, 570, 1986.

22. **Miller, F. N., Joshua, I. G., and Anderson, G. L.**, Quantitation of vasodilator-induced macromolecular leakage by in vivo fluorescent microscopy, *Microvasc. Res.*, 24, 56, 1982.

23. **Svensjö, E., Arfors, K.-E., Raymond, R. M., and Grega, G. J.**, Morphological and physiologocal correlation of bradykinin-induced macromolecular efflux. *Am. J. Physiol.*, 236, H600, 1979.

24. **Hulström, D. and Svensjö, E.**, Simultaneous fluorescence and electron microscopical detection of bradykinin-induced macromolecular leakage, *Bibl. Anat.*, 15, 466, 1977.

25. **Hultström, D. and Svensjö, E.**, Intravital and electron microscopic study of bradykinin-induced vascular permeability changes using FITC-dextran as a tracer, *J. Pathol.*, 129, 125, 1979.

26. **Joris, I., Majno, G., Ryan, G. B.**, Endothelial contraction in vivo: a study in the rat mesentery, *Virchows Arch. B*, 12, 73, 1972.

27. **Buonassisi, V. and Venter, J. C.**, Hormone and neurotransmitter receptors in an established vascular endothelial cell line, *Proc. Natl. Acad. Sci. U.S.A.*, 73, 1612, 1976.

28. **Buonassisi, V. and Colburn, P.**, Hormone and surface receptors in vascular endothelium, in *Advances in Microcirculation*, Vol. 9, Altura, B. M., Ed., S. Karger, Basel, 1980, 76.

29. **Heltianu, C., Simionescu, M., and Simionescu, N.**, Histamine receptors of the microvascular endothelium revealed in situ with a histamine-ferritin conjugate: characteristic high-affinity binding sites in venules, *J. Cell Biol.*, 93, 357, 1982.

30. **Ryan, U.**, The endothelial cell surface and response to injury, *Fed. Proc.*, 45, 101, 1986.

31. **Kozlowski, T., Raymond, R. M., Korthuis, R. J., Wang, C. Y., Grega, G. J., Robinson, N. E., and Scott, J. B.**, Microvascular protein efflux: interaction of histamine and H_1 receptors, *Proc. Soc. Exp. Biol. Med.*, 166, 263, 1981.

32. **Guth, P. H. and Hirabayashi, K.**, The effect of histamine on microvascular permeability in the muscularis externa of the rat small intestine, *Microvasc. Res.*, 25, 322, 1983.

33. **Grega, G. J. and Svensjö, E.**, Pharmacology of water and macromolecular permeability in the forelimb of the dog, in *Edema*, Staub, N. C. and Taylor, A. E., Eds., Raven Press, New York, 1984, 405.

34. **Becker, C. G. and Nachman, R. L.**, Contractile proteins of endothelial cells, platelets, and smooth muscle, *Am. J. Pathol.*, 71, 1, 1973.

35. **Herman, I., Pollard, T., and Wong, A.**, Contractile proteins in endothelial cells, *Ann. N.Y. Acad. Sci.*, 401, 50, 1982.

36. **Di Bona, D. R. and Schafer, J. A.,** Cellular transport phenomena, in *Edema,* Staub, N. C. and Taylor, A. E., Eds., Raven Press, New York, 1984, 61.
37. **Simionescu, N.,** Cellular aspects of transcapillary exchange, *Physiol. Rev.,* 63, 1536, 1983.
38. **Drenckhahn, D.,** Cell motility and cytoplasmic filaments in vascular endothelium, *Prog. Appl. Microcirc.,* 1, 53, 1983.
39. **Laposata, M., Dovnarsky, D. K., and Shin, H. S.,** Thrombin-induced gap formation in confluent endothelial cell monolayers in vitro, *Blood,* 62, 549, 1983.
40. **De Clerck, F., De Brabander, M., Neels, H., and Van De Velde, V.,** Direct evidence for the contractile capacity of endothelial cells, *Thromb. Res.,* 23, 505, 1981.
41. **Shepro, D. and Hechtman, H. B.,** Endothelial serotonin uptake and mediation of prostanoid secretion and stress fiber formation, *Fed. Proc.,* 44, 2616, 1985.
42. **Liddell, R. H. A., Scott, A. R. W., Simpson, J. G.,** Histamine-induced changes in the endothelium of postcapillary venules: effects of chelating agents and cytochalasin B, *Bibl. Anat.,* 20, 109, 1981.
43. **Crone, C.,** Modulation of solute permeability in microvascular endothelium, *Fed. Proc.,* 45, 77, 1986.
44. **Mayhan, W. G. and Joyner, W. L.,** The effect of altering the external calcium concentration and a calcium channel blocker, verapamil, on microvascular leaky sites and dextran clearance in the hamster cheek pouch, *Microvasc. Res.,* 28, 159, 1984.
45. **Olesen, S.-P.,** A calcium-dependent reversible permeability increase in microvessels in frog brain, induced by serotonin, *J. Physiol.,* 361, 103, 1985.
46. **Cerijido, M., Meza, I., and Martinez-Palomo, A.,** Occluding junctions in cultured epithelial monolayers, *Am. J. Physiol.,* 240, C96, 1981.
47. **Tosteson, M. T. and Tosteson, D. C.,** Melittin forms channels in lipid bilayers, *Biophys. J.,* 36, 109, 1981.
48. **Fox, J., Galey, F., and Wayland, H.,** Action of histamine on the mesenteric microvasculature, *Microvasc. Res.,* 19, 108, 1980.
49. **Adamski, S. W., Jeraj, K., Langone, J. J., and Grega, G. J.,** Effect of suffusion flow rate on the formation and disappearance of mediator-stimulated vascular macromolecular leakage sites and plasma to suffusate tracer fluxes, *Microvasc. Res.,* submitted.
50. **Svensjö, E. and Grega, G. J.,** Evidence for endothelial cell-mediated regulation of macromolecular permeability by postcapillary venules, *Fed. Proc.,* 45, 89, 1986.
51. **Svensjö, E. and Joyner, W. L.,** The effects of intermittent and continuous stimulation of microvessels in the cheek pouch of hamsters with histamine and bradykinin on the development of leaky sites, *Microcirc. Endothel. Lymphat.,* 1, 381, 1984.
52. **Svensjö, E. and Roempke, K.,** Microvascular aspects of oedema formation and its inhibition by β_2-receptor stimulants and some other anti-inflammatory drugs, in *Progress in Microcirculation Research II,* Courtice, F. C., Garlick, D. G., and Perry, M. A., Eds., Committee in Postgraduate Medical Education, University of New South Wales, Sydney, 1984, 449.
53. **Svensjö, E.,** Bradykinin and prostaglandin E_1-, E_2-, and F_{2a}-induced macromolecular leakage in the hamster cheek pouch, *Prostaglandins Med.,* 1, 397, 1978.
54. **Ohya, Y. and Guth, P. H.,** Potentiation of histamine-induced microvascular permeability by prostaglandin E_2 in rat mesentery, *Microcirc. Endothel. Lymphat.,* 2, 331, 1985.
55. **Williams, T. J.,** Vascular responses and their suppression: vasodilation and edema, in *Handbook of Inflammation, Vol. 5, The Pharmacology of Inflammation,* Bonta, I. L., Bray, M. A., and Parnham, M. J., Eds., Elsevier, Amsterdam, 1985, 61.
56. **Grega, G. J., Kline, R. L., Dobbins, D. E., and Haddy, F. J.,** Mechanisms of edema formation by histamine administered locally into canine forelimbs, *Am. J. Physiol.,* 223, 1165, 1972.
57. **Kline, R. L., Scott, J. B., Haddy, F. J., and Grega, G. J.,** Mechanism of edema formation by bradykinin in the canine forelimb, *Am. J. Physiol.,* 225, 1051, 1973.
58. **Haddy, F. J., Scott, J. B., Grega, G. J.,** Effects of histamine upon lymph protein concentration and flow in the dog forelimb, *Am. J. Physiol.,* 223, 1172, 1972.
59. **Kline, R. L., Sak, C. P., Haddy, F. J., and Grega, G. J.,** Pressure-dependent factors in edema formation in canine forelimbs, *J. Pharmacol. Exp. Ther.,* 193, 452, 1975.
60. **Haddy, F. J., Scott, J. B., and Grega, G. J.,** Peripheral circulation: fluid transfer across the microvascular membrane, in *Cardiovascular Physiology II,* Vol. 9, Guyton, A. C. and Cowley, A. W., Eds., Baltimore University Press, Baltimore, 1976, 63.
61. **Haddy, F. J. and Grega, G. J.,** Effects of bradykinin on skin lymph flow and protein concentration in the dog forelimb, *Acta Physiol. Lat. Am.,* 24, 469, 1974.
62. **Grega, G. J., Svensjö, E., and Adamski, S. W.,** Evidence that inflammatory mediators promote edema formation primarily due to a direct pressure-independent action on the microvascular membrane, in *Progress in Microcirculation Research II,* Courtice, F. C., Garlick, D. G., and Perry, M. A., Eds., Committee in Postgraduate Medical Education, University of New South Wales, Sydney, 1984, 413.

63. **Grega, G. J., Kline, R. L., Dobbins, D. E., and Haddy, F. J.,** Mechanisms of edema formation by histamine administered locally into canine forelimbs, *Am. J. Physiol.,* 223, 1165, 1972.

64. **Merrill, G. F., Kline, R. L., Haddy, F. J., and Grega, G. J.,** Effects of locally infused serotonin on canine forelimb weight and segmental vascular resistances, *J. Pharmacol. Exp. Ther.,* 189, 140, 1974.

65. **O'Neill, J. T., Haddy, F. J., and Grega, G. J.,** The effect of norepinephrine on lymph flow and edema formation in the canine forelimb, *Am. J. Physiol.,* 243, H575, 1982.

66. **Kjellmer, I. and Odelram, H.,** The effect of some physiologocal dilators on the vascular bed of skeletal muscle, *Acta Physiol. Scand.,* 63, 94, 1965.

67. **McNamee, J. E. and Grodins, F. S.,** Effect of histamine on microvasculature of isolated dog gracillis muscle, *Am. J. Physiol.,* 229, 119, 1975.

68. **Baker, C. H.,** Nonhemodynamic effects of histamine on gracilis muscle capillary permeability, *J. Pharmacol. Exp. Ther.,* 211, 672, 1979.

69. **Korthuis, R. J., Wang, C. Y., and Scott, J. B.,** Transient effects of histamine on microvascular fluid movement, *Microvasc. Res.,* 23, 316, 1982.

70. **Korthuis, R. J., Wang, C. Y., and Spielman, W. S.,** Transient effects of histamine on the capillary filtration coefficient, *Microvasc. Res.,* 28, 322, 1984.

71. **Maron, M. B.,** Differential effects of histamine on protein permeability in dog lung and forelimb, *Am. J. Physiol.,* 242, H565, 1982.

72. **Grega, G. J., Dobbins, D. E., Scott, J. B., and Haddy, F. J.,** Effects of histamine and increased venous pressure on transmicrovascular protein transport, *Microvasc. Res.,* 18, 95, 1979.

73. **Grega, G. J., Adamski, S. W., and Dobbins, D. E.,** Physiological and pharmacological evidence for the regulation of permeability, *Fed. Proc.,* 45, 96, 1986.

74. **Amelang, E., Prasad, C. M., Raymond, R. M., and Grega, G. J.,** Interactions among inflammatory mediators on edema formation in the canine forelimb, *Circ. Res.,* 49, 298, 1981.

75. **Prasad, C. M., Adamski, S. W., Svensjö, E., and Grega, G. J.,** Pharmacological modification of the edema produced by combined infusions of prostaglandin E$_1$ and bradykinin in canine forelimbs, *J. Pharmacol. Exp. Ther.,* 220, 293, 1982.

76. **Rippe, B. and Grega, G. J.,** Effects of isoprenaline and cooling on histamine induced changes of capillary permeability in the rat hindquarter vascular bed, *Acta Physiol. Scand.,* 103, 252, 1978.

77. **Rippe, B., Kamiya, A., and Folkow, B.,** Transcapillary passage of albumin, effects of tissue cooling and of increases in filtration and plasma colloid osmotic pressure, *Acta Physiol. Scand.,* 105, 171, 1979.

78. **Rippe, B., Kamiya, A., and Folkow, B.,** Is capillary micropinocytosis of any significance for the transcapillary transfer of plasma proteins, *Acta Physiol. Scand.,* 100, 258, 1963.

79. **Grega, G. J., Svensjö, E., and Haddy, F. J.,** Macromolecular permeability of the microvascular membrane: physiological and pharmacological regulation, *Microcirculation,* 1, 325, 1982.

80. **Grega, G. J., Adamski, S. W., and Svensjö, E.,** Is there evidence for venular large junctional gap formation in inflammation? *Microcirc. Endothel. Lymphat,* 2, 211, 1985.

81. **Persson, C. G. A., Erjefält, I., Grega, G. J., and Svensjö, E.,** The role of β-receptor agonists in the inhibition of pulmonary edema, in *Mechanisms of Lung Microvascular Injury,* Vol. 384, Malik, A. B. and Staub, N. C., Eds., New York Academy of Science, New York, 1982, 544.

82. **Persson, C. G. A.,** Leakage of macromolecules from the tracheobronchial microcirculation, *Am. Rev. Respir. Dis.,* 135, 571, 1987.

83. **Persson, C. G. A. and Svensjö, E.,** Vascular responses and their suppression: drugs interfering with venular permeability, *Handbook of Inflammation, Vol. 5, The Pharmacology of Inflammation,* Bonta, I. L., Bray, M. A., and Parnham, M. J., Eds., Elsevier, Amsterdam, 1985, 61.

84. **Miller, G. L., Kline, R. L., Haddy, F. J., and Grega, G. J.,** Effects of prolonged ischemia on canine forelimb weight, pressures, blood flows, and vascular resistances following relief of ischemia, *Proc. Soc. Exp. Biol. Med.,* 149, 581, 1975.

85. **Diana, J. N. and Laughlin, M. H.,** Effect of ischemia on capillary pressure and equivalent pore radius in capillaries of the isolated dog hindlimb, *Circ. Res.,* 35, 77, 1974.

86. **Svensjö, E.,** to be published.

87. **Joyner, W. L., Svensjö, E., and Arfors, K.-E.,** Simultaneous measurement of macromolecular leakage and arteriolar blood flow as altered by PGE$_1$ and β$_2$-receptor stimulant in the hamster cheek pouch, *Microvasc. Res.,* 18, 301, 1979.

88. **Maciejko, J. J., Dobbins, D. E., Marciniak, D. L., Haddy, F. J., Scott, J. B., and Grega, G. J.,** The effect of histamine infused into the left ventricle of the heart on forelimb lymph flow and protein efflux in anesthetized dogs, *Fed. Proc.,* 35, 1976.

89. **Marciniak, D. L., Maciejko, J. J., Dobbins, D. E., Scott, J. B., Haddy, F. J., and Grega, G. J.,** The failure of locally administered histamine to increase lymph flow and protein efflux in the presence of norepinephrine in forelimbs of anesthetized dogs, *Fed. Proc.,* 35, 1976.

90. **Marciniak, D. L., Dobbins, D. E., Maciejko, J. J., Scott, J. B., Haddy, F. J., and Grega, G. J.,** Effects of systemically infused histamine on transvascular fluid and protein transfer, *Am. J. Physiol.,* 233, H148, 1977.

91. **Marciniak, D. L., Dobbins, D. E., Maciejko, J. J., Scott, J. B., Haddy, F. J., and Grega, G. J.,** Antagonism of histamine edema formation by catecholamines, *Am. J. Physiol.,* 234, H180, 1978.

92. **Maciejko, J. J., Marciniak, D. L., Gersabeck, E. F., and Grega, G. J.,** Effects of locally and systemically infused bradykinin on transvascular fluid and protein transfer in the canine forelimb, *J. Pharmacol. Exp. Ther.,* 205, 221, 1978.

93. **Adamski, S. W., Svensjö, E., and Grega, G. J.,** Effects of AVP and DDAVP on histamine-induced increases in macromolecular permeability in the hamster cheek pouch, *Microcirc. Endothel. Lymphat.,* 2, 41, 1985.

94. **Svensjö, E. and Roempke, K.,** Dose-related antipermeability effect of terbutaline and its inhibition by a selective β_2-receptor blocking agent, *Agents Actions,* 16(1/2), 19, 1985.

95. **Bottaro, D., Shepro, D., and Hechtman, H. B.,** Intimal and microvessel endothelial cell barrier function in vitro, *Fed. Proc.,* 45, 1150, 1986.

96. **Kim, J. K., Linas, S. L., and Schrier, R. W.,** Catecholamines and sodium transport in the kidney, *Pharmacol. Rev.,* 31, 169, 1979.

97. **Grega, G. J., Maciejko, J. J., Raymond, R. M., and Sak, D. P.,** The interrelationship among histamine, various vasoactive substances, and macromolecular permeability in the canine forelimb, *Circ. Res.,* 46, 264, 1980.

98. **Raymond, R. M., Jandhyala, B. S., and Grega, G. J.,** The interrelationship among bradykinin, various vasoactive substances, and macromolecular permeability in the canine forelimb, *Microvasc. Res.,* 19, 329, 1980.

99. **Dobbins, D. E., Solka, C. Y., Premen, A. J., Grega, G. J., and Dabney, J. M.,** Blockade of histamine and bradykinin-induced increases in lymph flow, protein concentration, and protein transport by terbutaline in vivo, *Microcirculation,* 2, 127, 1982.

100. **Adamski, S. W., Dobbins, D. E., Lokhandwala, M. F., and Grega, G. J.,** Effects of dopamine and vasopressin on histamine-induced increases in macromolecular permeability in the canine forelimb, *Microcirc. Endothel. Lymphat.,* 2, 27, 1985.

101. **Grega, G. J., Dobbins, D. E., Parker, P. E., and Haddy, F. J.,** Effects of intravenous histamine on forelimb weight and segmental vascular resistances, *Am. J. Physiol.,* 223, 353, 1972.

102. **Grega, G. J., Marciniak, D. L., Jandhyala, B. S., and Raymond, R. M.,** Effects of intravenously infused histamine on transvascular protein efflux following adrenergic receptor blockade, *Circ. Res.,* 47, 584, 1980.

103. **Adamski, S. W., Svensjö, E., Su, K., and Grega, G. J.,** Effects of captopril and propranolol on bradykinin-induced changes in vascular pressures, lymph total protein concentration and weight in canine forelimbs, *Microvasc. Res.,* 25, 307, 1983.

Clinical Perspectives

Chapter 36

ENDOTHELIAL CHANGES ASSOCIATED WITH HIGH PULMONARY BLOOD FLOW AND PRESSURE

Marlene Rabinovitch

TABLE OF CONTENTS

I. INTRODUCTION

From our studies of patients with congenital heart defects and left-to-right shunts, (e.g., ventricular septal defect), there is evidence that endothelial cells may play a pivotal role in the pathogenesis of vascular disease associated with increased pulmonary blood flow and pressure. We have observed alterations in endothelial cell morphology and function which suggest a potential for abnormal interaction with blood elements (platelets and leukocytes) and with smooth muscle cells. The following chapter describes these features, puts forth a hypothesis, and discusses our current experimental approach using an in vitro system in which lamb pulmonary artery endothelial and smooth muscle cells from central and peripheral arteries are pulsated at high pressure.

II. PULMONARY VASCULAR CHANGES IN PATIENTS WITH HIGH PULMONARY BLOOD FLOW AND PRESSURE

In patients with congenital heart defects and intra- or extracardiac left-to-right shunts, increased pulmonary blood flow and pressure are associated with progressive pulmonary vascular changes (Figures 1 and 2). The earliest abnormalities have been identified by morphometric analysis of the lung at post-mortem and reflect a process of altered growth and remodeling of the pulmonary arteries.[1,2] They correlate with increasingly abnormal hemodynamics of the pulmonary circulation and can be graded accordingly.[3] They have come to be known as A, B, Cs of pulmonary vascular disease. *Grade A* is characterized by abnormal muscularization of arteries that are small and peripheral in location and normally nonmuscular. We have learned from the studies of Meyrick and Reid[4] that this feature is due to the "precocious" differentiation to mature smooth muscle of precursor cells, the pericyte in the normally nonmuscular region of the artery, and the intermediate cell in the partially muscular region (Figure 3). The stimulus which induces this change is unknown. However, peripheral arteries become muscular as they increase in size throughout childhood,[5] so it is possible that the wide pulmonary pulse pressure produced by a congenital heart defect with a left-to-right shunt "stretches" the peripheral artery "simulating" growth. The close proximity between the endothelial cell and the underlying precursor cell begs us to speculate that the former may "sense" the stretch and by some metabolic message induce differentiation of the latter. Patients with grade A changes have increased pulmonary blood flow and pulse pressure, but mean arterial pressure is generally normal.

Grade B, there is medial hypertrophy of normally muscular arteries. This is produced by an increase in both the size and number of smooth muscle cells and in the intercellular matrix proteins, collagen, elastin, and ground substance. What induces smooth muscle hypertrophy and proliferation is difficult to ascertain, but our recent observations on ultra-structural analysis of lung biopsy tissue have given us some clues. It seems that as medial hypertrophy develops, the elastic lamina which normally separates endothelial and smooth muscle cells breaks down.[6] Whether this is due to endothelial and smooth muscle cells breaks down.[6] Whether this is due to endothelial release of elastase or neutrophil release of elastase due to altered interaction with endothelial cells is uncertain. The result is a closer proximity between endothelial and smooth muscle cells (Figure 4), and this might facilitate transfer of endothelial derived growth factors[9] or platelet derived growth factors.[10] Patients with grade B changes have increased pulmonary blood flow and increased mean pulmonary pressure.

Grade C is characterized by a decrease in the number of peripheral arteries relative to the number of alveoli which is normal. The loss of arteries in children probably reflects a failure of normal growth. It is conceivable that arterial "loss" may also be due to resorption as has been described in a patient with idiopathic pulmonary hypertension.[11] where small

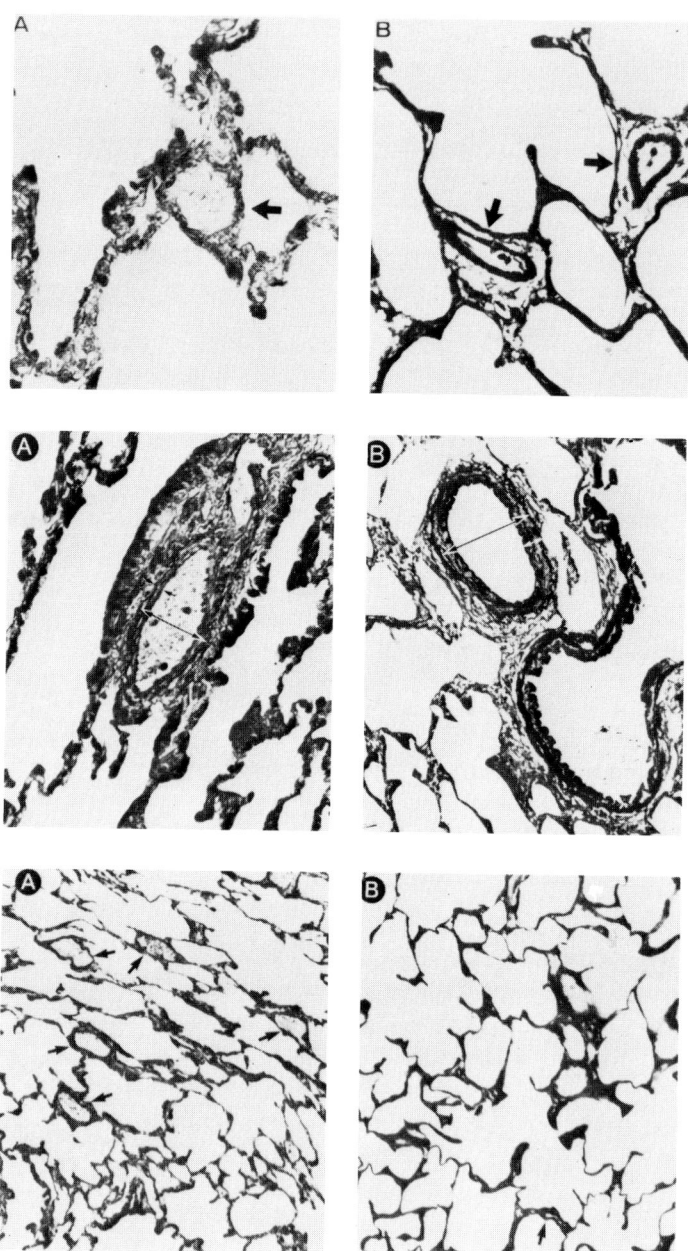

FIGURE 1. Morphometric features on lung biopsy tissue from: (left panels) 2-year-old patient with a ventricular septal defect and normal pulmonary artery pressure; and (right panels) a 2-year-old with a defect of the atrioventricular canal and increased pulmonary artery pressure and resistance. (Top A) Alveolar wall arteries (arrows) nonmuscularized, and (top B) surrounded by a complete muscularized coat. (Magnification × 175 (Center A) Artery accompanying respiratory bronchiolus (RB) with wall thickness only slightly increased, and (center B) with wall thickness greatly increased. Arrows denote external diameter and medial width. (Magnification × 70.) (Bottom A) An abundance of small arteries (arrows) relative to alveoli, and (bottom B) only one small artery in a similar microscopic field. (Elastin Van Gieson stain.) (From Rabinovitch, M., Haworth, S. G., Castaneda, A. R., Nadas, A. S., and Reid, L., *Circulation,* 58, 1107, 1978. With permission.)

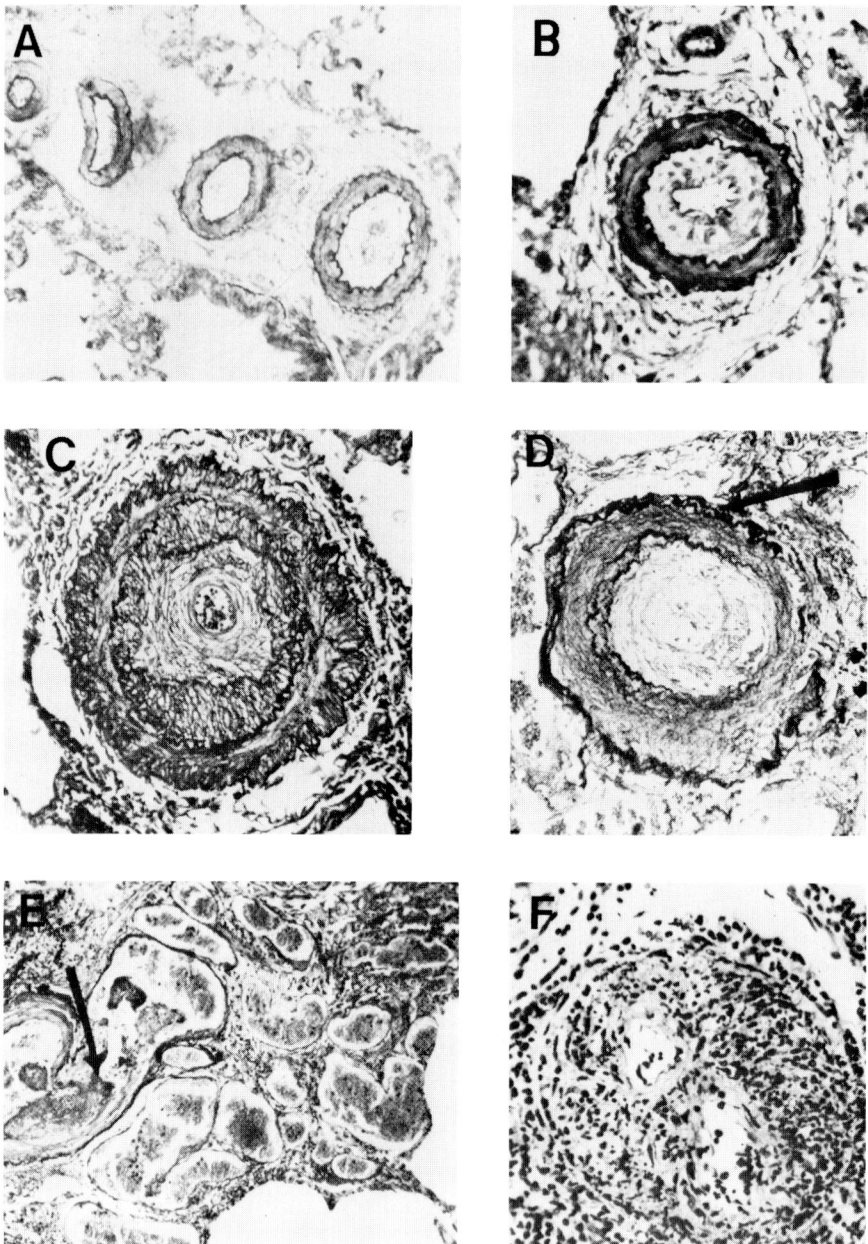

FIGURE 2.   Heath Edwards classification of pulmonary vascular changes. (A) Grade I: medial hypertrophy. (Elastin Van Gieson stain [EVG] magnification × 150.) (B) Grade II: cellular intimal proliferation in an abnormally muscular artery. (EVG; magnification × 250.) (C) Grade III: occlusive changes. Media is thickened due to fasciculi of longitudinal muscle, and vessel is all but occluded by fibroelastic tissue. (EVG; magnification × 150.) (D) Grade IV: dilatation. Vessel is dilated, and media is abnormally thin (arrow). Lumen is occluded by fibrous tissue. (EVG; magnification × 150.) (E) Grade V: plexiform lesion. There is cellular intimal proliferation (arrow); clustered around are numerous thin walled vessels which terminate as capillaries in the alveolar wall. (EVG: magnification × 95.) (F) Grade VI: acute necrotizing arteritis. A severe reactive acute inflammatory exudate is seen through all layers of the vessel. (Hematoxylin eosin stain; EVG; magnification × 250). (From Wagenvoort, C. A., Heath, D., and Edwards, J. E., *The Pathology of the Pulmonary Vasculature*, Charles C Thomas, Springfield, Ill., 1964. With permission.)

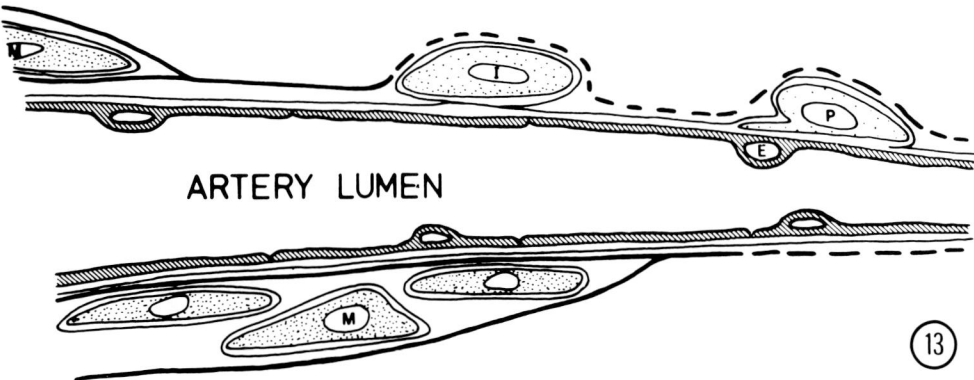

MUSCULAR   PARTIALLY   NON-MUSCULAR
                    MUSCULAR   & CAPILLARY

ARTERY LUMEN

FIGURE 3.   Diagrammatic representation of the cells in the wall of the distal part of a pulmonary artery. The smooth muscle cells (M) of the medial muscular coat are surrounded by a discrete basement membrane and are situated between both an internal and external elastic lamina (thick black lines). In the nonmuscular region of the partially muscular artery, the "intermediate" cell (I) is seen. This cell is surrounded by a basement membrane that fuses, in regions, with that of the endothelial cell (E) and is situated internal to the single fragmented internal elastic lamina (dashed lines). In the wall of the nonmuscular artery and alveolar capillary, the pericyte (P) is found. This cell is ensheathed by a basement membrane which is continuous with and thereby shares the basement membrane of the associated artery, and like the intermediate cell, it is situated internal to the single elastic lamina. (From Meyrick, B. and Reid, L., *Anat. Rec.*, 193(1), 71, 1979. With permission.)

arteries with pale swollen endothelial cells were seen. Patients with grade C generally have a disproportionate increase in pulmonary artery pressure relative to flow resulting in increased pulmonary vascular resistance.

The features of altered growth and development precede the changes, classified by Heath and Edwards,[12] i.e., intimal hyperplasia-cellular, fibrous, and occlusive, dilatation complexes, angiomatoid formation, and fibrinoid necrosis (Figure 2). The mechanism which leads to these further changes is unknown. Nor is it understood why the rate of their progression varies with the type of shunt lesion and even among individuals with the same lesion. The occlusive changes are associated with increased and abnormally distributed connective tissue proteins, collagen, elastin, and ground substance in the subendothelium and media, and are similar to those observed in individuals with idiopathic pulmonary hypertension.

We have established structural criteria on lung biopsy tissue which predict the presence and to some extent the severity of heightened pulmonary vascular reactivity in the early postoperative period following repair of the congenital heart defect as well as fixed or progressive pulmonary hypertension at later follow-up.[13] We could not, however, determine from light-microscopic analyses, which of the patients with only medial hypertrophy of pulmonary arteries would exhibit heightened pulmonary vascular reactivity or "pulmonary hypertensive crises"[14] in the early postoperative period. Nor could we identify the rare patient with only medial hypertrophy who would have pulmonary hypertension in the postoperative period that was severe and resistant to vasodilator therapy. We speculated that the abnormal postoperative pulmonary hemodynamics could be the result of altered metabolic function of pulmonary vascular endothelium[15] with abnormal production of vasoactive substances or altered interaction of the endothelium with blood elements (platelets or leukocytes)

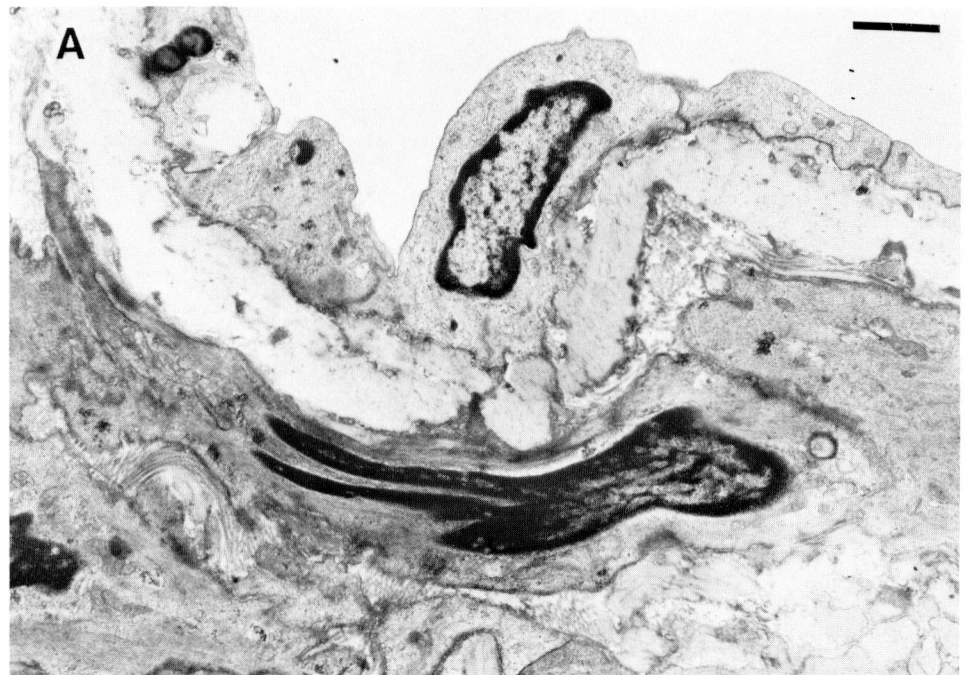

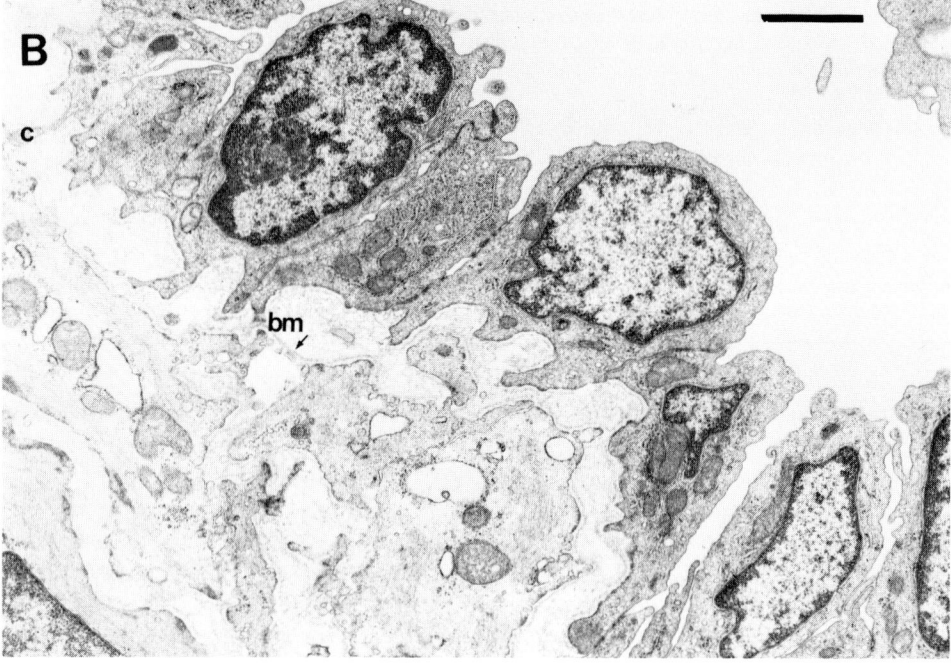

FIGURE 4.   (A) Section of a pulmonary artery 92 μm in external diameter in a patient with normal pulmonary artery pressure shows an intact internal elastic lamina. (B) Section from a pulmonary artery 108 μm in external diameter in a patient with increased pulmonary blood flow and pressure. Microfibrillar material is present in the subendothelium but no true internal elastic lamina. The endothelial and smooth muscle cells are separated by only a thick basement membrane (bm). A myoendothelial contact is seen (c). Bar = 1 μm in both. (From Rabinovitch, M. et al., *Lab. Invest.*, 55, 632, 1986. With permission.)

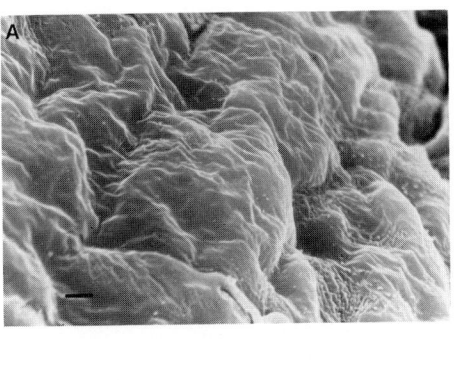
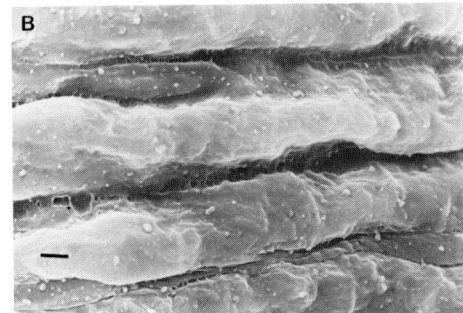
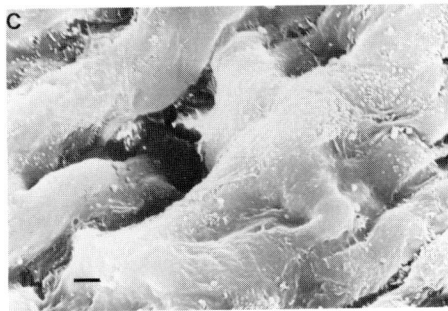
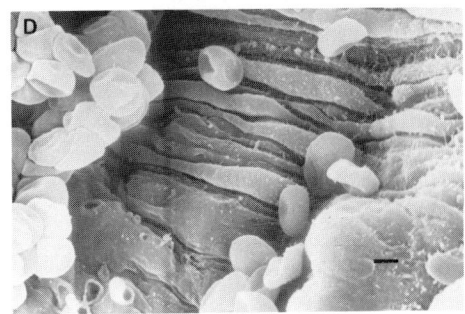

FIGURE 5. Scanning electron photomicographs depicting four typical surface patterns observed in pulmonary arteries of patients with congenital heart defects. (A) "Crinkled" and (B) "corduroy" seen in patients with normal pulmonary artery pressure. (C) "Cable" seen in patients with increased pulmonary blood flow and pressure. (D) "Chenille" observed in patients with advanced vascular disease (occlusion and dilatation). Bar = 2 μm in all. (From Rabinovitch, M. et al., *Lab. Invest.*, 55, 632, 1986. With permission.)

that were postcardiopulmonary bypass and therefore would more readily release vasoconstrictors such as thromboxane[16] or leukotrienes.[17]

To explore this hypothesis, we applied scanning electron microscopy to lung biopsy tissue obtained at intracardiac repair from patients with congenital heart defects associated with increased pulmonary blood flow and pressure, to determine whether there were changes in the endothelial surface characteristics of the pulmonary arteries which might suggest a potential for abnormal interaction with blood elements. We applied transmission electron microscopy to detect whether there were changes in the concentration of the intracytoplasmic components of pulmonary vascular endothelial cells which might reflect altered metabolic function.[6]

III. ENDOTHELIAL CHANGES ASSOCIATED WITH HIGH PULMONARY BLOOD FLOW AND PRESSURE

In patients with normal or near-normal pulmonary arteries on light microscopy, scanning electron microscopy revealed endothelial surfaces which were "crinkled" when the vessels were dilated or which were comprised of "narrow", uniform, corduroy-like ridges when the vessels were constricted (Figures 5A and B). In contrast, patients with high pulmonary blood flow and pressure and medial hypertrophy of the pulmonary arteries on light microscopy, had endothelial surfaces composed of thick, deep, intertwined "cable-like" ridges (Figure 5C). Also, there was an almost 200% increase in the areal density of microvilli. Occasionally, marked "blebbing" of the endothelial surface was apparent, associated with

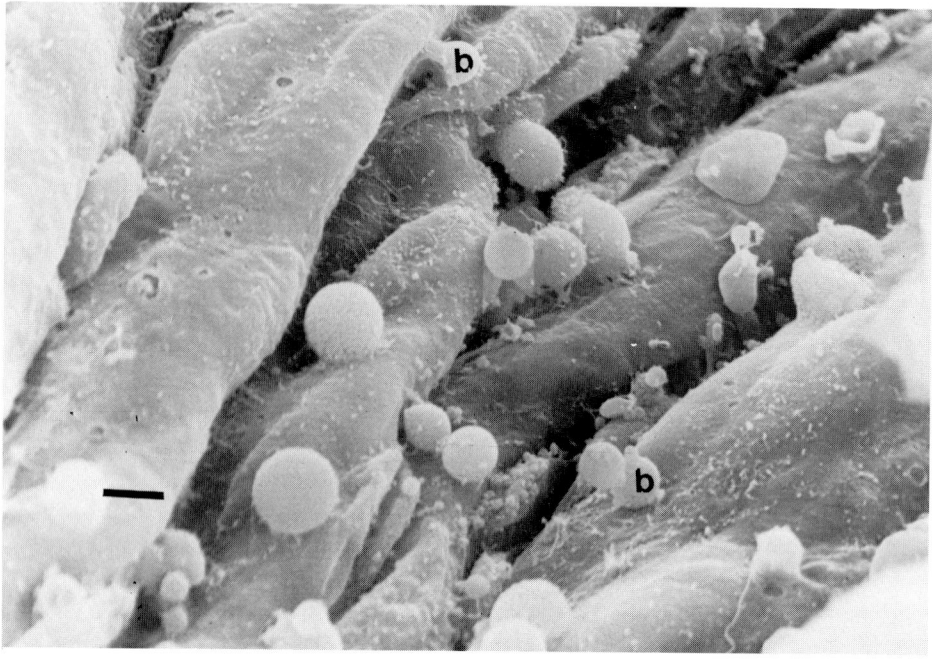

FIGURE 6.   A scanning electron photomicograph of a "cable" endothelial surface with numerous blebs (b) and evidence of fragmented blood cells. Bar = 2 μm. (From Rabinovitch, M. et al., *Lab. Invest.*, 55, 632, 1986. With permission.)

fragments of blood elements (Figure 6). Patients with the highest areal density of microvilli also had the highest concentration of adherent platelets and leukocytes. In patients with cellular intimal hyperplasia, the endothelial surface was similar in that it was "cable-like", but more irregularly so, in that occasionally the ridges seemed flatted down. Also, in any given artery, the increased areal density of microvilli was less uniformly seen. Patients with more advanced pulmonary vascular disease on light microscopy, i.e., occlusive intimal hyperplasia and dilatation, had arteries with "chenille-like" endothelial surface in which high convoluted ridges (in obstructed regions) alternated with relatively flat areas in dilated zones. Even in the flat areas, the endothelial cells were uneven in size and twisted (Figure 5D). Only patches of increased microvilli were seen.

On transmission electron microscopic analysis, pulmonary artery endothelial cells from patients with high pulmonary blood flow and pressure had a higher volume density of microfilament bundles and rough endoplasmic reticulum than in patients with normal pulmonary blood flow and pressure (Figures 7A and B). When advanced pulmonary vascular changes were observed on light microscopy (occlusion and dilatation), there was evidence of endothelial cell degeneration with concentric membranous bodies (Figure 7C).

A high concentration of microvilli could increase the lumenal surface area of the endothelium and this may allow for better exchange of metabolites under conditions of high flow.[15] The increase in microfilament bundles in endothelial cells of patients with high pulmonary blood flow and pressure probably serves to keep the endothelium well anchored to the subendothelium, and may influence vascular tone. The increase in rough endoplasmic reticulum suggests a generalized state of heightened metabolic activity. The proteins being synthesized might include the von Willebrand factor, cytoskeletal elements forming microfilament bundles, or enzymes related to the production or inactivation of vasoactive substances.[16]

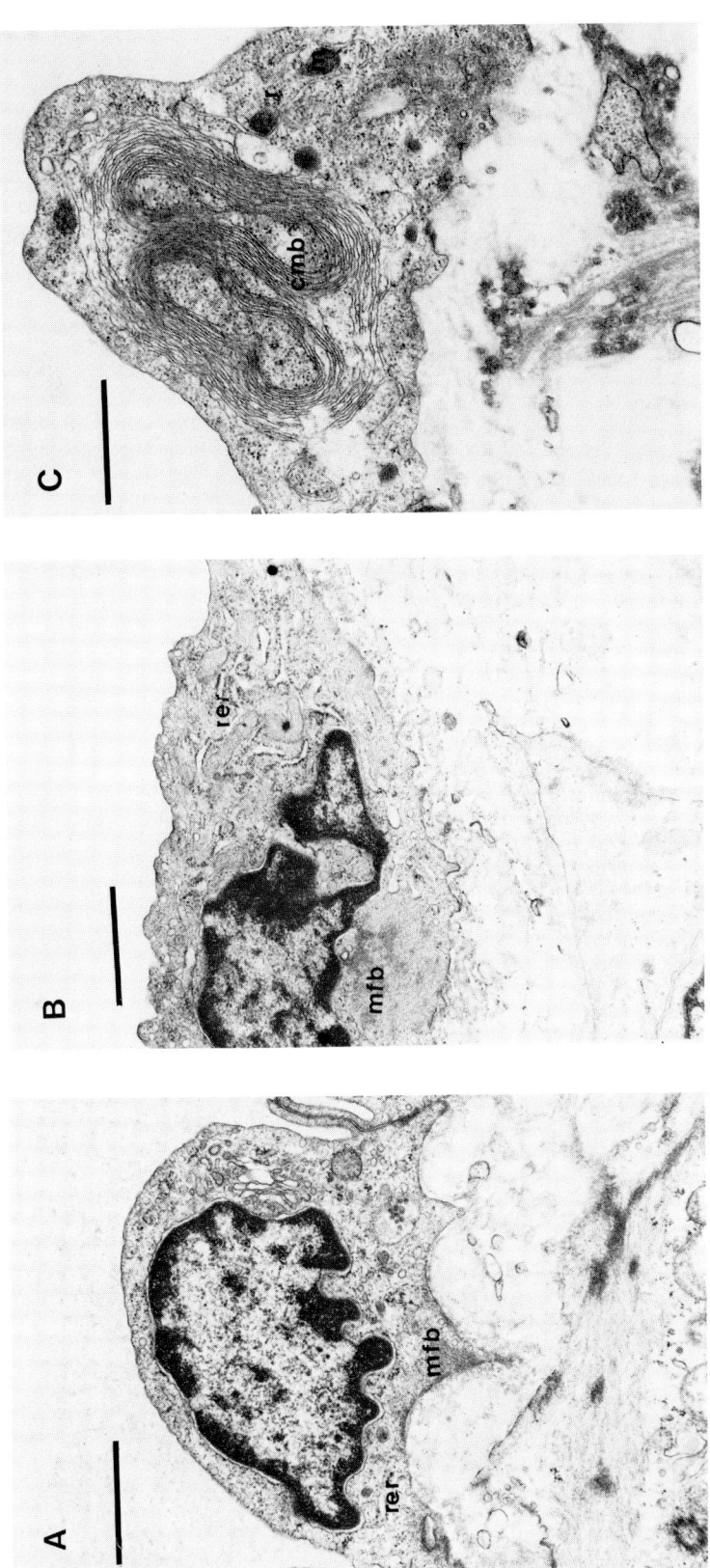

FIGURE 7.   Transmission electron photomicrographs. (A) An endothelial cell from a pulmonary artery 320 μm in external diameter in a patient with normal pulmonary artery pressure. (B) An endothelial cell from a pulmonary artery 240 μm in external diameter from a patient with increased pulmonary blood flow and pressure. Observe the increased rough endoplasmic reticulum (rer) and microfilament bundles (mfb) in the latter patient. (C) Transmission electron photomicrograph of an endothelial cell from a 316-μm external diameter pulmonary artery in a patient with advanced vascular disease; a concentric smooth membranous body (cmb) is apparently suggesting a degenerative process; m = mitochondria, r = ribosomes. Bar = 1 μm in both. (From Rabinovitch, M. et al., *Lab. Invest.*, 55, 632, 1986. With permission.)

IV. FUNCTIONAL CHANGES IN PULMONARY ARTERY ENDOTHELIUM

We determined whether pulmonary artery endothelial production of the von Willebrand factor (vWF)[19] is increased in patients whose endothelial abnormalities were associated with high pulmonary blood flow and pressure.[20] If so, then this may promote platelet adhesion and aggregation with occlusion of small vessels and worsening of the pulmonary hypertension, explaining both heightened pulmonary vascular reactivity and contributing to the progression of pulmonary vascular disease. We measured the antigenic (vWF:Ag) and biological (vWF:rist) activity of vWF in plasma and assessed vWF:Ag directly by an immunoperoxidase stain applied to lung biopsy tissue.

The majority of patients with congenital heart defects and pulmonary hypertension had increased production of vWF:Ag, judged by high circulating levels and by immunostain (Figure 8), but rarely with an increase in vWF:rist (biologic activity). Compatible with this discrepancy was a loss of high-molecular-weight forms determined by crossed immunoelectrophoresis and confirmed by multimeric analysis. There are two possible explanations. If the vWF being produced lacks high-molecular-weight forms, then there may actually be a propensity for decreased adherence of platelets to the endothelial surface, and this may serve as a protective mechanism. Alternatively, there may be ongoing consumption of the high-molecular-weight forms in platelet aggregates. This might be part of the mechanism of progression of pulmonary vascular disease or of the acute exacerbation of pulmonary hypertension in the postoperative period. Further studies are planned to help sort this out.

V. EXPERIMENTAL MODELS OF HIGH PULMONARY BLOOD FLOW AND PRESSURE

Experimental models of high pulmonary blood flow and pressure are invaluable in studying the functional abnormalities of altered endothelial cells and their role in the development of increased pulmonary vascular reactivity and in the pathogenesis of pulmonary vascular disease. In a collaborative study with Boucek et al. using their newborn lamb model with an experimentally created ventricular septal defect,[21] we have correlated altered pulmonary artery compliance with ultrastructural evidence of degradation of elastin. We will study further the possibility that this is related to endothelial release of elastase.

We have developed an in vitro method of pulsating pulmonary artery endothelial cells at high pressure to simulate the abnormal mechanical forces produced by a congenital heart defect with a left-to-right shunt.[22] We have harvested endothelial cells from both central pulmonary arteries[23] and from pulmonary microvessels[24] and have grown them to confluence onto the flexible polyvinylchloride membrane of a transducer dome. The dome is connected to a reservoir syringe and to a bellows stainless steel syringe filled with culture medium affixed to a pulsation generator. By varying the amplitude of the excursion of the generator syringe and the rate (using a microgearhead attachment), the cells can be pulsated at a given pressure and frequency (Figure 9). We elected first to pulsate endothelial cells for 48 hr, 100 times per minute at both high pressure 100/60 mmHg and low pressure 20/10 mmHg and to compare them ultrastructurally to cells nonpulsated. On scanning electron microscopy, the cells pulsated at high pressure remained confluent and on transmission electron microscopy, intercellular junctions were evident, intracytoplasmic organelles were intact and there was no evidence of cellular injury (Figure 10).

In our most recent and very preliminary studies, we have removed cell culture medium from pulsated central or microvessel pulmonary artery endothelial cells after 48 hr and incubated it with smooth muscle cells from the respective central or microvessel pulmonary artery.[25] The "conditioned medium" from endothelial cells pulsated at high pressure actually decreases smooth muscle incorporation of ^{3}H-thymidine and ^{14}C-leucine. This suggests that

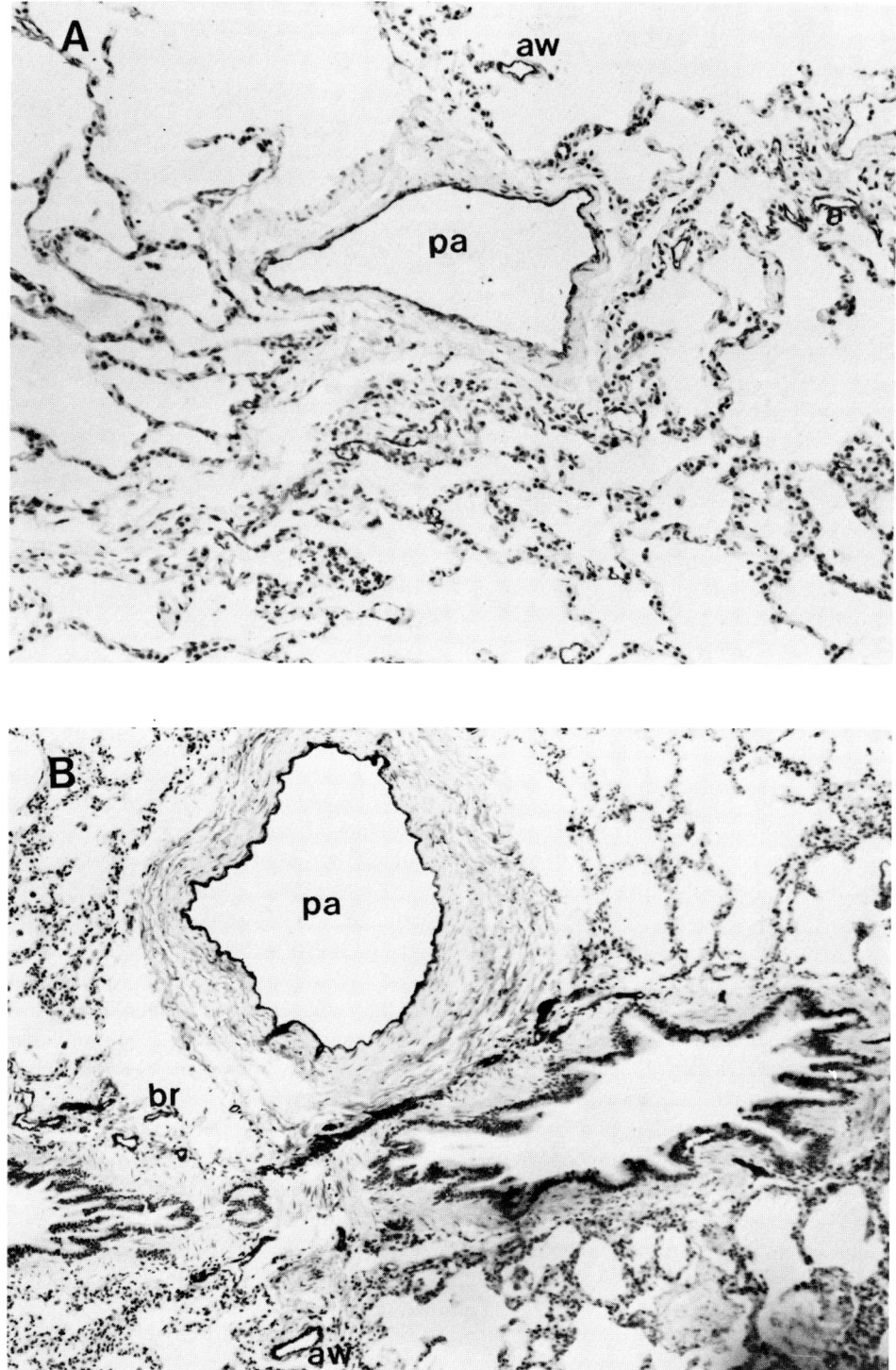

FIGURE 8.   Photomicrographs of vWF immunostaining. (A) A patient with normal pulmonary artery pressure, faint immunostain in preacinar (pa) and alveolar wall (aw) artery is observed, whereas in (B) a patient with increased pulmonary blood flow and pressure, dense immunostain of endothelium in pa, aw, and bronchial (br) artery is seen. (Magnification × 154.)

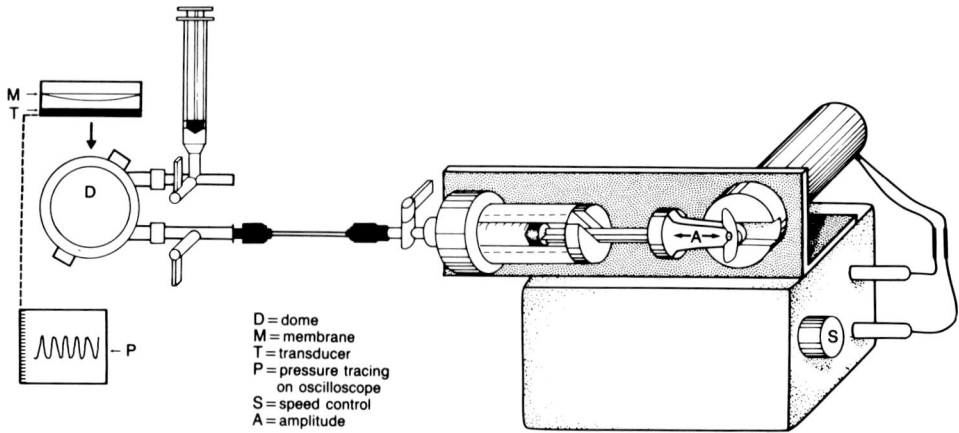

FIGURE 9. Diagrammatic representation of the pulsation system showing the generator pump and plunger (we now use stainless steel tubing), the reservoir syringe, and domes. Two domes are connected in series; the first is used to adjust the pressure to the appropriate range so that the endothelial cells on the second dome are not accidentally exposed to abnormally high pressure.

high-pressure pulsation causes release of a factor which inhibits DNA and protein synthesis. This inhibition may balance whatever other growth factor (? platelet derived) is stimulating smooth muscle hyperplasia and hypertrophy in vivo during conditions of increased pulmonary blood flow and pressure.

Functional abnormalities of endothelial cells may not only be important in "structural remodeling" of the pulmonary artery wall but in altering vascular tone[26,27] as well. We have shown in recent studies[28] that when smooth muscle cells are incubated in endothelial conditioned medium, they migrate to align in a uniform direction (Figure 11). The endothelial derived factor which determines smooth muscle alignment, and likely has an important bearing on the control of vascular tone, may be altered by pulsation at high pressure, and this will be addressed in future studies.

VI. A HYPOTHESIS TO CONSIDER

The following then is our working hypothesis for the pathogenesis of pulmonary vascular disease in the setting of high pulmonary blood flow and pressure (Figure 12). The endothelial cell plays a pivotal role. Mechanical forces produced by high pulmonary blood flow and pressure, alter its interactions with blood elements (platelets and leukocytes) and with smooth muscle cells. In the normally nonmuscular distal pulmonary artery, the endothelial cell may release a protease, which alters the subendothelial matrix and permits precursor cell differentiation to mature smooth muscle. Thus, there is abnormal muscularization of small arteries. In the normally muscular artery, the endothelial cell may release an elastase, and the degradation of the internal elastic lamina permits induction of smooth muscle hypertrophy and hyperplasia. The hypertrophied smooth muscle cells, together with the altered endothelial cells, synthesize increased connective tissue proteins, elastin, ground substance, and collagen, and this results in progressive decreased compliance of the artery and narrowing of the vascular lumen.

V. CONCLUSION

Clearly, much more work is necessary before these mechanisms are elucidated, and their role in the pathogenesis of pulmonary vascular disease established with certainty. Identifying

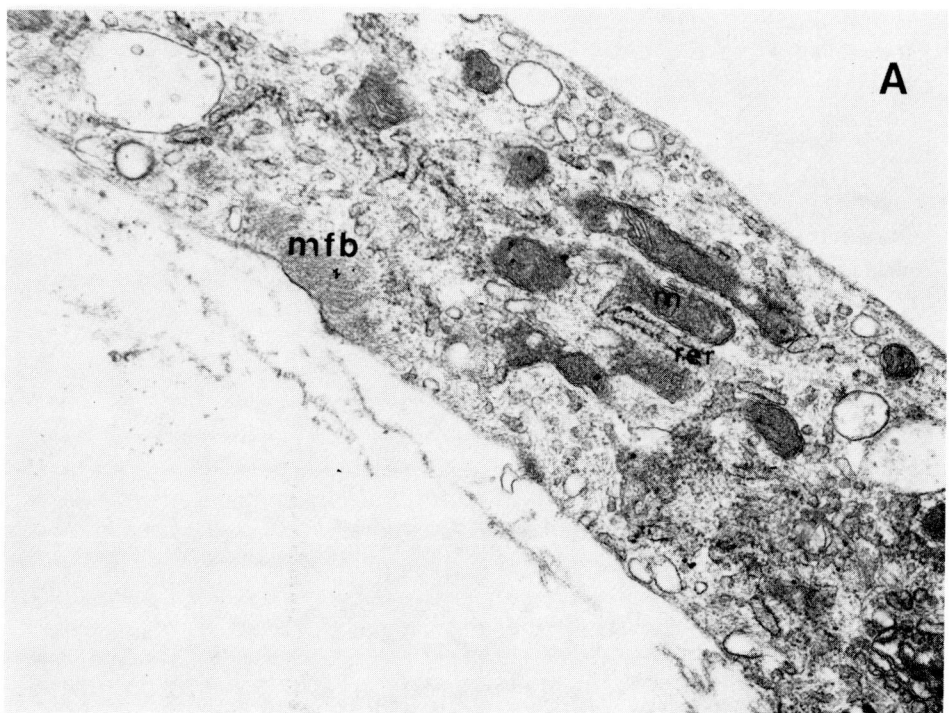

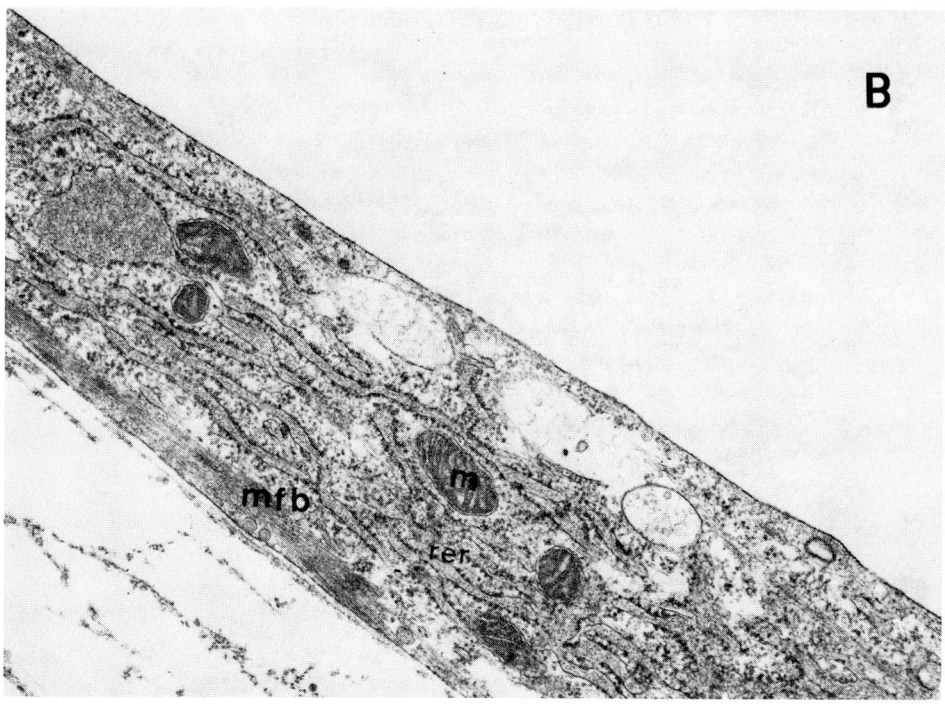

FIGURE 10.   Transmission electron photomicographs from a confluent monolayer stationary on a dome for 48 hr in (A), and from a confluent monolayer pulsated at 100/60 mmHg in (B). Note the increase in rough endoplasmic reticulum and microfilament bundles in (B); m = mitochondria. (Magnification × 25,000.)

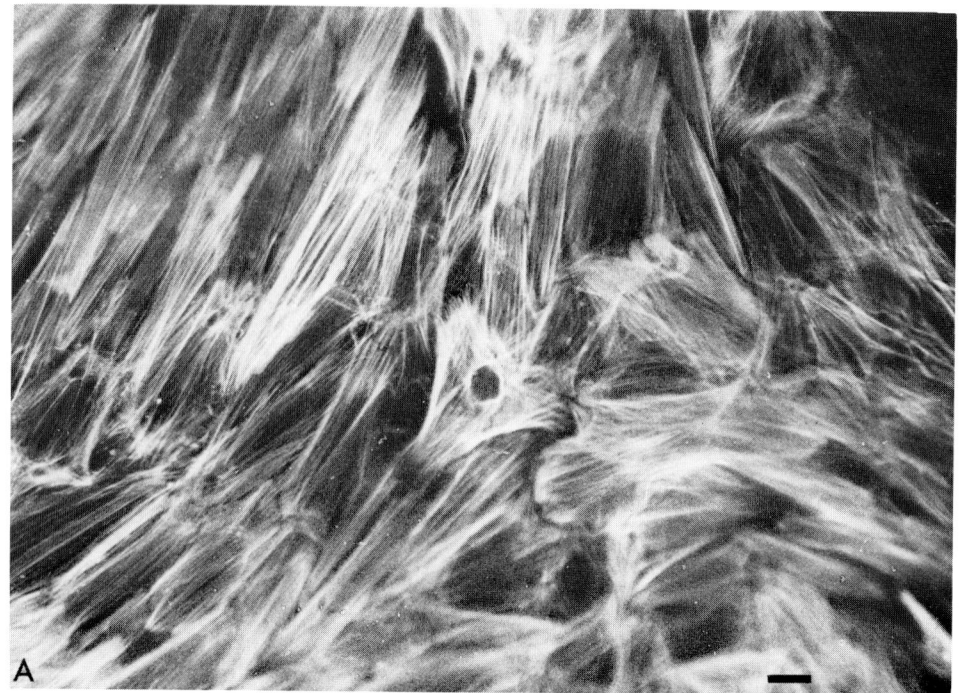

FIGURE 11.   Smooth muscle cells stained with rhodamine phalloidin to identify F-actin filaments. In (B), where smooth muscle cells have been incubated with endothelial conditioned medium, the cells show uniform direction of alignment when compared to the control (A).

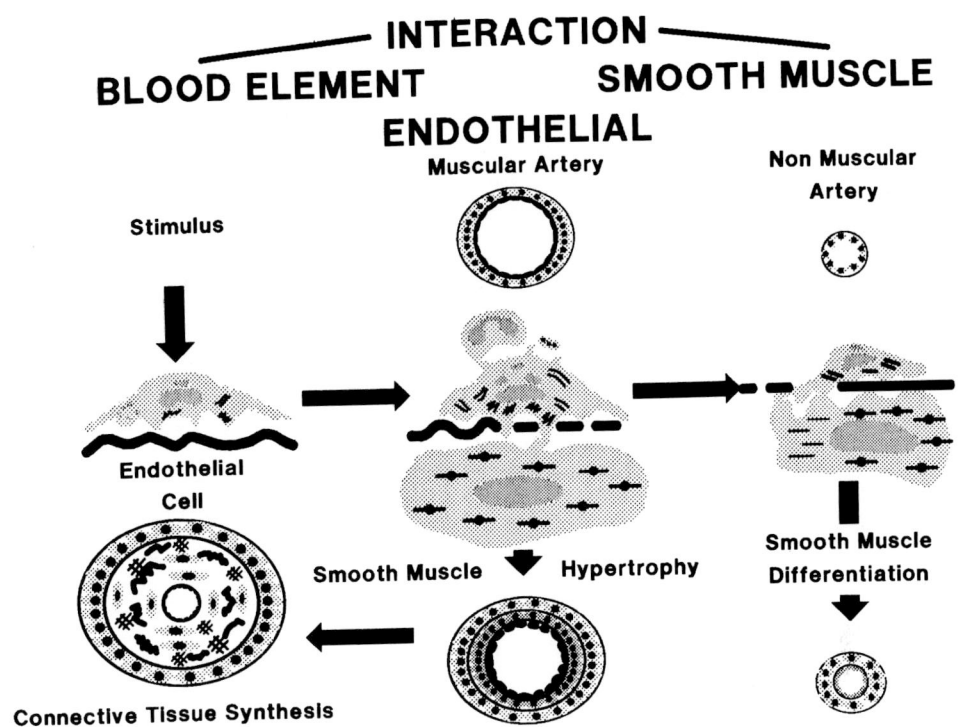

FIGURE 12. Schema of the mechanism of development of pulmonary vascular disease as outlined in the text. (From Wilson, W. Bothwell, T., and Rabinovitch, M. A., *Tissue Cell,* 19, 177, 1987. With permission.)

the key systems will lead to further research to uncover the molecular basis for the observed abnormalities. This will ultimately provide a means to identify patients at risk and to develop a more rational therapeutic approach to control heightened pulmonary vascular reactivity and to arrest and perhaps even reverse advanced disease.

REFERENCES

1. **Hislop, A., Haworth, S. G., and Reid, L.,** Quantitative structural analysis of pulmonary vessels in isolated ventricular septal defect, *Br. Heart J.,* 37, 1014, 1975.
2. **Haworth, S. G. and Reid, L.,** A morphometric study of regional variation in lung structure in infants with pulmonary hypertension and congenital heart defect. A justification of lung biopsy, *Br. Heart J.,* 40, 825, 1978.
3. **Rabinovitch, M., Haworth, S. G., Castaneda, A. R., Nadas, A. S., and Reid, L.,** Lung biopsy in congenital heart disease. A morphometric approach to pulmonary vascular disease, *Circulation,* 58, 1107, 1978.
4. **Meyrick, B. and Reid, L.,** Ultrastructural findings in lung biopsy material from children with congenital heart defects, *Am. J. Pathol.,* 101, 527, 1980.
5. **Hislop, A. and Reid, L. M.,** Pulmonary arterial development during childhood: branching pattern and structure, *Thorax,* 18, 129, 1973.
6. **Rabinovitch, M., Bothwell, T., Hayakawa, B. N., Williams, W. G., Trusler, G. A., Rowe, R. D., Olley, P. M., and Cutz, E.,** Pulmonary vascular endothelial abnormalities in patients with congenital heart defects and pulmonary hypertension. A correlation of light with scanning and transmission electron microscopy, *Lab Invest.,* 55, 632, 1986.
7. **Yamada, E., Hazama, F., Kataoka, H., Amano, S., Sashara, M., Kayembe, K., and Katayama, K.,** Elastase like enzyme in the aorta of spontaneously hypertensive rats, *Virchows Arch. Cell Pathol.,* 44, 241, 1983.

8. **Reilly, C. F. and Travis, J.,** The degradation of human lung elastin by neutrophil proteinases, *Biochim. Biophys. Acta,* 621, 147, 1980.

9. **Ross, R., Glomset, J., Kariya, B., and Harker, L.,** Platelet dependent serum factor that stimulates the proliferation of arterial smooth muscle cells in vitro, *Proc. Natl. Acad. Sci. U.S.A.,* 71, 1207, 1974.

10. **Gadjusek, C., Ross, R., and Schwartz, S. M.,** An endothelial cell-derived growth factor, *J. Cell Biol.,* 15, 467, 1980.

11. **Meyrick, B., Clarke, S. W., Symons, C., Woodgate, D. J., and Reid, L.,** Primary pulmonary hypertension. A case report including electron microscopic study, *Br. J. Dis. Chest,* 68, 11, 1974.

12. **Heath, D. and Edwards, J. E.,** The pathology of hypertensive pulmonary vascular disease, *Circulation,* 18, 533, 1958.

13. **Rabinovitch, M., Keane, J. F., Norwood, W. I., Castaneda, A. R., and Reid, L.,** Vascular findings in lung biopsy tissue correlated with pulmonary hemodynamic findings after repair of congenital heart defects, *Circulation,* 69, 655, 1984.

14. **Jones, O. D. H., Shore, D. F., Rigby, M. L., Leijala, M., Scallan, J., Shinebourne, E. A., Lincoln, J. C. R.,** The use of tolazoline hydrochloride as a pulmonary vasodilator in potentially fatal episodes of pulmonary vasoconstriction after cardiac surgery in children, *Circulation,* 64, II-134, 1981.

15. **Ryan, U. and Ryan, J. W.,** Electron microscopy of endothelium and endothelial components of the lungs. Correlation of structure and function, *Fed. Proc.,* 32, 1957, 1973.

16. **Addonizio, V. P., Jr., Smith, J. B., Strauss, J. F., III, Colman, R. W., and Edmunds, L. H., Jr.,** Thromboxane synthesis and platelet secretion during cardiopulmonary bypass with bubble oxygenation, *J. Thorac. Cardiovasc. Surg.,* 7, 91, 1980.

17. **Yokochi, K., Olley, P. M., Sideris, E., Hamilton, F., Huhtanen, D., and Coceani, F.,** Leukotriene D_4: a potent vasoconstrictor of the pulmonary and systemic circulations in the newborn lamb, in *Leukotrienes and Other Lipoxygenase Products,* Samuelson B., Ed., Raven Press, New York, 1982, 211.

18. **Ryan, U. S. and Ryan, J. W.,** Correlations between the fine structure of the alveolar capillary unit and its metabolic activity, in *Metabolic Function of the Lung,* Vane, J. R. and Bakhle, Y. S., Eds., in *Lung Biology in Health and Disease Series,* L'Enfant, C., Ed., Marcel Dekker, New York, 1977, 197.

19. **Hoyer, L. W.,** The factor VIII complex. Structure and function. *Blood,* 58, 1, 1981.

20. **Rabinovitch, M., Andrew, M., Thom, H., Trusler, G. A., Williams, W. G., Rowe, R. D., and Olley, P. M.,** Abnormal endothelial Factor VIII metabolism associated with pulmonary hypertension, *Circulation,* 74, 1043, 1987.

21. **Boucek, M. M., Chang, R., and Synhorst, D. P.,** Hemodynamic consequences of inotropic support with digoxin and amrinone in lambs with ventricular septal defect, *Pediatr. Res.,* 19, 887, 1985.

22. **Rabinovitch, M., Bothwell, T., Mullen M., and Hayakawa, B. N.,** Central and microvessel endothelial cells pulsated at high pressure, *Am. J. Physiol.: Cell Physiol.,* in press.

23. **Ryan, U., Clements, E., Habliston, D., and Ryan, J. W.,** Isolation and culture of pulmonary artery endothelial cells, *Tissue Cell,* 10, 535, 1978.

24. **Ryan, U. S., White, L. A., Lopez, M., and Ryan, J. W.,** Use of microcarriers to isolate and culture pulmonary microvascular endothelium, *Tissue Cell* 14, 597, 1982.

25. **Rabinovitch, M. and Bothwell, T.,** An inhibitor of pulmonary artery smooth muscle growth i released by endothelial cells pulsated at high pressure, *Fed. Proc.,* 46, 730, 1987.

26. **Furchgott, R. F.,** Role of endothelium in responses of vascular smooth muscle, *Circ. Res.,* 53, 557, 1983.

27. **DeMey, J. G. and Vanhoutte, P. M.,** Heterogeneous behaviour of the canine arterial and venous wall; importance of the endothelium, *Circ. Res.,* 51, 439, 1982.

28. **Wilson, W. L., Bothwell, T., and Rabinovitch, M.,** In vitro studies suggest a pulmonary artery endothelial factor may facilitate smooth muscle contraction, *Tissue Cell,* in press.

Chapter 37

VASCULAR ENDOTHELIUM AND ATHEROSCLEROSIS: A MULTIDISCIPLINARY APPROACH

Vladimir N. Smirnov, Vadim S. Repin, Vsevolod A. Tkachuk, and Eugene I. Chazov

TABLE OF CONTENTS

I. INTRODUCTION

Vascular wall cells play an important role in the atherosclerotic thickening of intima and the progression of plaques; all cell types which form the population of vascular cells finally become involved in this process. Nevertheless, the function of smooth muscle cells and macrophages in atherogenesis seems to be better understood than the role of endothelial cells (EC). Many decades ago several eminent pathologists suggested that the atherosclerotic process starts from changes in the lumen of the arteries. However, it is only recently that this hypothesis is acquiring experimental support.

The progress in the study of endothelium is explained by a number of factors. First, the development of techniques for growing vascular endothelium has broadened the possibilities for biochemical, immunological, and molecular biological experiments. Second, techniques for short- and long-term culturing of vascular endothelium from human tissues have been developed. And third, the methods of cellular biology allow the modeling of some features of vascular atherosclerosis in vitro. The complicated picture of morphological manifestations of this disease in vivo is now being reconstituted stage by stage in culture. Experiments with living cells permitted study of the sequence of morphological, biochemical, and functional responses in atherosclerosis.

In this paper an attempt is made to review the main trends in the study of vascular endothelium carried out in the U.S.S.R. Cardiology Research Center in Moscow. Data are presented which were collected in laboratories of the Center over the last several years on the peculiarities of morphology, metabolism, and function of endothelium with special attention to the study of "normal" and "atherosclerotic" human vessels.

It is known that human atherosclerosis has a number of features which are difficult to reproduce in animal models. One of the important trends in the study of endothelium from human autopsy material is related to the description of initial morphological and biochemical changes of EC which can be considered as early cellular signs of atherosclerosis. An important thought which underlies all observations is the morphological and functional heterogeneity of EC in normal and atherosclerotic human vessels. A part of this review is an attempt to illustrate this point.

The second direction in our studies of endothelium is related to understanding the role of cellular receptors and second messengers in normal EC and in endothelial damage *in situ*. Until recently the high probability of EC injury was often explained by the unique location of endothelium in the circulatory system. The endothelial lining is extremely susceptible to mechanical and hemodynamic forces and is the target for damaging agents circulating in blood (viruses, cytotoxins, cholesterol oxiderivatives, fatty acids, catecholamines, activated neutrophils and macrophages, etc.). However, it is now evident that the receptor machinery of EC is a more sensitive and effective mechanism for controlling the integrity of the endothelial layer. Understanding of the conditions that stimulate receptors or enzymes which control the intracellular level of second messengers is of both fundamental and applied value. These studies may help to identify those undesirable combinations of hormones or various vasoactive substances in the blood which are capable of directly potentiating cytotoxic effects of other substances on vascular endothelium.

The third direction in our studies relates to understanding the mechanisms which control organization of the normal endothelial monolayer and its changes in atherosclerosis. It becomes evident now that the size, shape, mutual orientation, and intercellular contacts in endothelium are not incidental and statistic events, but are precisely and dynamically regulated. By cooperative stimulation of the receptors or the systems of second messengers it is possible to reversibly change the shape and mutual orientation of cells in certain directions, to change the intercellular contacts, and to stimulate proliferation and hypertrophy of endothelium. The ability to model morphological changes of endothelium in culture would permit understanding of how this monolayer is organized and how the morphology of EC is related to their function.

The last direction in our experiments is related to the development of approaches to target the endothelial surface or the components of subendothelium. The aim of these studies is to develop a technique for local targeting of potential drugs to the discrete zones of the luminal surface of vessels.

The study of ''normal'' and ''atherosclerotic'' human endothelium at the U.S.S.R. Cardiology Research Center is a ''hot spot'' which initiates experiments involving specialists of various disciplines. We hope that this spontaneous cooperation of many specialists will be useful in deeper understanding of human atherosclerosis.

II. ENDOTHELIAL CELL PATTERNS IN NORMAL AND ATHEROSCLEROTIC REGIONS OF HUMAN AORTA *IN SITU*

Progress in the study of vascular endothelium *in situ* was associated with the development of the techniques for studying the cellular organization of endothelium *en face* using light microscopy. By analysis of the preparations stained with silver nitrate, the heterogeneity of endothelium in the thoracic region of human adult aorta by cell size and shape has been demonstrated.[1] It was noticed that large and giant cells are often seen in human aorta after 40 years of age, whereas among people younger than 30 years they are practically absent. It was also shown that endothelium in the adult aorta can form groups of morphologically homogeneous cells.

Later, polymorphism of EC in atherosclerotic lesions over fatty streaks and plaques in human aorta has been described.[2-4] Quantitative examination of morphological heteroteneity of endothelium in human vessels by scanning electron microscopy (SEM) revealed various types of monolayer cellular organization with cell density varying from 300 to 500 up to

2000 to 3000 cells per square millimeter. By morphometric analysis it was possible to classify EC by size into small, (up to 400 μm^2), medium (400 to 800 μm^2), large, (800 to 1200 μm^2), and giant cells (over 1200 μm^2).[3]

The findings which follow below were obtained with so-called total aortic preparations (intima-media) stained with silver nitrate used for light microscopy. The advantage of this technique is that it permits analysis of large luminal areas of an artery, local fluctuations in the luminal surface with parallel evaluation, and determination of EC shape and area.

This approach allowed us to find a number of specific features in the organization of the endothelial lining in zones without gross atherosclerotic changes, and to examine the endothelial layer over atherosclerotic plaques. Thoracic and abdominal regions of 14 aortas obtained at autopsy of sudden-death victims aged 39 to 74 years, and aortas of five children (age up to 3 years, death in the hospital) were used in the analysis. The autopsy was performed within 3 hr after death. In the adult aortas in the regions having no visible atherosclerosis, EC vary by shape and size (Figure 1A). Similar cells usually form cell clusters. The degree of clustering may vary between different aortas and within the same aorta by a wide range, from random heterogeneity to practically 100% clusters (Figure 1B).

Morphometric analysis of EC in the specimen partly shown in Figure 1A demonstrated that in high cell-density regions of the monolayer about 80% of cells are small and medium sized, with cell areas not exceeding 800 μm^2 (Figure 2A); the size of a significant number of the cells is even smaller than the average area of EC in child aorta (Figure 3). In the adjacent regions with lower cell density, the majority of cells (75%) are large and giant cells with areas varying from 800 to 3500 μm^2 and more (Figure 2B).

Calculating the percentage of the areas taken by the clusters of small and large cells, the number of EC in the clusters, the following data were collected. In the fragment of aorta, for example, the clusters of small cells include about 80% of the total EC population but occupy less than 40% of the corresponding lumen. The main part of the luminal surface (60%) is covered with endothelium consisting of large, giant, and unclustered small- and medium-sized-cells.

A similar study of the organization of endothelium over atherosclerotic plaques led to conclusions somewhat different from earlier investigations.[3]

Out of 63 plaques examined in the samples prepared from 14 donors, a homogeneous monolayer consisting of medium-sized cells over the central region of the atherosclerotic cap has been found in 53 cases (Figures 4 and 5A). As a rule, these plaques had a diameter not less than 1 cm, which is characteristic of "adult" plaques. The larger the plaque, the larger the area covered with homogeneous endothelium. All calcinated plaques were covered with homogeneous endothelium.

In over nine plaques (14% of cases) a heterogeneous monolayer involving unclustered EC was found (Figure 5B), and in only one case a homogeneous monolayer composed of large and giant cells was seen over the plaque (Figure 5C).

The periphery of all plaques examined was covered with endothelium highly polymorphous by cell size. Even in cases where the central region of the plaque cap was not homogeneous the polymorphism of EC on the periphery of the plaque was more pronounced. The maximal cell heterogeneity by size accompanied by clusterization of endothelium was found in the transition area from the margin of the plaque to lesion-free intima.

Thus, the peculiar characteristic of the adult aortic human endothelium is its polymorphism by size in the regions without visible atherosclerotic changes. The most pronounced manifestation of the heterogeneity is the clustering of EC. At present, we have no information related to the role of clusters in atherogenesis. It can not be ruled out, however, that endothelial clusters include cells possessing higher proliferative activity.[5] In any case, the cellular heterogeneity and clustering of endothelium found in lesion-free regions of adult human aorta deserve further investigations.

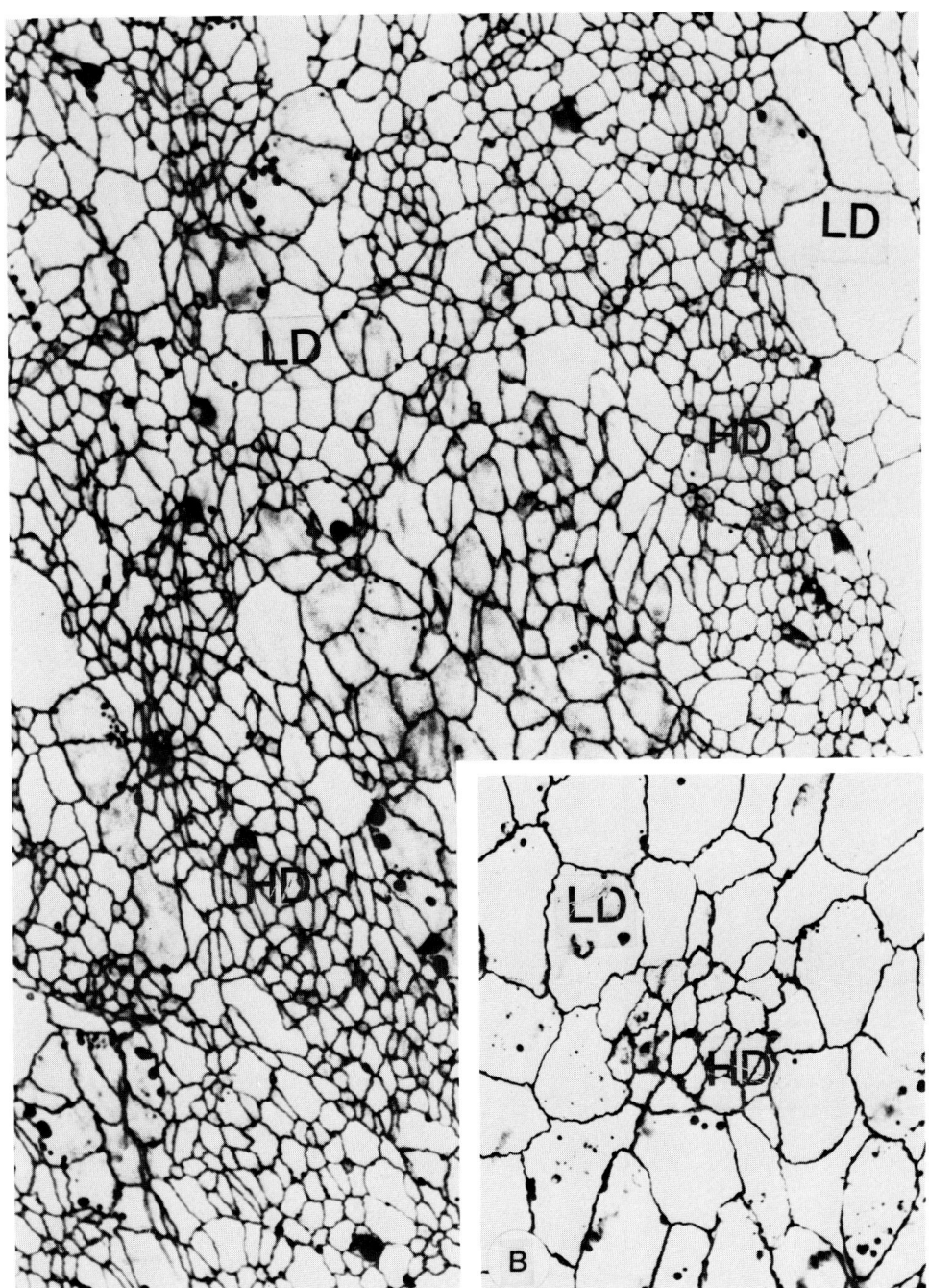

FIGURE 1. Human aorta endothelium *in situ*. (A) Heterogeneous monolayer of EC in unaffected zone (magnification × 150). (B) Cluster of homogeneous small size EC (magnification × 235). (HD), zones of high density of monolayer; (LD), zones of low density of monolayer. (Silver nitrate stain.)

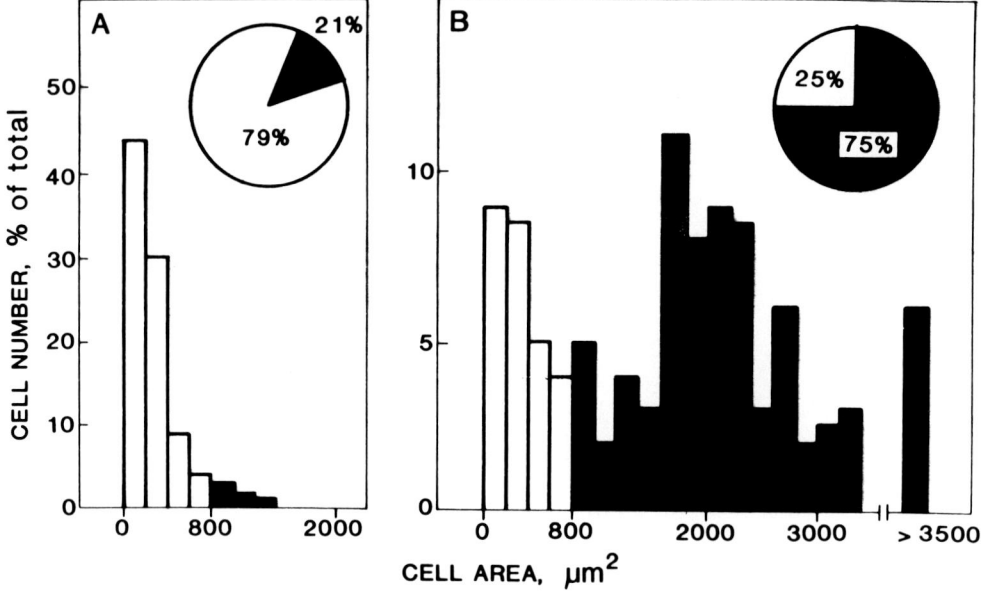

FIGURE 2.   Morphometric analysis of EC in zones of high (A) and low (B) density of monolayer. Distribution by area. White, small- and medium-sized EC (>800 μm²).

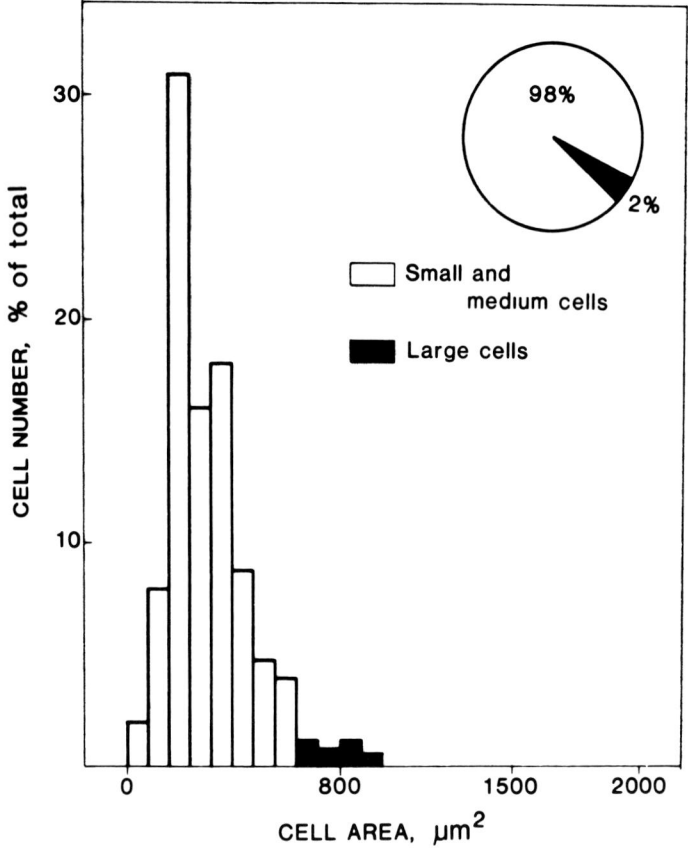

FIGURE 3.   Morphometric analysis of EC in child aorta. Distribution by area. See legend to Figure 2.

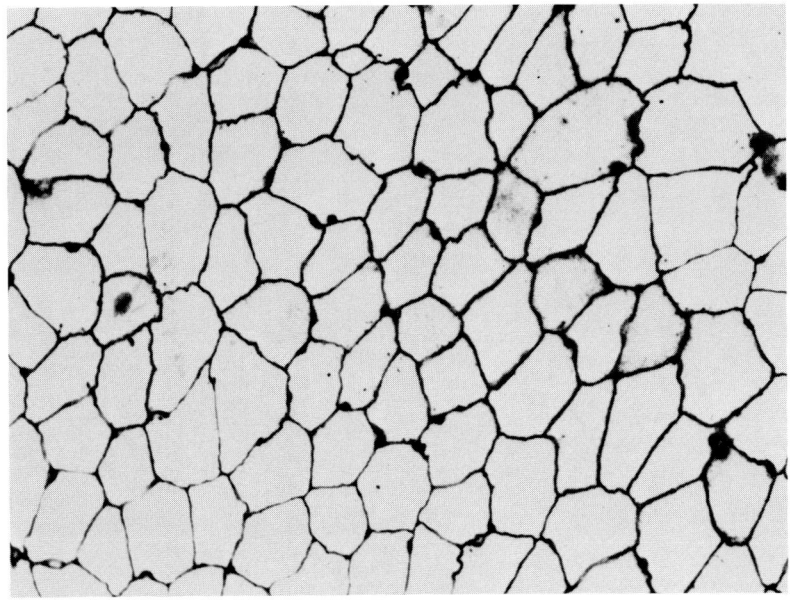

FIGURE 4.   Homogeneous endothelium over atherosclerotic plaque. (Silver nitrate stain; magnification × 300). See text.

III. SOME ULTRASTRUCTURAL CHARACTERISTICS OF ENDOTHELIUM IN ATHEROSCLEROTIC HUMAN AORTA

The comparative ultrastructural analysis of EC in vessel regions differing by the degree of atherosclerosis is one way to find out whether atherosclerosis in human vessels is related to the changes in the function of vascular endothelium.

The study of the ultrastructure of human aortic EC (men age 50 to 60 years) has shown that endothelial cells covering uncomplicated atheromatous plaques, in contrast to endothelium of other regions (noninvolved areas, fibrous plaques), have a different frequency of cells with primary cilium. It was found that up to 90% of EC over atheromatous plaques are ciliated, with the cilia protruding from their cell basal surface (Figure 6). Ciliated cells are normally rarely seen (20%) in noninvolved aortic regions or over fibrous plaques. In these cases, the cilium protrudes from the apical surface of the EC (Figure 6e). The majority of these endothelial cells in noninvolved regions are not ciliated (Figure 7, type I-IV). It is noteworthy that by a number of criteria about one half of unciliated cells have centrosomal organization (Figure 7, type III-IV) similar to that of ciliated cells (Figure 7, type V). The structure and centriolar localization are the same, thin filament bundles are located between centrioles, and the appendages of a mother centriole are in contact either with the plasma membrane or with the membranous vesicle (Figure 8). It is quite possible that types III and IV of the centrosome are, in fact, different stages in the formation of the primary cilium in EC; hence, most EC in noninvolved areas and over fibrous plaques are in various stages of cilliogenesis (Figure 9). Variations in the frequency of ciliated cells in different vessel regions may reflect the differences in life span of fully formed cilia in the cells. Over the surface of atheromatous plaques, a cilium in EC should exist as long as the plaque retains its histological characteristics used in morphological identification. A primary cilium (with a long life span) is thought to be formed in cells after the peak of proliferation followed by a period of lowered proliferative activity. This situation is normal for untransformed cells

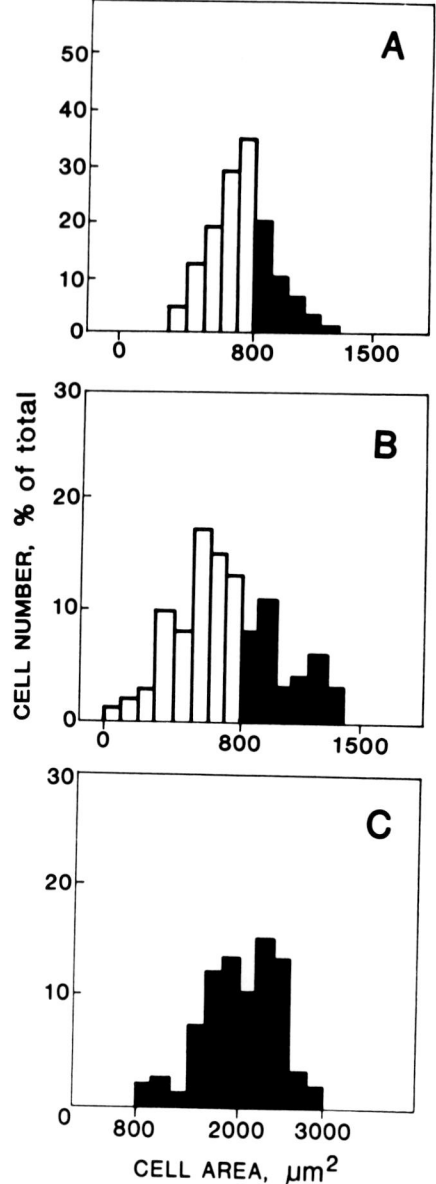

FIGURE 5.   Morphometric analysis of EC over
atherosclerotic plaque. Distribution by area. (A)
homogeneous monolayer of small- and medium-
sized EC; (B) heterogeneous monolayer of non-
clustered EC, (C) monolayer consisting of large
and giant EC.

when G1 block is induced by serum deficiency in the culture medium or when the cells
reach confluency.[6,7] According to some authors the cells behave in a similar way in vivo.[8]
It is thought that atheromatous plaque formation in the human aorta is accompanied (or
preceded) by the proliferation of corresponding EC which subsequently are capable of
forming the primary cilium. In other parts of the vessel (macroscopically normal regions
and fibrous plaques), the cells proliferate asynchronously and at any given moment are in
different stages of ciliogenesis.

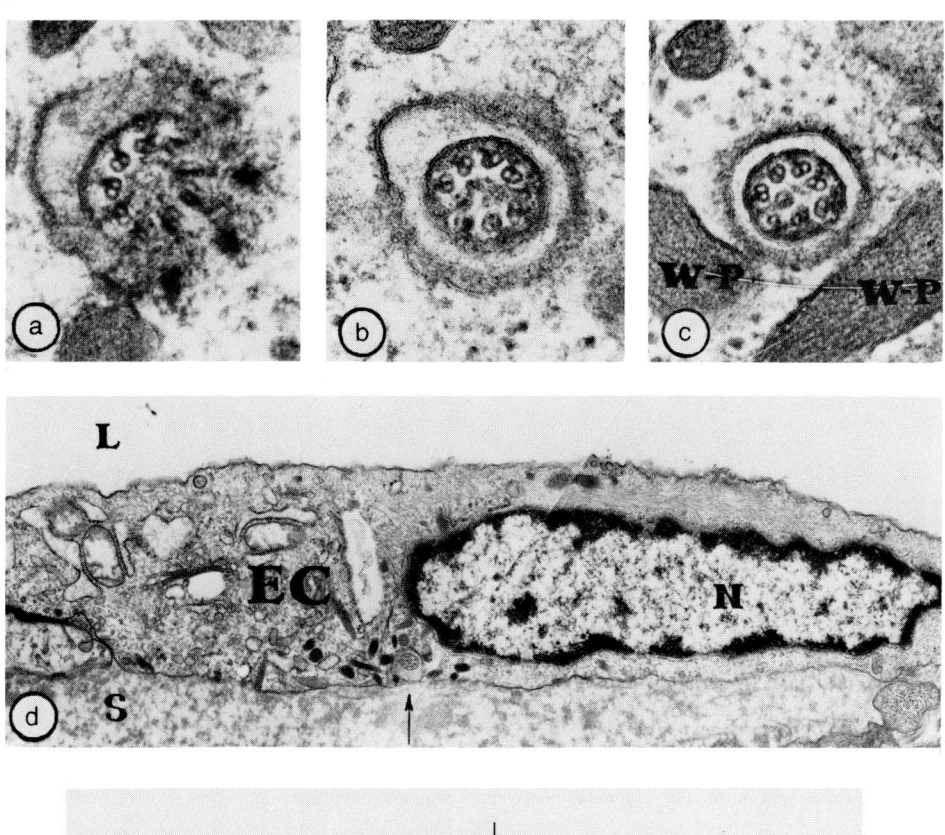

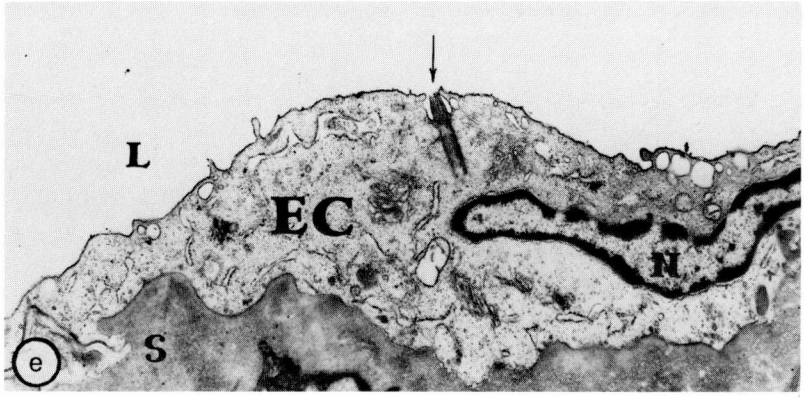

FIGURE 6. The primary cilia (arrows) of human aortic arch EC (man aged 60 years) in atheromatous plaque (a—d) and in a grossly normal region (e). (a—c) Serial transverse sections of a cilium L, lumen; S, subendothelial space; EC, endothelial cell; N, nucleus; W-P, Weibel-Palade bodies. (Magnification in (a—c × 83,000; d × 15,000; and e × 8,300.)

The difference in endothelial lining between atheromatous plaque and other regions of the aorta lies both in the frequency of ciliated cells and in the type of intracellular location of cilium distribution in EC. In macroscopically normal regions and fibrous plaques the mother centriole of EC more frequently contacts the apical cell surface. The mother centriole of the cells covering atheromatous plaques always contacts the basal cell membrane (Figure 9). Irrespective of their distribution within a cell, centrioles in all EC are always surrounded with Golgi membranes. The fact that the centrioles in EC contact the basal cellular membrane is of special significance. It is known that the centrioles in cells of epithelial layers in vivo

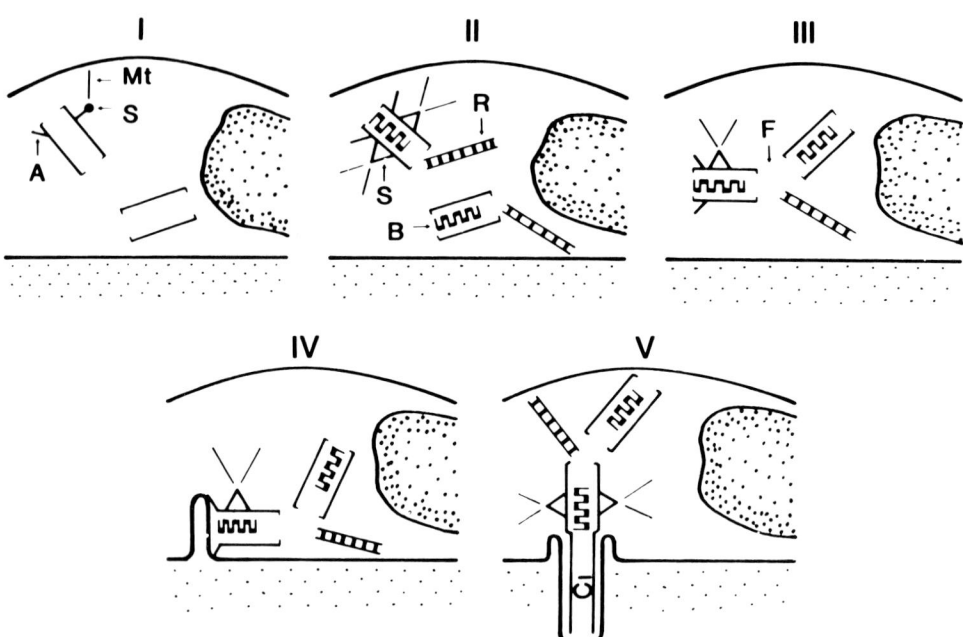

FIGURE 7.   Schematic representation of centrosomes of various forms (types I—V) in human aortic EC (90 cells were examined). Dimensional reconstruction by serial sections. Centrosomes of five types were differentiated according to: (1) cell centrioles disposition to each other or to the cell membrane; (2) details of centriole organization and centriolar microenvironment; (3) presence or absence of primary cilium. C, centriole; Ci, primary cilium; A, centriolar appendages; S, centriolar satellites; B, electron dense material in the lumen of centriole, amorphous bush; R, striated roots; Mt, microtubules; F, thin filament bundles.

with distinct polarity are, as a rule, located near the apical cell surface.[9-12] If a 9 + 0 cilium is formed in those cells, it is also found on the apical cell surface.[11,12] The functional significance of the apical location of centrioles in these cells is, at present, unknown. The interphase centrioles in the secretory cells are believed to play a role in directing the secretion by Golgi structures.[10] The intimate connection between microtubule-organizing center and the secretion is supported by the fact that the disruption of microtubules blocks intracellular transport of synthesized products and results in secretion disturbances.[13,14] The abnormal polymerization of microtubules is accompanied by dramatic changes in the localization of the Golgi apparatus.[15] It is likely that EC which differ in centriolar localization (at the apical or basal cell surface) have different types of Golgi apparatus function. These differences may be related to different directions of secretory flow or to changes in the intensity of the secretion in a certain direction, namely, to the lumen or to the subendothelium. This explanation obviously needs experimental support. Further studies in this area are of special significance for the understanding of cellular changes in atherosclerosis, for endothelial cell secretory activities (synthesis of varius collagens, glycocalyx, prostacyclin, etc.) which profoundly affect vessel wall metabolism.[16]

IV. PRIMARY CULTURE OF HUMAN AORTIC ENDOTHELIUM IN THE STUDY OF ATHEROSCLEROSIS

Further understanding of the nature of morphological and functional polymorphism found in human aortic endothelium was attempted using primary cultures of cells derived from normal and ''atherosclerotic'' human aorta.

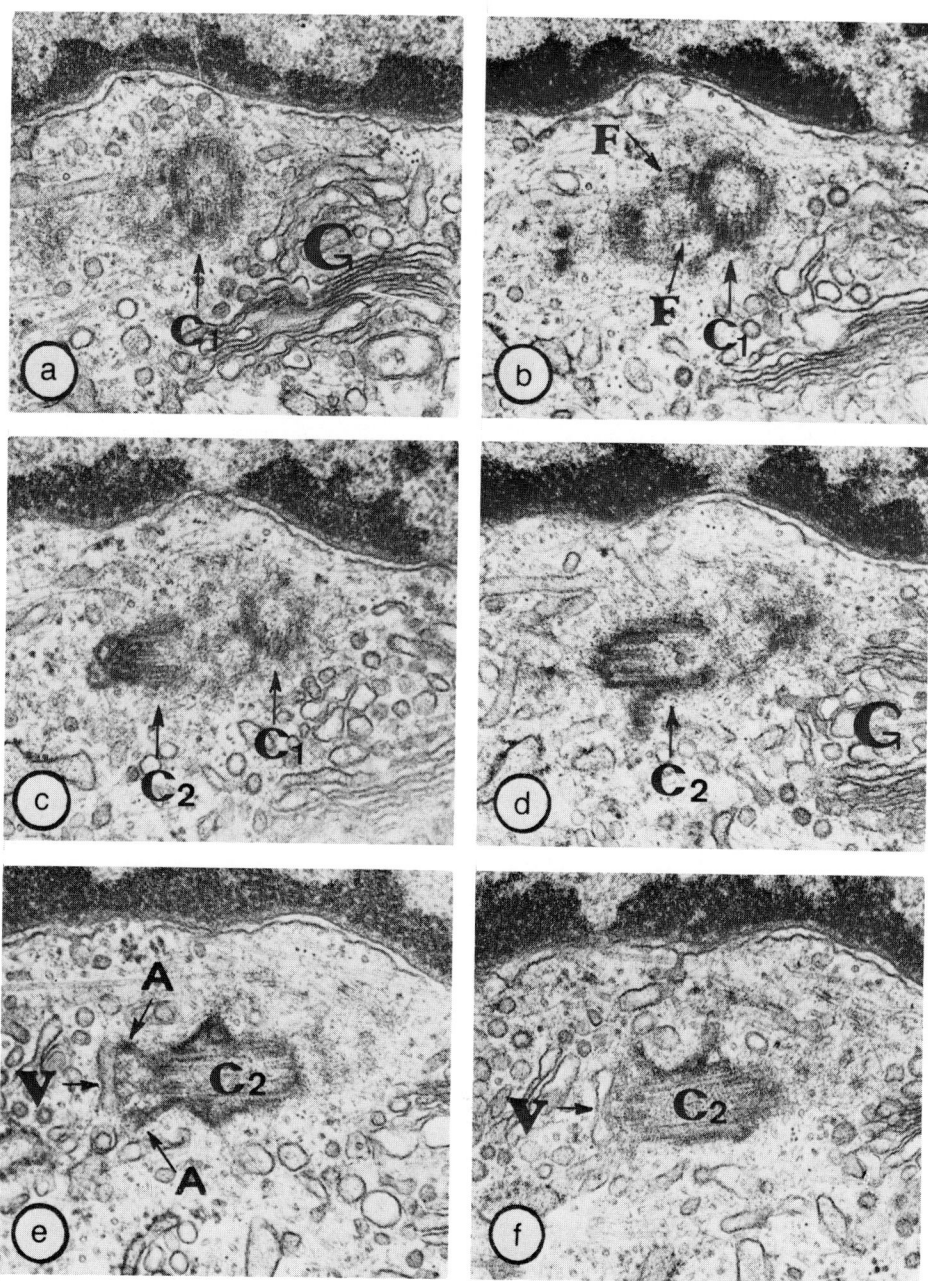

FIGURE 8.   Centrosome organization in EC from plaque-free region of human aorta arch (a—f, serial sections).
C_1, daughter centriole; C_2, mother centriole; F, thin filament bundles; V, membrane vesicle; G, Golgi apparatus.
(Magnification × 50000.)

A. Isolation of Endothelial Cells from Atherosclerotic Human Aorta

For isolation of EC, thoracic aortas were obtained at autopsies of males who died suddenly at ages from 20 to 70 years. The autopsies were performed within 3 hr after death. The type of atherosclerotic lesion was determined visually as described earlier.[17] Vessels without visible atherosclerotic changes or those where the combined area of fatty streaks and lipid spots did not exceed 10 to 15% of the surface were considered "normal". As "atheros-

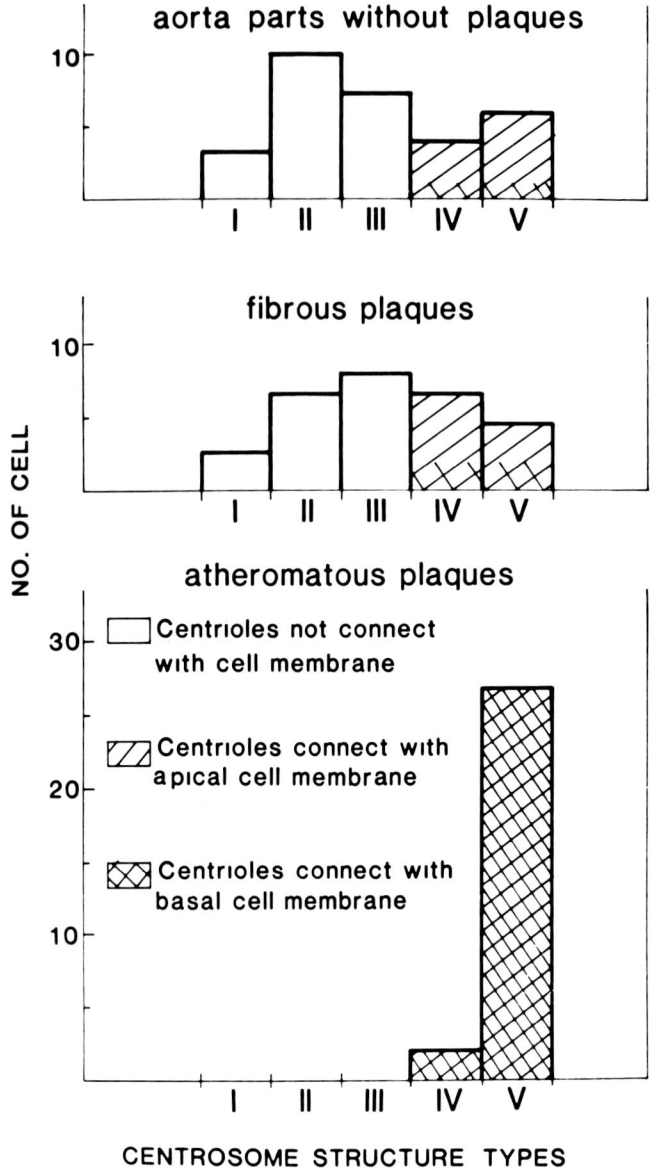

FIGURE 9. Distribution of human aortic EC with various centrosomes (I—IV types) in regions differing in the degree of atheroscerotic lesion. Centrosomes were examined in 90 EC (1) 30 cells from 9 plaque-free aorta pieces; (2) 30 cells from 5 fibrous plaques; and (3) 30 cells from 5 atheromatous plaques. All cells were taken at random.

clerotic'', those aortas were used where over 50% of the luminal surface was covered with fatty streaks, spots, and atherosclerotic plaques.

To isolate EC, we applied a previously described technique[18] with the exception that dispase (Boehringer, Mannheim GmbH, BRG) was substituted for collagenase. The use of dispase for isolation of EC has certain advantages over collagenase treatment. In the course of incubation of aortic samples in 0.15% dispase (Figure 10), a gradual increase in the number of released EC occurs; digestion comes to saturation by 60 to 90 min of incubation with the enzyme. Altogether about 5 to 8 $\times$ 10^4 cells per square centimeter of aortic surface

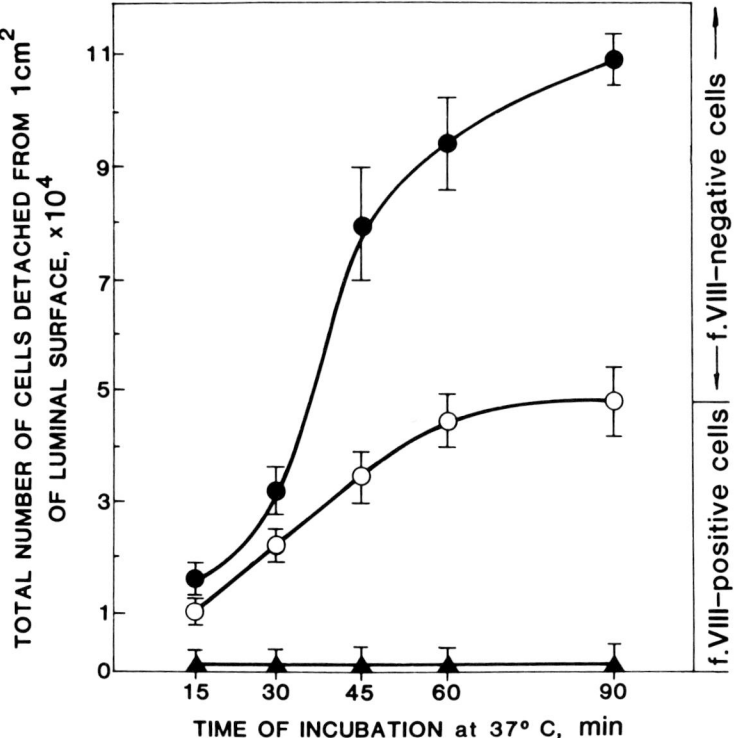

FIGURE 10. Harvesting of human aortic EC with collagenase (●) and dispase (○). A segment of aorta was incubated in medium 199 containing 0.1% collagenase type IV (Sigma) or 0.1% dispase (Boehringer). Enzyme-free medium 199 was assumed as control (▲).

can be isolated by this technique. Thus, the number of EC taken off by the dispase from the intimal surface of the vessel is roughly equal to the density of the endothelial monolayer *in situ*.[3,17] Cell suspensions obtained by this procedure include aggregates of 5 to 15 EC.

The content of subendothelial cells, as a rule, does not exceed 1 to 3% of all cells isolated. Viability of EC estimated by the trypan blue exclusion test exceeds 90 to 95%. Cells were seeded on plastic dishes at high densities comparable to the density of monolayer (5 to 8 × 10[4] cells per square centimeter). The attachment to plastic surface and spreading were about 70 to 80%. In 6 to 8 hr, the cultures were washed and the culture medium changed. The composition of the culture medium and the conditions for maintenance of cultures were described elsewhere.[3]

B. Growth of Human Aortic Endothelial Cells and their Morphology in Culture

Within the first 4 to 6 hr, EC become attached to the substrate and partially spread. The complete spreading of cells takes place within 24 hr after seeding. Colonies are formed consisting of 5 to 20 epitheloid cells. The further growth in culture occurs from these colonies by proliferation and migration of EC.

At this stage EC are elongated rather than polygonal in shape, and contain one or several nuclei with well defined nucleoli. In the colonies containing more cells, the central zone is covered with polygonal cells, and the periphery contains elongated EC (Figure 11A). Cells possess relatively homogeneous ectoplasm; cellular endoplasm has dense elements and inclusions. Within 2 to 3 days the morphology of the primary colonies does not essentially change.

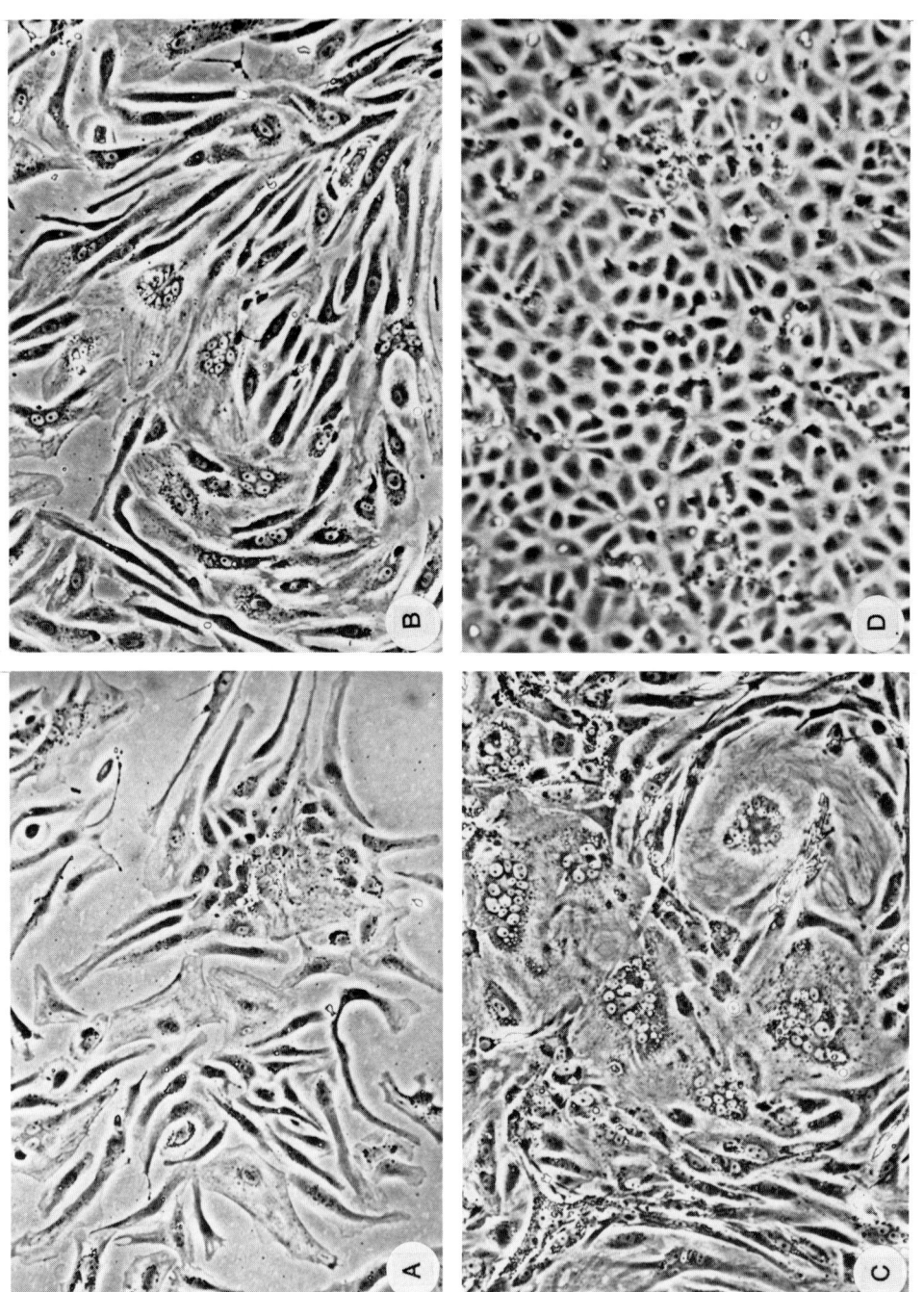

FIGURE 11.   Primary cultures of EC derived from atherosclerotic human aorta at different stages of growth (A—C) and EC from child aorta (D). (A) EC attachment and formation of primary colonies, 48 hr after seeding; (B) growing EC culture, 4 days after seeding; (C) monolayer of polymorphic EC, 7 days in culture; (D), monolayer of homogeneous EC from child aorta, 7 days in culture. (Phase contrast; magnification × 135.)

In the period from 3-4 to 5-6 days, active growth and migration of EC occur. Separate colonies fuse together, and cells cover most of the plastic surface. The shape of cells is elongated and sometimes spindle-like; cells may possess long processes. Cell nuclei at this stage are oval in shape, and the cytoplasmic inclusions are well delineated (Figure 11B).

On days 7 to 9, the formation of a compact monolayer is observed. Cells become polygonal, with oval nuclei and a well developed endoplasm and a somewhat more transparent ectoplasm. The formation of contacts along cellular perimeters becomes evident. The confluent monolayer is formed (Figure 11C). During long-term culturing of EC (10 to 15 days) multilayer cultures are never found. The migration of separate cells over the surface of the monolayer was never registered.

At all stages of growth, in addition to typical small mononucleated cells, large cells with several nuclei (from 2 to 10-15, and more rarely up to 20 to 30) of round or oval shape are found in culture. The nuclei are usually located in the central region of cells or form two to three groups, each containing several nuclei. The number of multinucleated cells containing two or more nuclei, varies in various cultures and may vary from 1-3 up to 20-30% of the total number of cells in the culture. Using high seeding densities we did not observe any increase in the number of multinucleated cells in the cultures. The percentage of multinucleated cells did not correlate with the age of donors; however, a significant correlation was found between the relative number of multinucleated cells in culture and the severity of atherosclerosis of the aorta used for cell isolation. Thus, in cultures isolated from "normal" aorta (age group 40 ± 14.2 years, N = 12), the percentage of multinucleated cells was about 7%. More than twofold growth in the number of multinucleated cells ($p < 0.05$) was found in the comparable age group of "atherosclerotic" aorta (59 ± 10 years, N = 14); the percentage of multinucleated cells in these cultures was about 14.9%.

The multinucleated cells were never seen in cultures of EC isolated from child aorta (Figure 11D), aorta from donors aged younger than 20 years, and human umbilical vein endothelium.

Using indirect immunofluorescence with antibodies against Factor-VIII antigen, very distinct specific staining of the perinuclear zone of the endoplasm was always seen in mononucleated as well as in multinucleated cells (Figure 12). Factor-VIII antigen was present in more than 95% of cells at all stages of growth.

Factor VIII negative cells were represented by smooth muscle cells. They had characteristic morphology and stained with antibodies to smooth muscle myosin.[19]

A significant number of EC isolated from "atherosclerotic" human aorta contained lipid inclusions stained with oil red O (Figure 13). The percentage of cells with lipid inclusions varied in different isolations and amounted to 10-20 up to 80-100% cells. In endothelial cultures derived from child and "normal" adult aorta, cells with lipid inclusions were either not found or their content did not exceed 1 to 5% of total cell population. In these cells lipid inclusions appeared on the 3rd to 5th day after plating, when cells were maximally spread. We never noticed the formation of lipid inclusions in cells which did not contain lipids *in situ*.

C. Ultrastructure of Human Aortic Endothelium in Culture

For transmission electron microscopy (TEM) we used cultures of EC isolated and grown from "normal" and "atherosclerotic" human aorta (age 40 to 55 years). The cells were grown under standard conditions until confluency was reached (7 to 9 days after seeding). The ultrastructure of typical mononucleated and multinucleated EC was examined by standard electron microscopy. No significant differences in the ultrastructural organization of these two cell types have been noticed.

In culture, EC grow as monolayer and form dense junctions (Figures 14 and 15A). The surface of the cells is usually smooth or has single microvilli. On apical as well as on basal

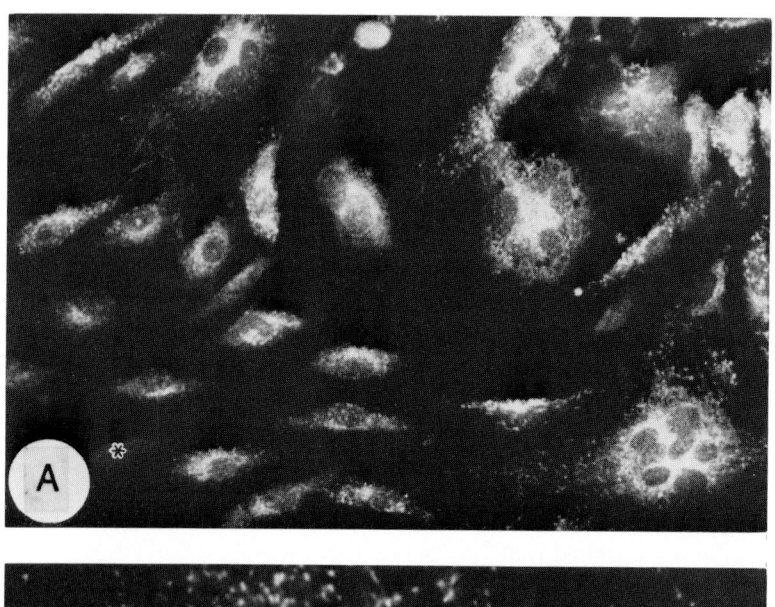

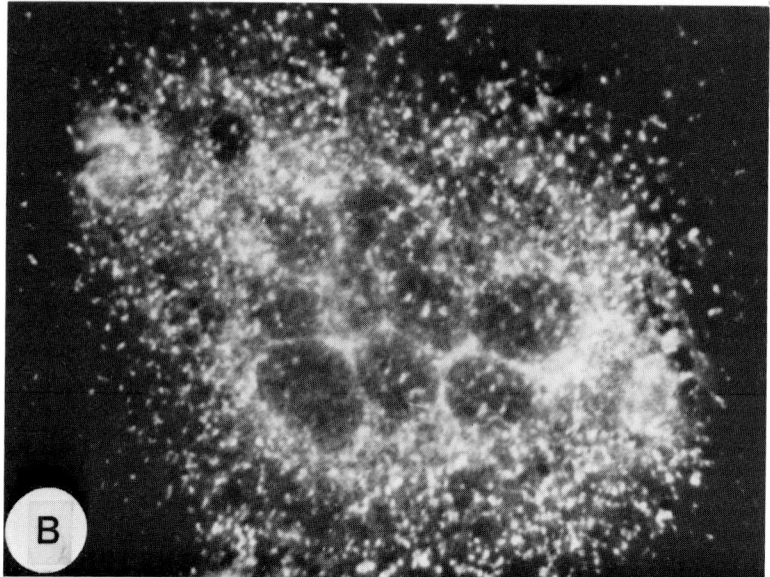

FIGURE 12.   Localization of Factor VIII antigen in EC from human atherosclerotic aorta. (A) Monolayer of aortic endothelium, Factor VIII antigen-negative cell (*). (Magnification × 200.) (B) Giant multinuclear EC. (Magnification × 500.) Cell cultures were treated with rabbit antiserum against human Factor VIII antigen and FITC-goat antibodies against rabbit IgG conjugates. Control staining was performed by replacing primary antisera either with the serum from nonimmune rabbits or DPBS. The control staining was negative.

surfaces of cells, numerous micropinocytotic vesicles of 60 to 80 nm and coated vesicles are seen (Figure 15B). Round and oval nuclei are centrally located in the cell. Condensed chromatin is localized in the vicinity of the nuclear membrane. The nucleoli (varying from one to three per cell) are usually large, with well delineated granular components (Figure 15C). Cells have well developed Golgi complexes consisting of the system of membrane cavernae and vesicles (Figure 15D). In multinucleated cells the Golgi complex (one or several) is usually located in the perinuclear zone. The majority of the cytoplasm is occupied by rough and smooth endoplasmic reticulum (Figure 15D). Ribosomes not bound to mem-

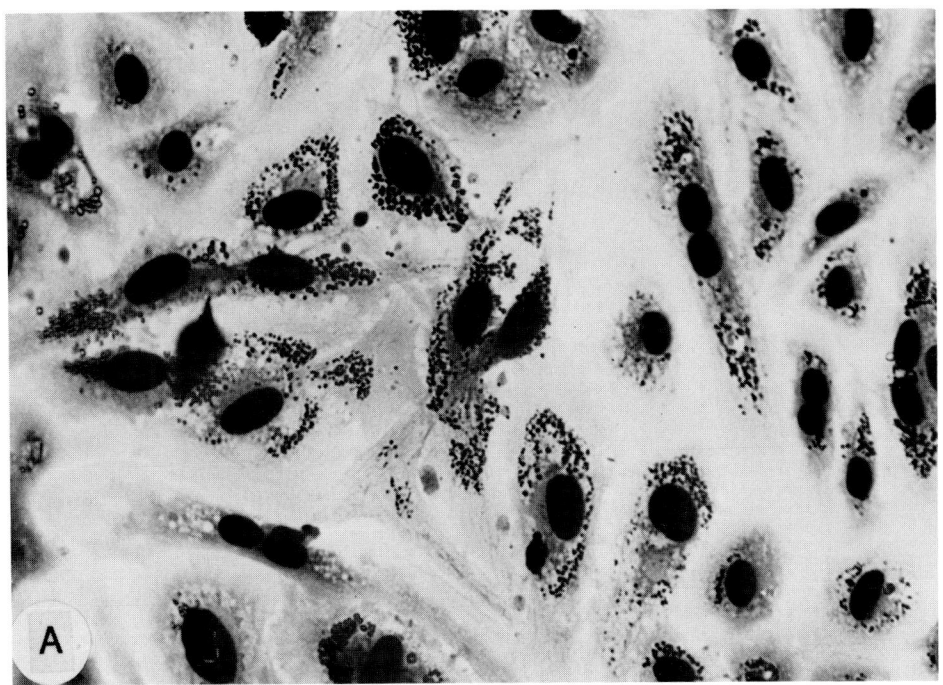

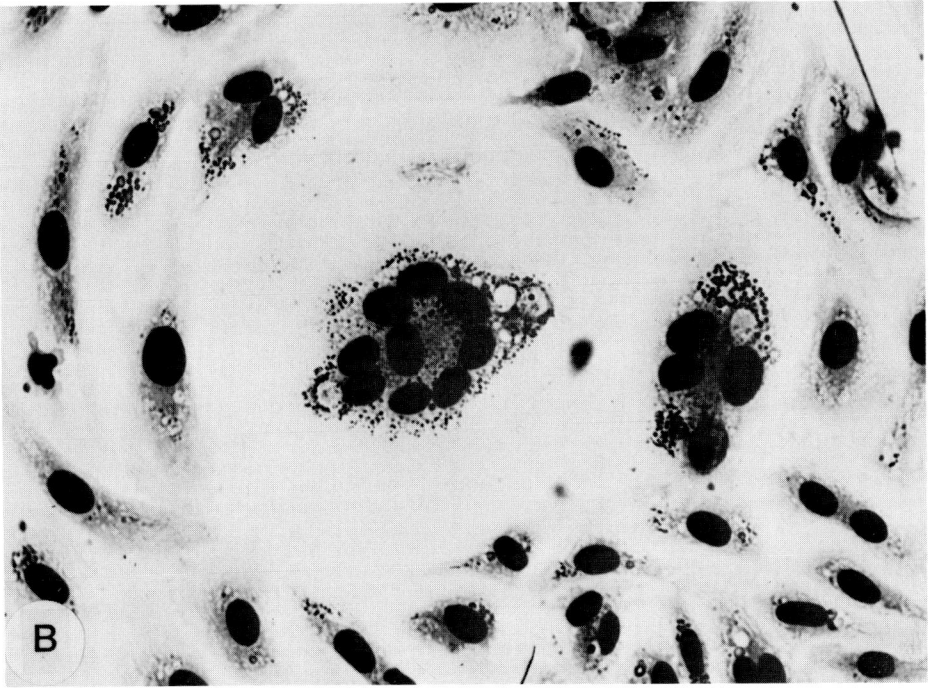

FIGURE 13.   Lipid inclusions in aortic EC on the 5th day of culturing. (A) Mononuclear EC with lipid inclusions; (B) giant multinuclear EC with lipid inclusions. (Light microscopy; oil-red-o-hematoxylin staining; magnification × 300.

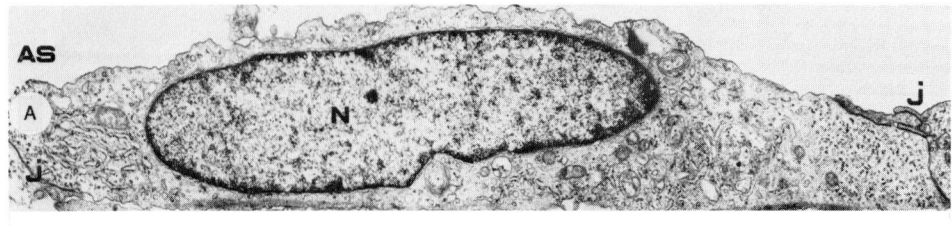

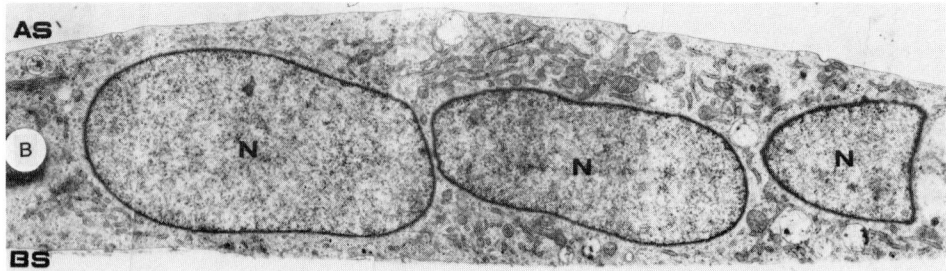

FIGURE 14. TEM of mononuclear (A) and multinuclear (B) EC from atherosclerotic human aorta in primary culture. The section was cut perpendicular to the culture plastic surface (7 days in culture). BS, basal cell surface; AS, apical cell surface; N, nucleus; J, a junction with neighboring cell. (Magnification in × 1,000; in B × 6,900.)

branes are organized in polysomes consisting of three to eight ribosomes. Mitochondria are branched and have an irregular elongated shape. Their cristae are located in the longitudinal as well as in the cross-sectional direction, and the matrix between cristae has medium electron density (Figure 15D). In the cytoplasm of cells, large numbers of microtubules and microfilaments are found. In the basal part of EC microfilaments form bundles of parallel fibrils with dense bodies. Cytoplasm contains lysosomes, multivesicular bodies, myelin figures, and phagocytotic vacuoles of various size (Figure 15E). Multinuclear EC contain, as a rule, several groups of centrioles (Figure 15F). Almost all EC analyzed contain lipid inclusions of different sizes (Figure 15D and E). Under basal membrane regions not involved in the formation of contacts with the substrate, the presence of amorphous extracellular matrix has been noticed.

In addition to the above-mentioned organelles and inclusions, mononucleated and multinucleated cells contain Weibel-Palade bodies which are represented by longitudinally oriented microtubules surrounded by a membrane (Figure 15G). Most often Weibel-Palade bodies were found in the region of the Golgi complex or in the vicinity of the cellular membrane. All ultrastructural characteristics of EC described above indicate not only high viability of the cells, but also high functional activity of EC in culture.[19]

Thus, the primary cultures of EC from human aorta can be used as an adequate model to study physiological and functional characteristics of normal human vascular endothelium and their rearrangements caused by atherosclerosis.

D. Heterogeneity of Human Aortic Endothelial Cells: Biparametric Analysis of Cellular DNA and Protein Content by Flow Cytometry

The increase in the number of nuclei per cell and increase in cell size are usually accompanied by polyploidy. At the same time, the elevation in the amount of DNA per cell is either not paralleled by the increase in the protein content[21] or results in the hypertrophy of a cell.[22] However, the relationships between protein content and ploidy are not clear.

To answer this question, flow cytometry was used.[23,24] Cells were fixed and stained with

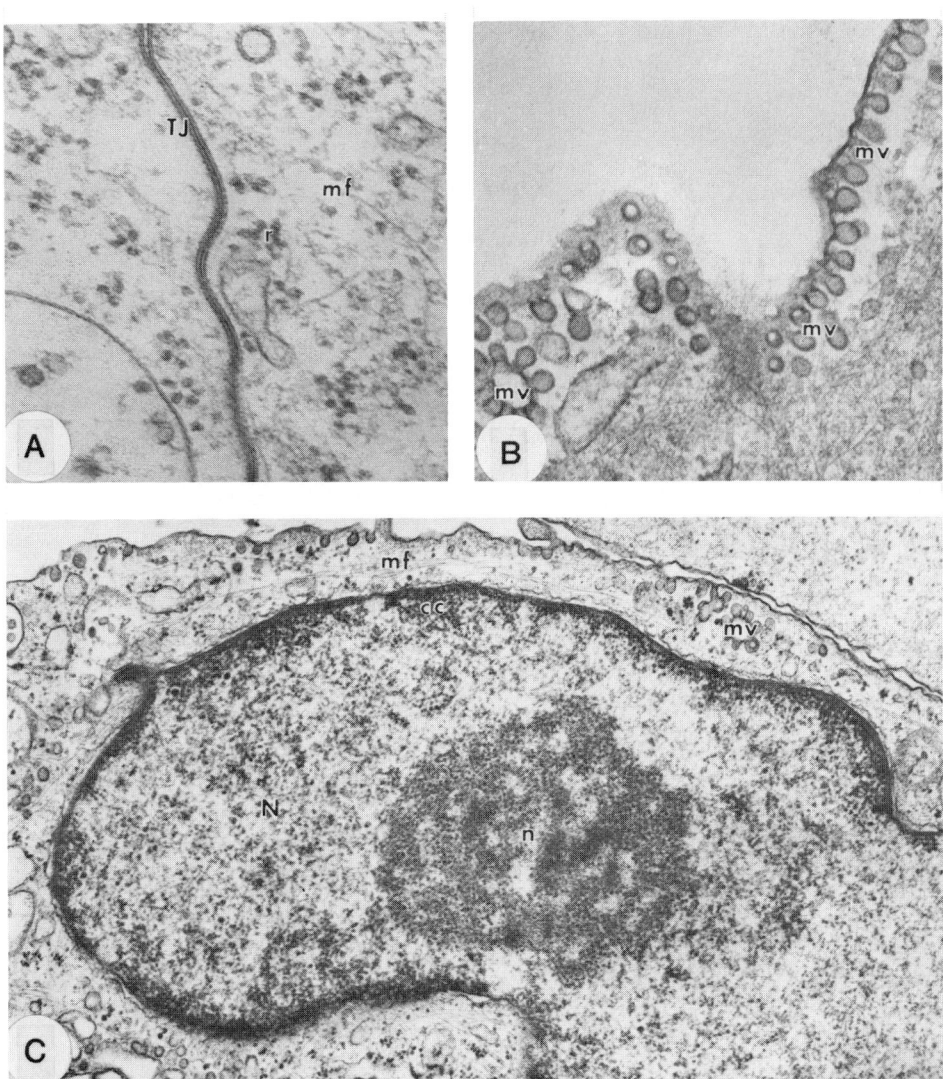

FIGURE 15. TEM of EC from atherosclerotic human aorta on the 7th day in primary culture. (A) Intercellular junctional area. (Magnification × 76,500.) (B) A part of the apical cell surface. (Magnification × 39,000.) (C) fragment of the nucleus. (Magnification × 19,000.) (D, E) Fragments of cytoplasm, (Magnification in D × 28,000; in E × 22,000.) (F) Centriolar region of a multinuclear EC. (Magnification × 22,000.) (G) Weibel-Palade bodies. (Magnification × 47,000.) N, nucleus; tj, tight junction; mf, microfilaments; r, ribosomes, mv, micropinocytic vesicles; n, nucleolus; cc, condensed chromatin; G, Golgi complex; RER, rough endoplasmic reticulum; SER, smooth endoplasmic reticulum; l, lipid inclusions, mt, microtubules; c, centrioles; m, mitochondria; Ly, lysosomes; Mlb, myelin-like bodies; WPb, Weibel-Palade bodies.

propidium iodide and fluoresceinisothiocyanate 12 to 16 hr after seeding. Three-dimensional distribution of cells by DNA and protein content was generated by the computer and analyzed (Figure 16).

Figure 17 shows the number of cells in each zone of biparametric distribution of cell population for human umbilical vein, "normal" and "atherosclerotic" human aortic endothelium. It is seen that a significant part of EC population in all vessels are the diploid cells with medium protein content. Tetraploid cells are found in the zones corresponding to cells with high and very high protein content, and hypertetraploid cells are found in the

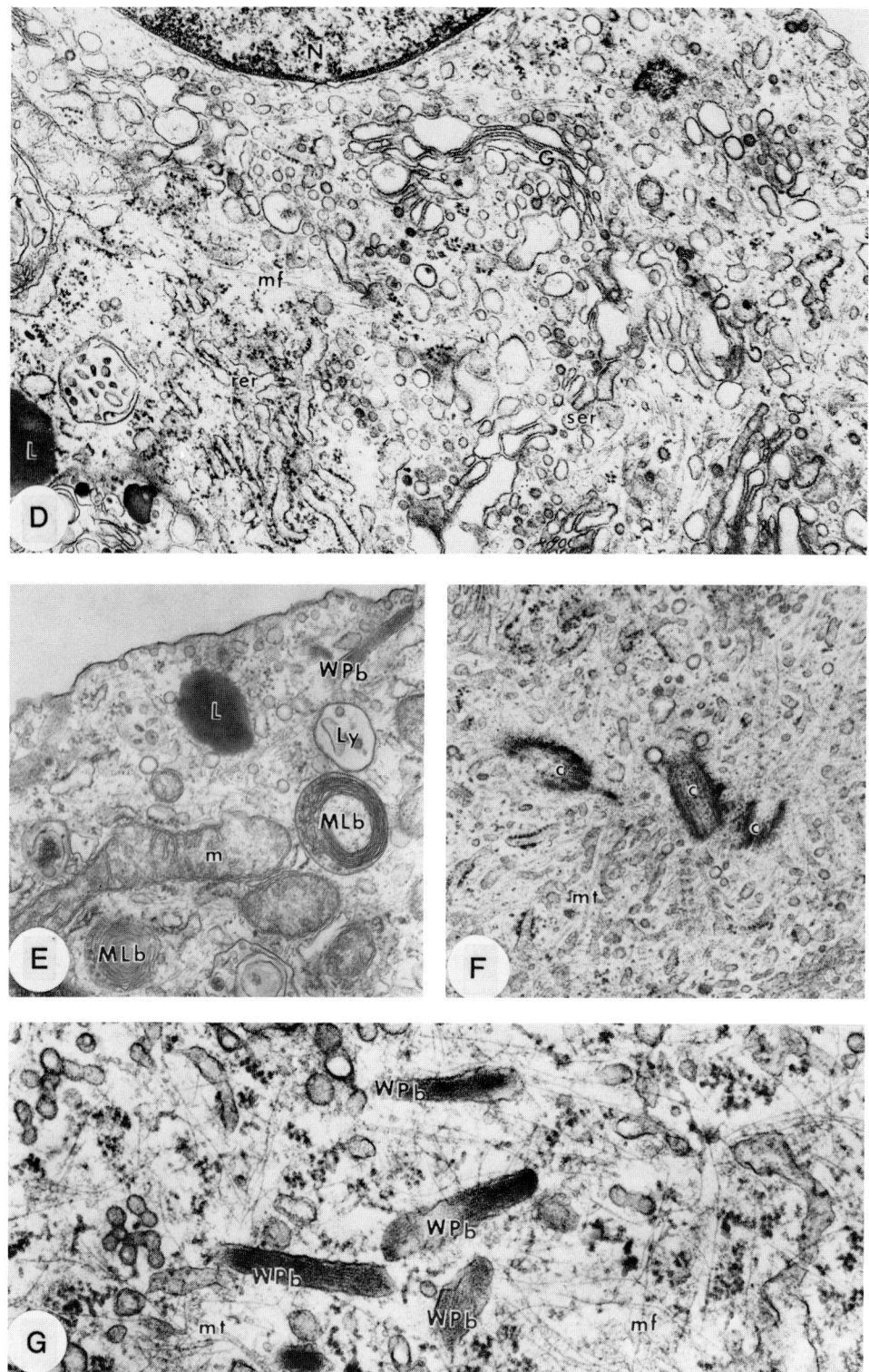

FIGURE 15D—G

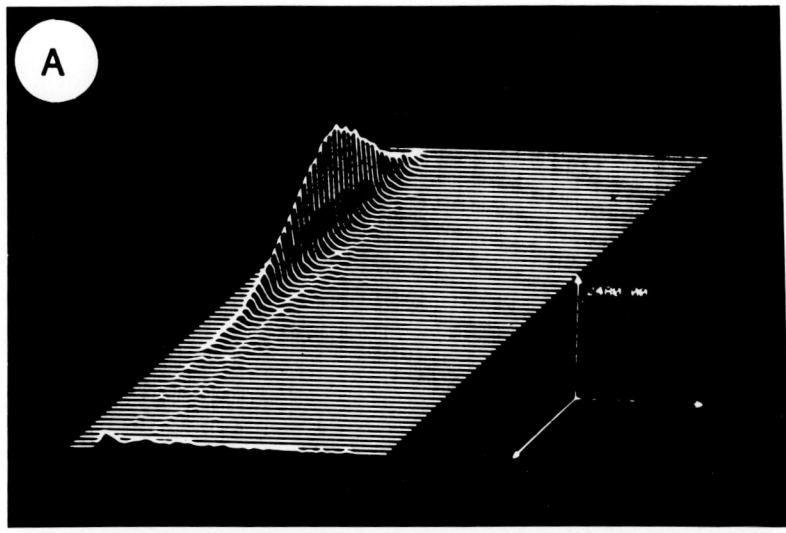

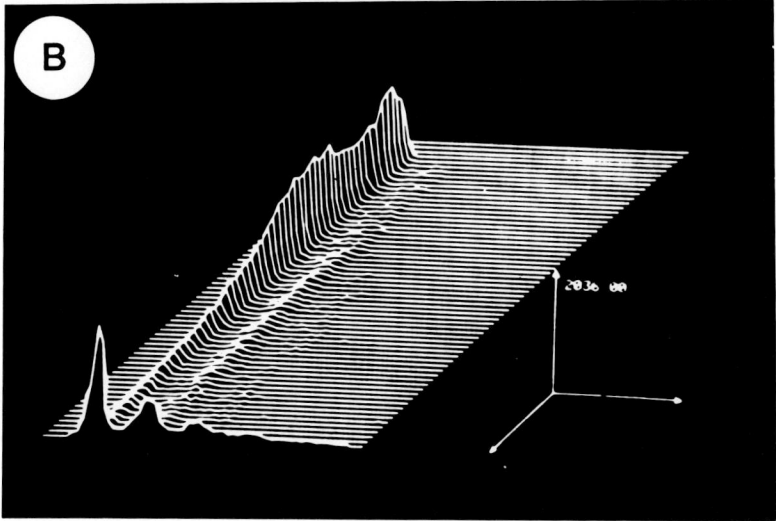

FIGURE 16. Cell distribution according to DNA-protein level in EC isolated from human atherosclerotic aorta or umbilical vein. Fluorescence analysis was performed on a cell sorter FASC-II (Becton-Dickinson). The excitation wavelength of argon ion laser (Spectraphysics 164-05) was 488 nm with a 400-mW output. The optical system was arranged with two filters: BP-520 (Ditric Optics) for green (absciss) and LP-580 (Ditric Optics) for red (ordinate) fluorescence. In a histogram for atherosclerotic EC, a subpopulation of cells with low protein content was evident. These cells were identified as lymphocytes and cut planimetrically from the analysis.

region with very high protein content. Thus, for endothelium the increase in the ploidy is accompanied by the increase in the cellular protein content. Nevertheless, in "normal", as well as in "atherosclerotic" aorta, the significant part (up to 43%) of hypertrophied cells are, in fact, diploid cells (Figure 17).

Upon analyzing the histograms of DNA distribution, it as found that for human umbilical endothelium, "normal" and "atherosclerotic" aortic endothelium the percentage of diploid cells was 93, 81, and 69%, respectively. The percentage of cells with DNA content more than 2C was increased in "atherosclerotic" aorta (31%) compared to "normal" vessels by 1.6-fold ($p < 0.05$; Figure 18).

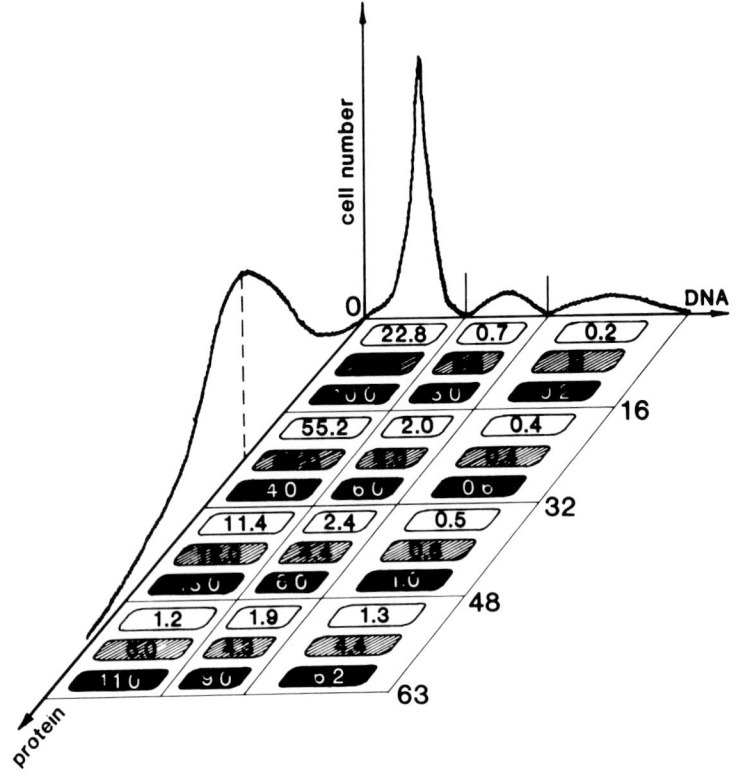

FIGURE 17.   The mean cell number in each of 12 subgroups according to DNA-
protein content for EC isolated from grossly normal, atherosclerotic human aorta, or
umbilical vein. Biparameter distribution was divided in the following way: along the
DNA axis three subpopulations, corresponding to 2C, 4C, and more than 4C were
defined. The position of 2C cells on the histogram was determined independently
using propidium-iodine-stained human lymphocytes. Cells were subdivided into four
classes according to green fluorescence level: low (0—16 channels), medium (17—
32), high (33—48), and very high (49—64) protein content. The mode of distribution
was located in the medium-range class (17—32 channels).

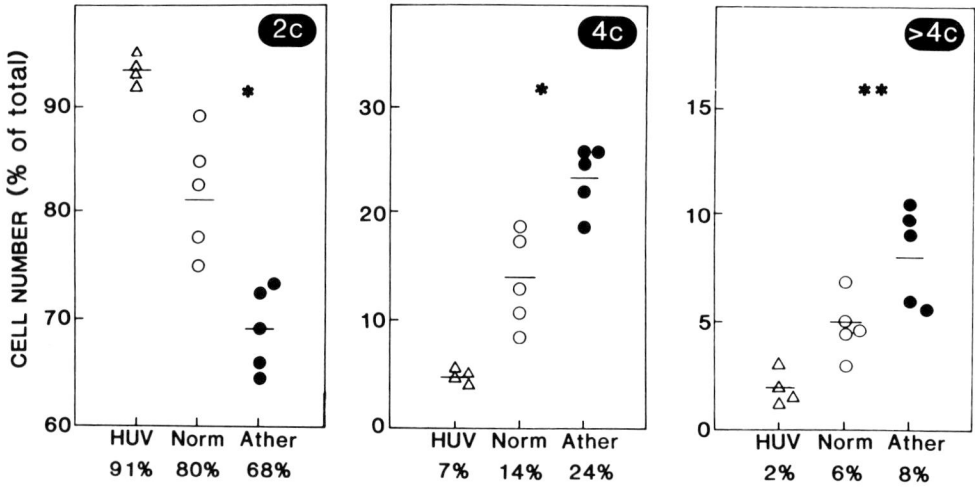

FIGURE 18.   Distribution of EC isolated from grossly normal (Norm) or atherosclerotic (Ather) human aorta, or
umbilical vein (HUV) according to DNA-content. A histogram for each subpopulation was obtained projecting
two-dimension distribution on the red fluorescence axis.

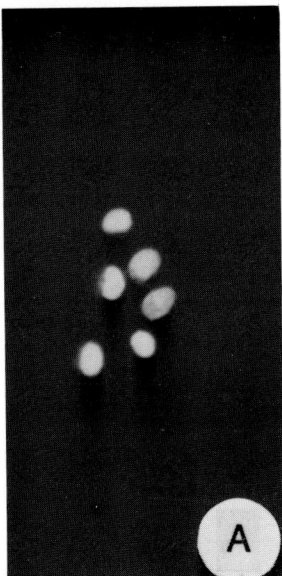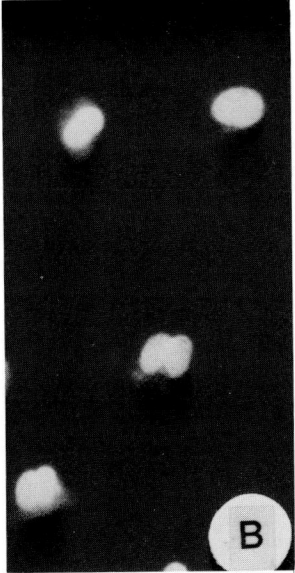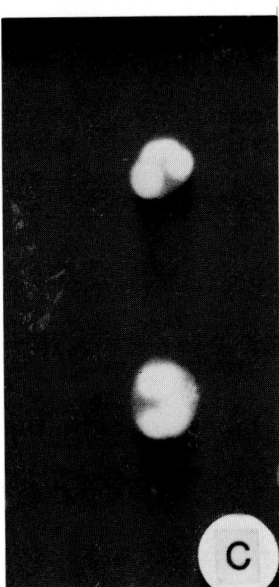

FIGURE 19. Subpopulations of 2C (A), 4C (B), and >4C (C) cells sorted out by a FACS-II cell sorter. In the 2C cell subpopulation, only mononuclear cells are present, whereas in 4C both mononuclear and binuclear cells and in the >4C subpopulation mononuclear, binuclear, and polynuclear cells are present. (Magnification × 1000.)

By sorting out cells differing in protein and DNA content, it was found that the subpopulation of diploid cells is represented by mononucleated cells; tetraploid cells included mononucleated and binuclear cells; hypertetraploid cells were represented by mononucleated, binuclear, and multinucleated cells (Figure 19).

The formation of multinucleated cells in EC culture was noticed after the interaction of endothelium with HTLV-1 virus,[25] known to occur in cell aging,[26] and a number of damaging agents.[27] The mechanisms of formation of multinucleated polyploid cells as a result of atherosclerosis are still obscure.

The histogram of the distribution of EC by protein content is continuous and asymmetric. If in the case of human umbilical endothelium the distribution approached normal (coefficient of asymmetry As = 0.3), for the endothelium of atheroscerotic aorta this distribution was highly assymmetric (As = 1.1) due to an increase in the percentage of cells with high protein content. The variation coefficient, which reflects the heterogeneity of a distribution, was equal to 47, 58, and 72% for human umbilical endothelium, "normal", and "atheroscerotic" aorta, respectively. Thus, significant difference is seen between "normal" and "atherosclerotic" aorta by this criterion (p <0.05).

Calculating the number of cells in corresponding parts of the histogram, it was shown (Figure 20) that the percentage of hypertrophied cells in human umbilical endothelium is about 4%, whereas in the endothelium of "atherosclerotic" aorta this value is 27%. EC from normal aorta can be divided into two groups differing by the number of hypertrophied cells: (1) with a relatively low percentage of hypertrophied cells (not more than 15%) compared to "atherosclerotic" aorta (p < 0.05); and (2) with high content of these cells not significantly different from their amount in "atherosclerotic" aorta. The relative number of hyperploid cells in both groups is similar and significantly lower than in "atherosclerotic" aorta. The appearance of hypertrophic EC may be related to higher synthetic activity in the regions of an elevated vessel permeability in experimental atherosclerosis,[27] under conditions of hypoxia,[28] or when nicotin is present in the culture of bovine aortic endothelium.[29] It was also found that human aortic endotheliocytes can selectively accumulate plasma proteins.[30]

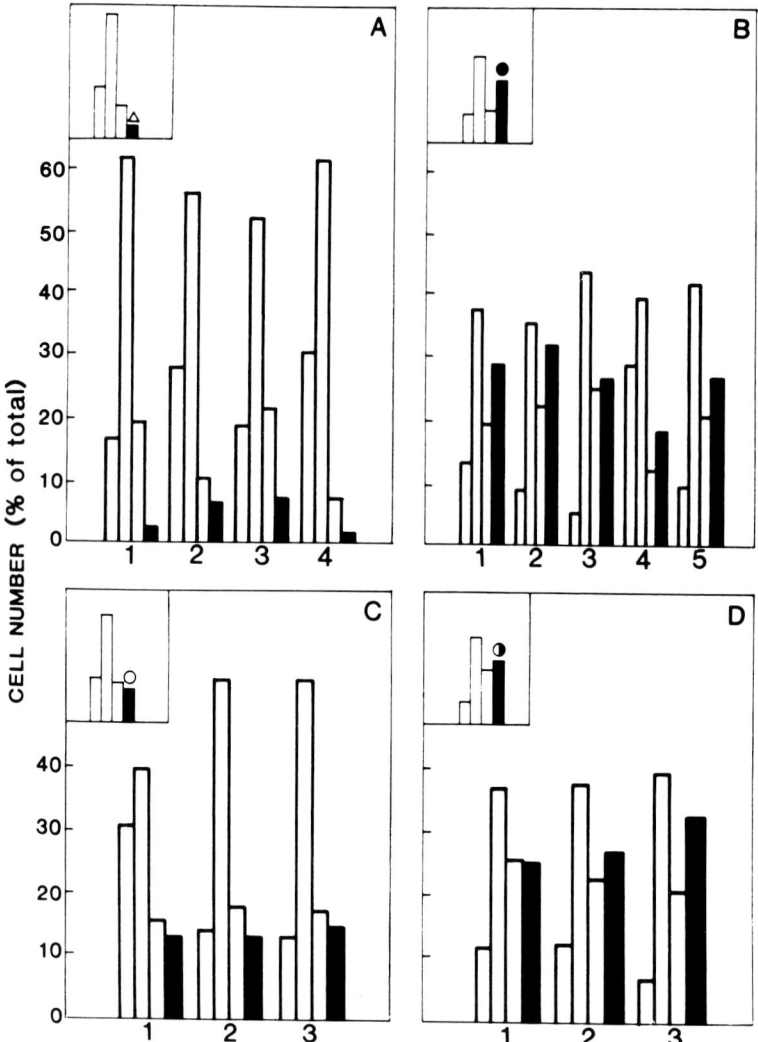

FIGURE 20.   Distribution of EC isolated from human umbilical vein (A), grossly normal (B), or atherosclerotic (C,D) human aorta according to protein content. A histogram for each subpopulation was obtained projecting biparameter distribution on the green fluorescence axis. Insert, mean cell number in each relevant region of histogram. ($\triangle$,●), P <0.01; (●,○), P <0.05; (●,◑), P = NS; (◑,$\triangle$), P <0.05.

Earlier it was shown that media thickening in experimental hypertension is due to hypertrophy and polyploidization of smooth muscle cells without their proliferation.[31] As is shown above, hypertrophic polyploid and diploid EC are present in adult human aorta, but in addition, the number of these cells grows in atherosclerosis. It is likely that common mechanisms of polyploidization and hypertrophy exist in vascular wall cells. The mechanisms of polyploid and hypertrophic cell formation in human arteries and their involvement in atherogenesis require further studies.

V. HUMAN ENDOTHELIUM AND LOW DENSITY LIPOPROTEINS (LDL)

A. LDL Receptors and Transport of Native and Modified LDL in Perfused Human Arteries, Identification of Modified LDL in Subendothelium of Human Arteries

The excessive accumulation of cholesterol and its esters in the vascular wall is one of the

Table 1
ACCUMULATION OF NATIVE, REDUCTIVELY METHYLATED, AND MDA-TREATED LDL IN PERFUSED HUMAN ARTERIES

	Uninvolved region	Lipid streak	Plaque
	^{125}I-LDL accumulation (pg of protein/mm^2)		
Native LDL	219 ± 24 (14)	289 ± 32 (13)	269 ± 25 (11)
Methyl LDL	140 ± 20 (10)	189 ± 32 (10)	180 ± 30 (9)
MDA-LDL	101 ± 12 (15)	186 ± 31 (14)	232 ± 22 (13)

Note: Segments of human carotid or iliac arteries were perfused in a recirculation system with 3 to 5 mℓ of medium 199 containing 10% lipoprotein-deficient serum and 5 μg/mℓ of radioactive native, reductively methylated, or MDA-modified ^{125}I-LDL for 1 hr at 37°C. After washing with cold medium 199, the vessels were cut into standard pieces. ^{125}I-LDL incorporation was assessed by the measurements of the radioactivity of the samples. At least 15 pieces were examined from both an uninvolved region and an atherosclerotic lesion. The number of studied arteries is given in parentheses. Each value represents mean ± SD.

most typical features of the atherosclerotic lesions. According to present views, the likely explanation for the accumulation of lipids in the vascular wall is the excessive accumulation of LDL. The function of EC in the regulation of LDL flow into the vascular wall is still poorly understood. A number of experimental facts suggest that endothelium can perform a barrier function, limiting LDL transport into the vessel wall.[32,33] According to other authors[34] it is thought that EC can actively transport LDL into the arterial wall. It is not yet clear what the role of receptors is for native and modified LDL in this process.[36,37]

Recently new experimental results have shown that uptake of LDL in the atherosclerotic lesions of arteries is elevated.[37-40] The perfusion of human arteries with LDL also demonstrates an increase in the incorporation of LDL in the regions of fatty streaks and atherosclerotic plaques (Table 1). In the lesion area not only native but also LDL modified by methylation or malone dialdehyde (MDA) are incorporated at higher rates compared to control areas. It is known that modified LDL can specifically bind to receptors for native LDL.[41,42] It is likely that the increase in the incorporation of LDL in atherosclerotic lesions is due to nonspecific receptor-independent mechanisms.

It is worth noting that the incorporation of native LDL in lesions and control areas was always higher than incorporation of modified LDL. This suggests the existence of a specific mechanism by which incorporation of LDL into the vascular wall takes place. This idea is supported by the results shown in Table 2. When arteries were perfused with ^{125}I-LDL-containing medium in the presence of excess unlabeled native or methylated LDL, the incorporation of ^{125}I-LDL was lower in lesion and control areas by 60% only when native LDL was present in excess. When methylated unlabeled LDL was added, the incorporation of ^{125}I-LDL was not affected. Thus, the incorporation of LDL into perfused human artery may be partially due to the interaction of LDL with a limited number of specific binding sites, possibly LDL receptors.

The involvement of LDL receptors in LDL incorporation into arterial wall is confirmed by the data on lipoprotein incorporation into normal and denuded arteries (Table 3). It is evident that the removal of endothelium results in the drop of the incorporation of native LDL into normal and lesioned areas of the arteries, but the incorporation of modified LDL is not affected. At the same time, vascular EC, as well as macrophages, have so-called "scavenger-receptors" that recognize chemically modified negatively charged LDL pro-

Table 2
**EFFECT OF THE EXCESS OF UNLABELED NATIVE
OR REDUCTIVELY METHYLATED LDL ON ^{125}I-LDL
ACCUMULATION IN PERFUSED HUMAN ARTERIES**

	Uninvolved region	Lipid streak	Plaque
	^{125}I-LDL accumulation (pg of protein/mm^2)		
^{125}I-LDL	240 ± 62 (8)	543 ± 229 (6)	469 ± 105 (6)
Native LDL	112 ± 23 (7)	176 ± 30 (7)	172 ± 46 (4)
Methyl LDL	269 ± 62 (10)	371 ± 124 (9)	298 ± 120 (6)

Note: Segments of human carotid or iliac arteries were perfused in a recirculation system with 3 to 5 mℓ of medium 199 containing 10% lipoprotein-deficient serum and 5 μg/mℓ of radioactive ^{125}I-LDL and 200
μg/mℓ of unlabeled native or reductively methylated LDL for 1 hr at
37°C. After washing with cold medium 199, the vessels were cut into
standard pieces. ^{125}I-LDL incorporation was assessed by the measurements of the radioactivity of the samples. At least 15 pieces were
examined from both an uninvolved region and an atherosclerotic lesion.
The number of arteries under experiment is given in parentheses. Each
value represents mean ± SD.

Table 3
**EFFECT OF DEENDOTHELIALIZATION ON ^{125}I-LDL
ACCUMULATION IN PERFUSED HUMAN ARTERIES**

	Normal endothelium		Denuded vessels	
	Uninvolved region	Lipid streak	Uninvolved region	Lipid streak
LDL	201 ± 31 (12)	280 ± 41 (6)	143 ± 14 (26)	182 ± 18 (4)
Methyl LDL	151 ± 11 (7)	214 ± 31 (5)	130 ± 13 (10)	170 ± 18 (8)
MDA-LDL	101 ± 12 (15)	186 ± 31 (14)	195 ± 31 (18)	186 ± 86 (11)

Note: Segments of human carotid or iliac arteries were perfused in a recirculation system with 3 to
5 mℓ of medium 199 containing 10% lipoprotein-deficient serum and 5 μg/mℓ native, reductively
methylated, or MDA-modified ^{125}I-LDL for 1 hr at 37°C. The vessels were denuded with a
balloon catheter. Intact and denuded arteries were perfused in parallel in the same experiment.
After washing with cold medium 199, the vessels were cut into standard pieces. ^{125}I-LDL
incorporation was assessed by the measurements of the radioactivity of the samples. At least
15 pieces were examined from both uninvolved region and atherosclerotic lesion. The number
of studied arteries is given in parentheses. Each value represents mean ± SD.

teins.[47] In contrast to classical LDL receptors, the activity of "scavenger-receptors" does
not depend on the mechanism of "up" and "down" regulation;[48] the incubation of macrophages with modified LDL results in the accumulation of intracellular lipids and the
formation of foam cells.[49] Besides, in culture, endothelial cells are capable of transforming
LDL into a form which can be taken up by macrophages and degraded via high-affinity
saturated pathways through receptors to modified LDL.[44] These facts suggest that "scavenger" receptors may play an important role in atherogenesis.

In studying the uptake and the degradation of native LDL and malone dialdehyde modified
LDL (MDA-LDL) in primary cultures of polymorphic endothelium using an antibody against
native LDL (8G4) and the PAP technique, the diffuse staining of human intima of noninvolved areas of human artery was found (Plate 2).*

* Plate 2 appears after page 180.

In order to identify modified LDL in the human arterial wall, monoclonal antibody (12E7) was raised, which interacts specifically with MDA-LDL and does not recognize native LDL, proteins of delipidated human serum, or MDA-modified albumin.

The staining with antibody 12E7 (using the PAP technique) of cryostat sections prepared from normal and lesioned areas of human arteries demonstrated the absence of specific binding sites for MDA-LDL in the intima of noninvolved human arteries (Plate 2B). When the lesioned areas (fibrous plaque) were stained with these antibodies, the staining of intimal cells was observed (Plate 2C and D).

These results demonstrate the existence of specific binding for antibodies to MDA-LDL in the intimal cells only in the lesioned areas of human arteries. The ability of endotheliocytes to induce modification of LDL in vitro is known.[44] If similar modification takes place after LDL interaction with EC, native LDL are transformed into modified LDL and can be "taken up" by subendothelial cells, for example, by macrophages.[45]

The question which still remains open is what the other cell types in the intimal cell population are which are capable, in addition to EC, of modifying LDL or taking up modified LDL in atherogenesis. Monoclonal antibodies against modified LDL are convenient tools to study the fate of MDA-LDL and may be useful in the future to answer this question.

B. Metabolism of Lipoproteins in Human Endothelial Cells: the Dependence on Age and Relation to Atherosclerosis

All EC examined in culture were shown to have specific receptors for LDL.[46] It was found that the uptake and the degradation of native, as well as MDA-LDL labeled with iodine-125, are indeed high-affinity and saturable processes with the maximum saturation at 15 to 30 μg of protein per milliliter. (Figure 22). The degradation and association of native LDL and MDA-LDL were inhibited only by homologous LDL (Figure 21). The activity of receptors to native LDL grows after preincubation of cells in medium containing delipidated human serum and diminishes after preincubation with both native and MDA-LDL. The association of ^{125}I MDA-LDL does not depend on the mechanism of "up and down" regulation, whereas the degradation of ^{125}I-MDA-LDL was "down" regulated after preincubation with unlabeled MDA-LDL and stayed unchanged after preincubation with native LDL or delipidated serum (Figure 23). These data demonstrate that both types of receptors are present on the same cell. This conclusion is also confirmed by results of experiments on internalization of native LDL and MDA-LDL labeled with the fluorescent probe 1,1-dioctadecyl-3,3,3,3-tetramethylindocarbocyanine-perchloride (Figure 23).

A comparison of association and degradation of ^{125}I-labeled native LDL and ^{125}I-labeled MDA-LDL was made between EC isolated from "atherosclerotic" and visually unchanged aorta from donors of comparable age. The association of native LDL and MDA-LDL varied, and no statistically significant differences were noticed between endothelium from "normal" and "atherosclerotic" aorta. Degradation of native LDL and MDA-LDL was essentially lower in endothelium from "atherosclerotic" aorta (by 1.7- and 4.6-fold, respectively. Figure 24). The ratio of association of ^{125}I-MDA-LDL to the association of ^{125}I-native LDL for EC from "normal" and "atherosclerotic" aorta was equal to 0.7 ± 0.2 and 0.8 ± 0.3, respectively. The ratio of MDA-LDL degradation to the degradation of native LDL in endothelium derived from "normal" and "atherosclerotic" vessels was 0.5 ± 0.2 and 0.2 ± 0.1 (p <0.05).

The analysis of age-related changes in LDL metabolism (Figure 25) demonstrates that the rate of the metabolism of native LDL goes down with age, and the metabolism of MDA-LDL does not correlate with age. This conclusion is also supported by the inverse correlation between the age of the donor and the rate of association and degradation of native LDL (r = − 0.63, p <0.05 and r = −0.72, p <0.025, respectively). For association and degradation of MDA-LDL, this correlation was not observed (r = −0.24 and R = −0.26).

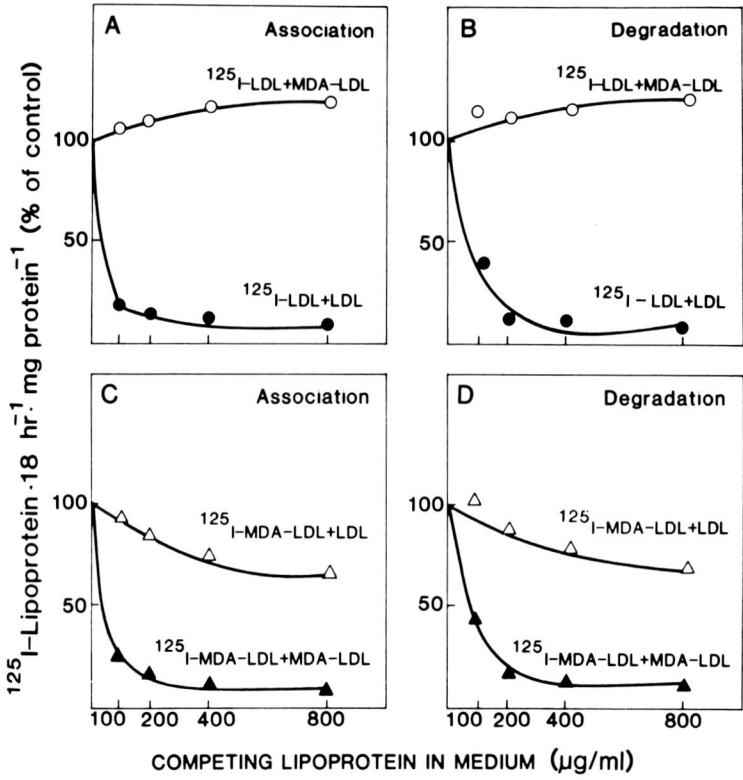

FIGURE 21.   Association (A, C) and degradation (B, D) of native ^{125}I-LDL (○,●) and ^{125}I-MDA-LDL (△,▲) in primary culture of heterogeneous endothelium from adult human aorta in excess of unlabeled native LDL and MDA-LDL.

In cellular LDL metabolism, the key point is the balance between the rates of uptake and degradation. This balance of lipoprotein metabolism is best reflected by the "ability" of cells to degrade LDL. This parameter may be calculated by measuring the ratio of LDL degraded by a cell to the total LDL present at a given moment within a cell. This analysis of the lipoprotein metabolism in EC isolated from "normal" and "atherosclerotic" human aorta demonstrates (Figure 26) that the ability of the EC to degrade native LDL stays unchanged. Thus, the metabolism of native LDL by EC in atherosclertic lesion remains in balance. On the other hand, the ability of these cells to degrade MDA-LDL is significantly lowered compared to endothelium from "normal" aorta.

Further studies of native and modified LDL were carried out in primary cultures of human aortic endothelium by indirect immunofluorescence in an attempt to analyze binding sites for native and MDA-LDL.[50] in addition, lipoproteins labeled with fluorescent stain 3,3 dioctadecylindocarbocyanine (DIL) were used.[51,52] To control the specificity of immunofluorescent technique, incubation of cells with lipoproteins or antibodies against LDL was omitted; the specificity of the uptake of DIL-labeled LDL and MDA-LDL was determined by simultaneous incubation of cells in the presence of a 20-fold excess of unlabeled lipoproteins. Cells were examined with a fluorescent microscope. The binding sites for native LDL were seen as separate granules which were discretely distributed over the surface of cells (Figure 27A). The accumulation of labeled LDL in coated pits was also found in other cell types, i.e., vascular smooth muscle cells and human skin fibroblasts.[53] In primary cultures of adult human endothelium some cells did not show LDL binding sites (Figure 27B). The number of such cells (both mononucleated and multinucleated) in each culture varied from

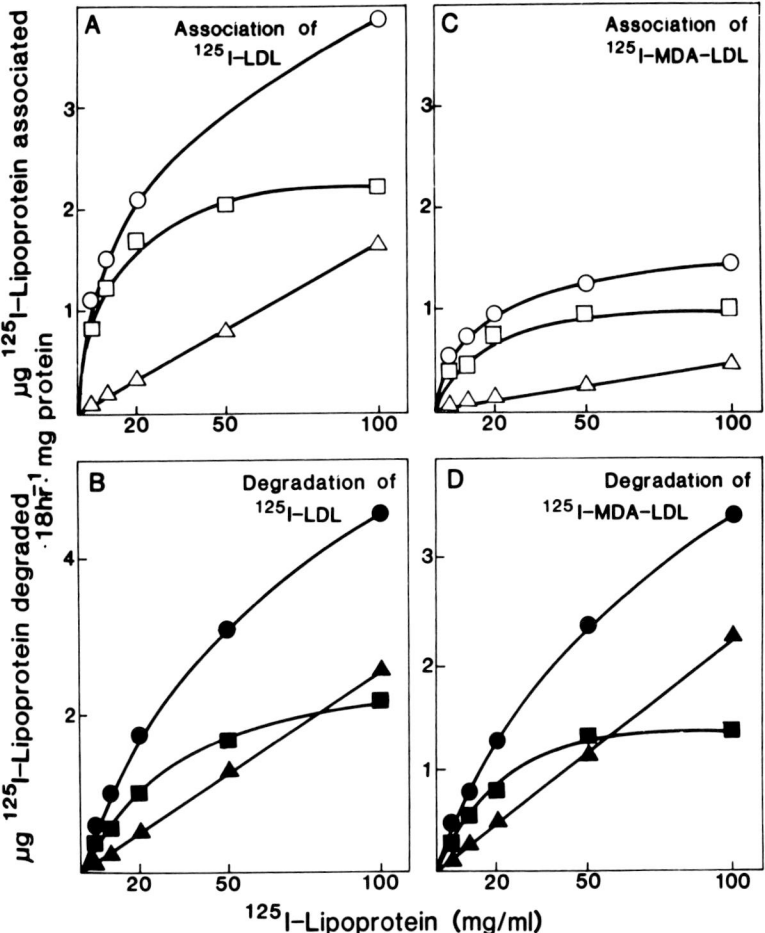

FIGURE 22. Association (A, C) and degradation (B, D) of native ^{125}I-LDL (A, B) and ^{125}I-MDA-LDL (C, D) in primary culture of heterogeneous endothelium from adult human aorta in the absence of unlabeled lipoprotein (O), in a 20-fold excess of unlabeled lipoprotein (△), and specific interaction (□).

12 to 26%. The heterogeneity in the functional activity of the receptors to LDL was also evident in experiments on incubation of cultured EC with DIL-LDL. Different aortic EC took up DIL-LDL with varying intensity producing specific fluorescence around nuclei (Figure 28A).

In contrast to native LDL, the binding sites for MDA-LDL in EC could not be visualized, though the incubation of DIL-labeled MDA-LDL indeed resulted in the staining of almost all cells of the culture (Figure 28B). It is thought that the receptors to MDA-LDL in EC do not aggregate on the cell surface, but rather are distributed evenly on the surface, as is observed in macrophages.[54,55]

Thus, in endothelium isolated from "atherosclerotic aorta" the rate of LDL degradation is lower than in control nonlesioned vessels. This may be explained by lower activity of lysosomal enzymes, and, in particular, of cholesterolesterase, which occurs in atherosclerosis.[56] High activity of "scavenger receptors" and unregulated uptake of LDL may result in the accumulation of intracellular cholesterol esters and the formation of foam cells. It cannot be ruled out that exocytosis of LDL-containing endocytotic vesicles (retroendocytosis) not fused to lysosomes may take place,[57] which can finally result in the accumulation of LDL in subendothelial extracellular space.

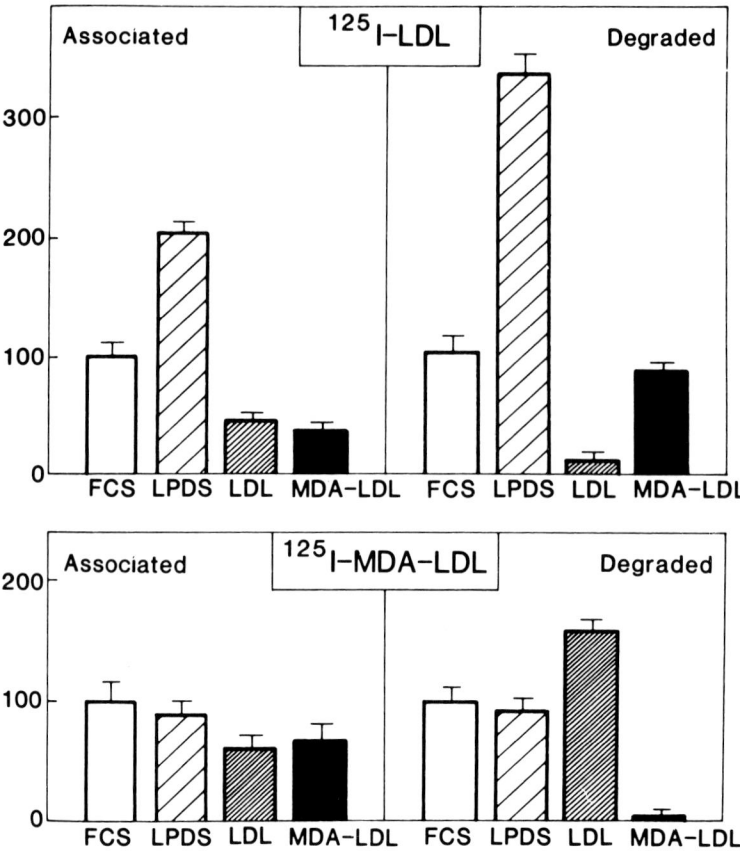

FIGURE 23.   Effect of preincubation conditions (FCS, 10% fetal calf serum; LPDS, 10% lipoprotein deficiency serum) on association and degradation of native ^{125}I-LDL and ^{125}I-MDA-LDL. Primary culture of heterogeneous endothelium from adult human aorta.

VI. RECEPTORS AND IONIC CHANNELS IN THE ENDOTHELIAL CELL MEMBRANE

A. Histamine and Ionic Permeability of Endothelial Cells

Using the patch-clamp technique,[58] we studied electrophysiological characteristics of human umbilical vein endothelium and human pulmonary artery endothelium cultured for 1 to 10 days. It was found that the resting potential for these cells averaged −30 mV. Volt-ampere characteristics (VAC) of the membrane were close to linear in the broad range of membrane potentials (Figure 29). This type of VAC behavior demonstrates that in contrast to muscle and nerve cells, the membrane of EC is likely to have only a low level of expression of potential-activated ionic channels.

The input resistance of cells was dependent on cell size and on the availability of intercellular contacts. Isolated cells had R = 3 to 8 GOm and R of cells of similar size, but at confluency, varied from 50 to 600 MOm. Thus, intercellular contacts are likely to play an essential role in the regulation of permeability of EC.

EC are known to have receptors to such hormones and mediators as histamine, acetylcholine, insulin, and catecholamines.[59] It is also known that in many cell types these substances increase the permeability of membranes for ions.[60-62] We studied the effect of histamine on the permeability of membranes from human umbilical vein endothelium. In whole cells in approximately 10% of the cells studied, histamine (10 μ*M*) reversibly elevated membrane

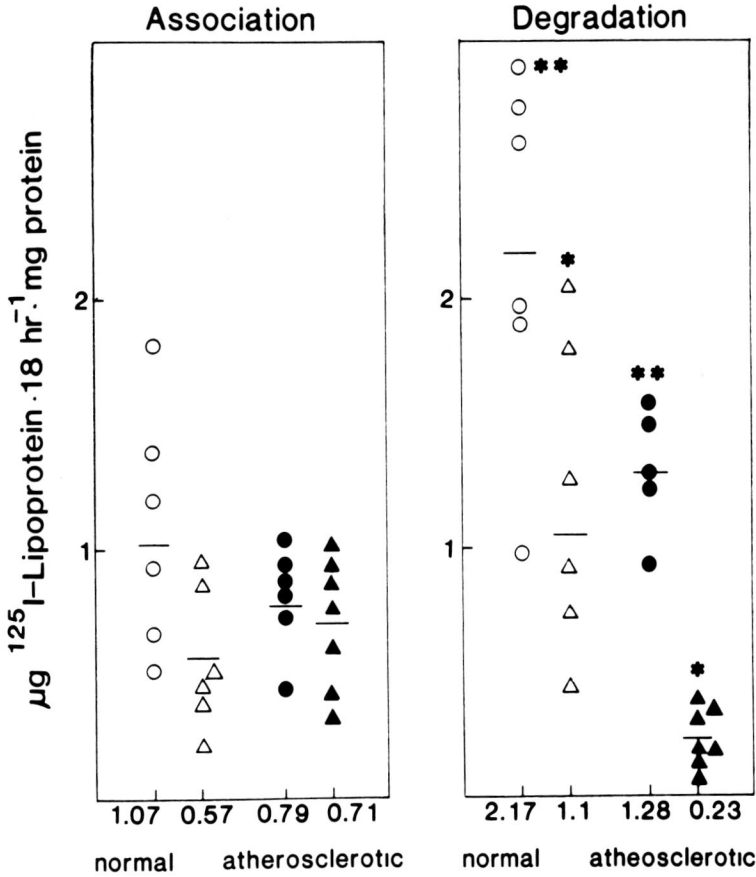

FIGURE 24. Association and degradation of native ^{125}I-LDL and ^{125}I-MDA-LDL in cultures of EC derived from grossly normal (○,△) and atherosclerotic (●,▲) adult human aorta.

conductance by three- to tenfold. (Figure 29). For histamine-activated currents VAC were close to linear, and the reversion potential (E) was close to O mV (Figure 29). The effect of histamine developed relatively slowly (3 to 5 min). The washout was also slow; complete restoration of the conductance required 10 to 15 min. Registration in the cell-attached mode showed that the addition of 10 μM histamine to the washing solution results in the activation of two groups of ionic channels shown in Figure 30 A and B. The conductance of both types was equal to about 20 pCm; however, the reversion potential in one population of channels was close to E_K, and in the other to E_{Cl}.

Histamine was effective only in the presence of Ca^{2+} in the outer solution (Figure 31). In calcium-free solution and in the absence of histamine the membrane was silent (upper track). At 6 min after addition of 10 μM histamine to the calcium-free solution, the function of channels was still not observed (the medium track). In several seconds after the addition of calcium into this solution (final concentration 1 mM), the ionic channels with conductance of about 20 pSm were activated (Figure 31, lower track).

Thus, it is thought that either Ca^{2+} is required to activate histamine receptors or that histamine causes the entrance of calcium ions into the cell which, in turn, activates Ca^{2+}-dependent ionic channels. The latter suggestion is more likely and is in agreement with earlier findings made on EC from guinea pig aorta, where calcium ions are required for manifestation of histamine effect[58] and, in the presence of this compound, the entrance of

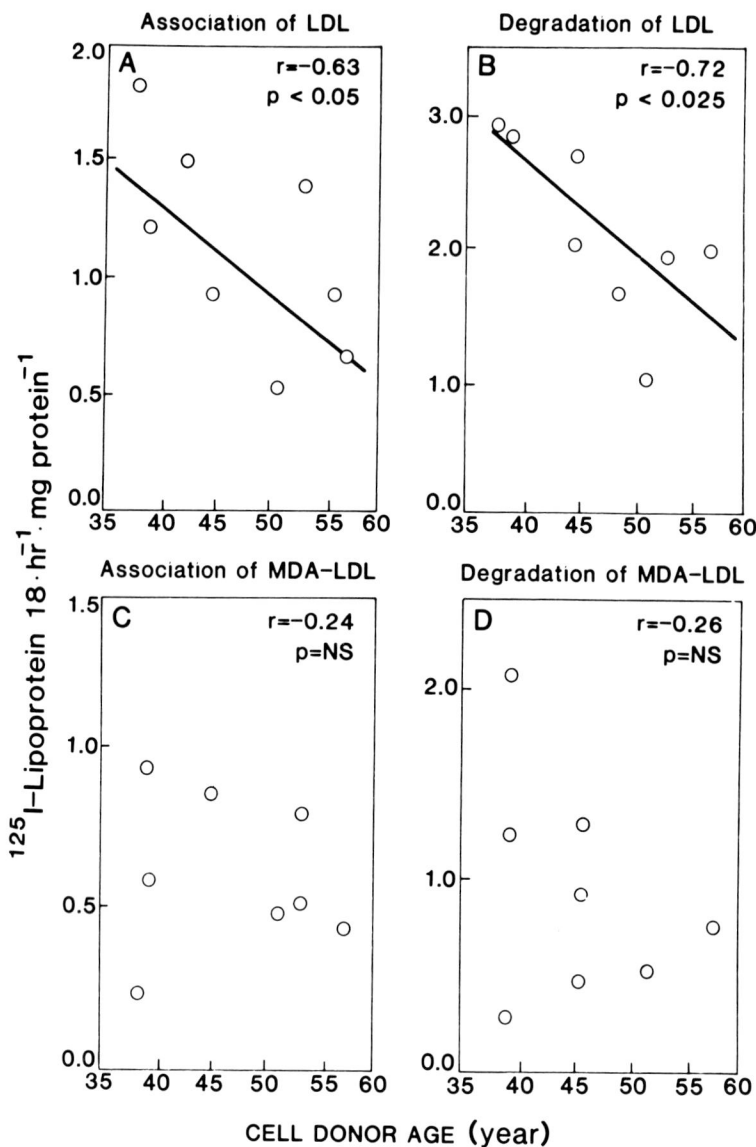

FIGURE 25. Associaton (A, C) and degradation (B, D) of ^{125}I-LDL (A, B) and ^{125}I-MDA-LDL (C, D) as a function of donor age.

calcium inside the cell was significantly increased.[63] This suggestion was confirmed by our studies of calcium ionophore A23187. The addition of 1 μM A23187 resulted in the activation of single channels with amplitude and kinetic characteristics close to those described earlier in response to histamine (Figure 32).

The experiments on isolated membrane fragments in an inside-out mode directly confirmed the existence of Ca^{2+}-activated channels in the membranes of vascular human EC. At concentrations of free calcium on the internal side of the membrane more than 1 μM, we observed single channel currents with conductance in symmetric potassium solution equal to 20 to 25 pSm. The probability of their activation increased with the elevation of "intracellular" calcium. Two types of Ca^{2+}-activated channels were registered. For the first type, the decrease in K^+ concentration resulted in a drop of elementary conductance and a shift

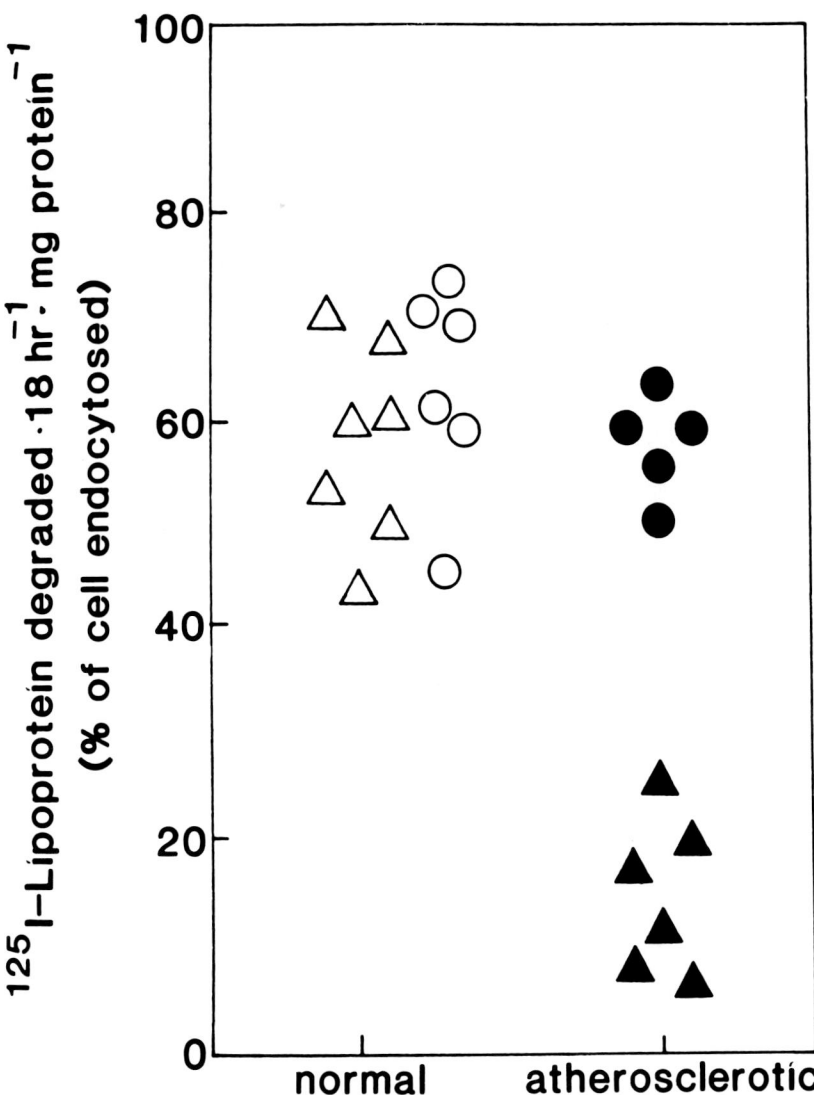

FIGURE 26.   A comparison of EC ability to utilize native ^{125}I-LDL and ^{125}I-MDA-LDL in primary culture of EC derived from grossly normal (○,△) and atherosclerotic (●,▲) human aorta.

of E. For the other channel, the substitution of Na^+ for K^+ did not result in changes in conductance or E.

B. Histamine-Induced Metabolism of Phosphoinositides

At present it is known that stimulating H1 receptors by histamine activates the synthesis and the secretion of prostacyclin in cultures of human umbilical vein endothelium.[64,65] In addition, EC secrete small amounts of thromboxane in response to the hormones which cause the relaxation of vascular smooth muscle cells.[66,67] However, the physiological significance of this phenomenon remains unclear.

It is known that H1 subtype of histamine receptor is not coupled to adenylate cyclase.[68] The role of calcium as second messenger in this process has been suggested[69] and the possible coupling of H1-receptor to phosphoinositide metabolism.

To check this idea, EC were incubated with myo-[2-^{3}H]inositol followed by incubation

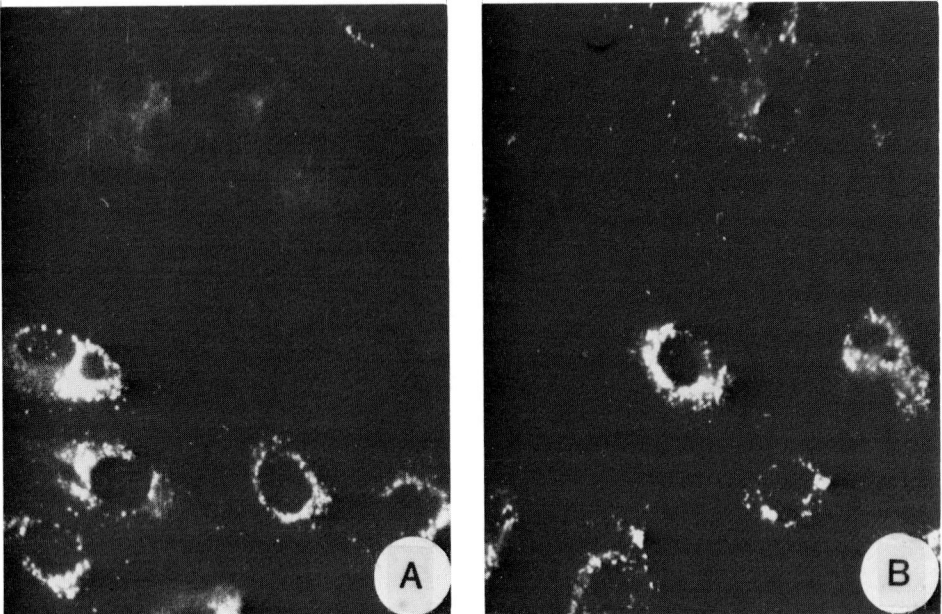

FIGURE 27.   Distribution of the binding sites for native LDL in primary human aortic EC culture. Note mononuclear (A) and polynuclear (B) cells without the binding sites for LDL. Indirect immunofluorescence. (Magnification in A × 400; in B × 320.)

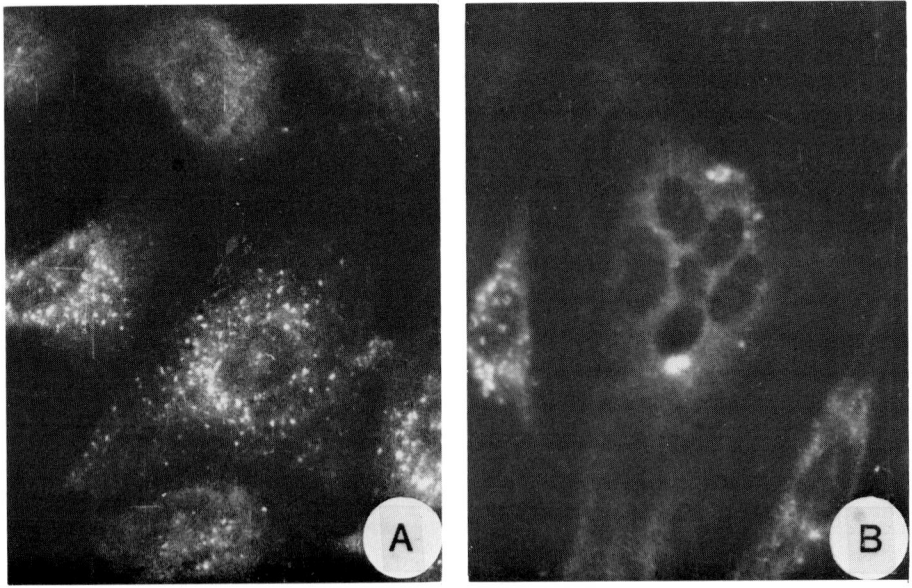

FIGURE 28.   Uptake of labeled DIL-LDL and DIL-MDA-LDL by human aortic EC in primary culture. Incubation with DIL-LDL (A) and DIL-MDA-LDL (B) for 5 hr at 37°C. (Rhodamine filter; magnification in A × 250; in B × 200.)

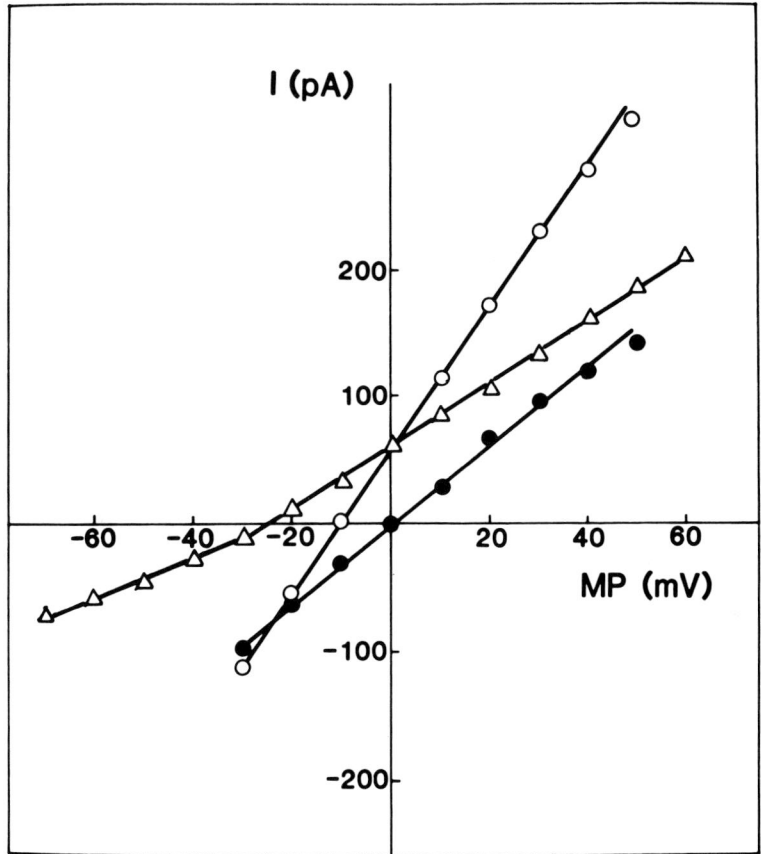

FIGURE 29.   Histamine-induced currents in the whole-cell record mode (○); I—V relationship in the absence of histamine (●); —V relationship 10 min. after addition of 10
mM histamine (△), differential I—V relationship, characterizing the real effect of histamine.
Cells were cultured on cover glass. For patch-clamp experiments, the cells were placed in
Petri dishes and maintained at room temperature. The bath solution contained (in mM):
NaCl, 140; KCl, 4.5; $CaCl_2$, 1; $MgCl_2$, 2; HEPES (pH 7.4), 10. The patch pipettes were
filled with "intracellular" solution containing (in mM): KCl, 140; $MgCl_2$, 2; HEPES (pH
7.2); EGTA, 1.1; $CaCl_2$, 0.55 (pCa 7). The resistance of the pipettes ranged between 3 to
10 MOm. Currents were recorded using an EPC-5 amplifier (List Electronic, FRG) on a
tape recorder and analyzed with a microcomputer. No analog compensation for series
resistance was applied (the error was usually <5 mV).

with 50 mM LiCl. The effect of histamine was evaluated on the synthesis of prostacyclin,
thromboxane, and on the formation of the metabolic products of phosphoinositodes: gly-
cerophosphoinositol (GroInsP), inositol-1-phosphate (InsP), inositol-1,4-biphosphate ($InsP_2$),
and inositol-1,4,5-trisphosphate ($InsP_3$) as described.[69]

The addition of histamine resulted in the increase of all four metabolic products of
phosphoinositides with an activation constant of 1.2 to 1.5 μM. Half-maximal effects of
histamine on the synthesis of thromboxane and prostacyclin were observed at a concentration
4.2 to 5.2 μM.

The effect of histamine (10 μM) on the formation of phosphoinositide metabolic products
and on the secretion of thromboxane and prostacyclin was completely blocked by pyrilamine
(10 μM), a known antagonist of H1 histaminergic receptors, whereas cimetidine, the antagonist of H2 receptors, at the same concentration had no effect.

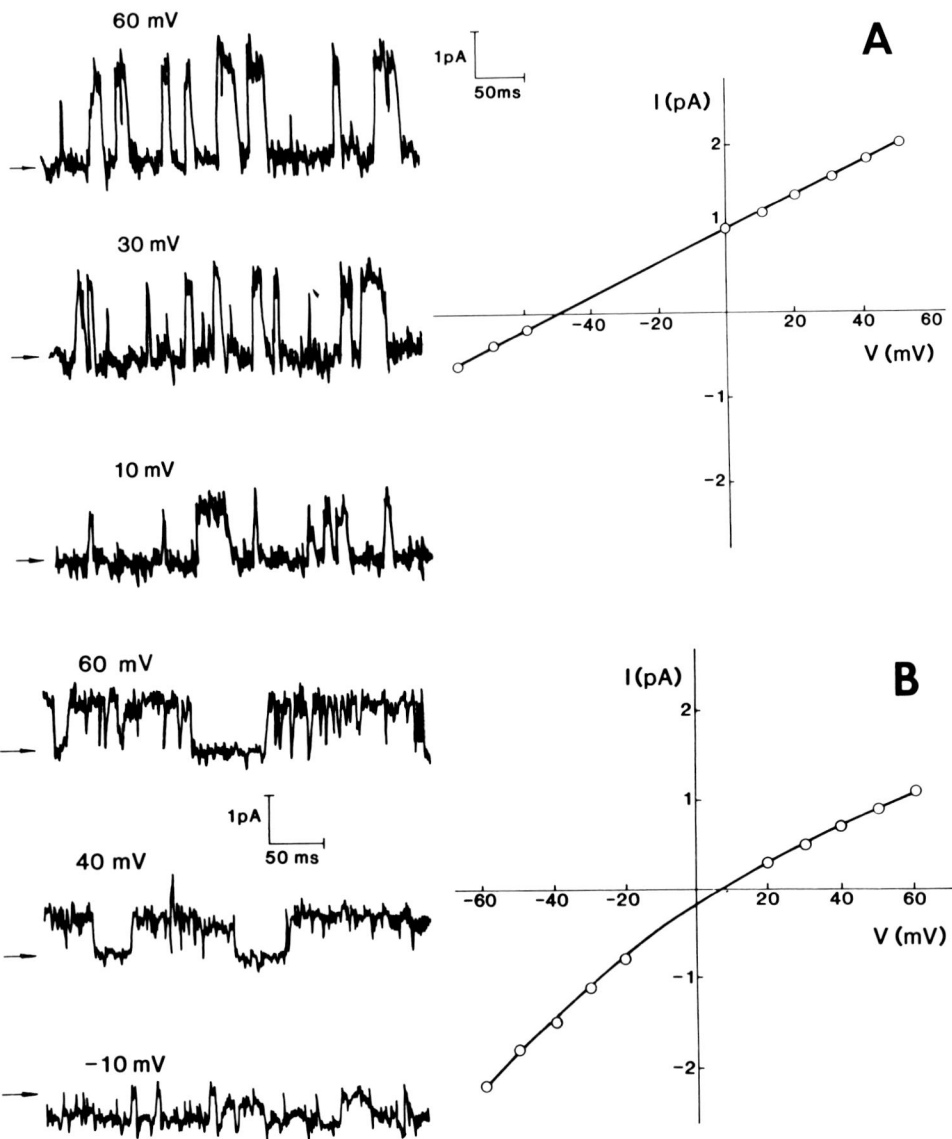

FIGURE 30.   Histamine-induced single-channel current in cell-attached configuration. Left, single-channel currents at different membrane potentials. Pipette potentials are indicated above the traces. Close state of channels is indicated with an arrow. Right, single-channel I—V relationship. The slope conductance is about 20 pS. (A) Currents possessing presumably K$^+$ selectivity; (B) currents possessing presumably Cl$^-$ selectivity.

Kinetic measurements of histamine effects on the catabolism of phosphoinositides and on the secretion of thromboxane and prostacyclin allowed study of the mechanism of its effect on EC. As seen in Figure 33, the InsP$_2$ and InsP$_3$ start to increase immediately after initiation of the reaction and reach maximum (threefold above the control) in 2 min after addition of histamine (10 μM). The elevation of GroInsP and InsP was noticeable in only 1 min after the initiation of the reaction and then grew linearly till saturation (four- to fivefold above the control) was reached in 20 min.

The time dependence of the secretion of thromboxane and prostacyclin is shown in Figure 34. The increase in the amount of these prostanoids in culture medium was registered in 1 to 2 min after the beginning of incubation of cells with histamine and reached maximum in

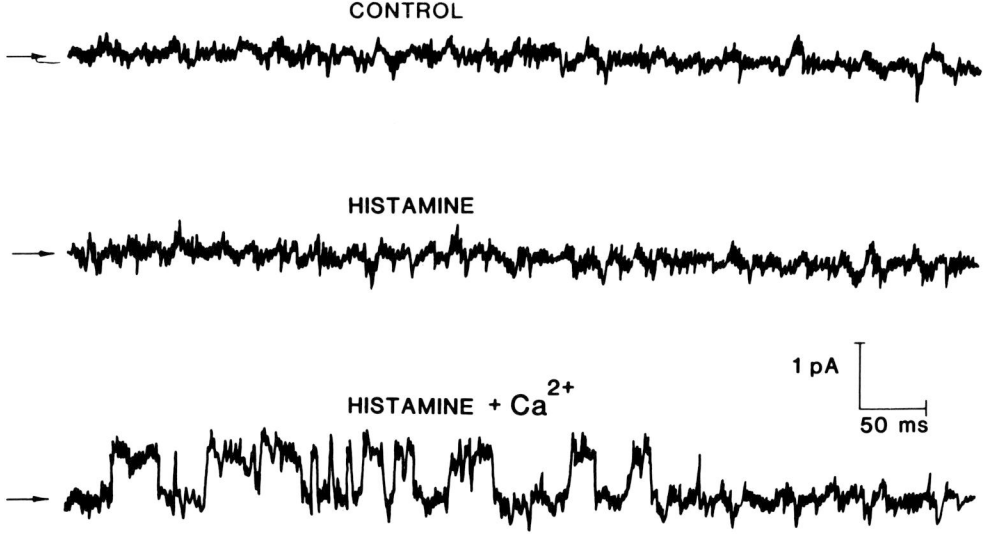

FIGURE 31. Ca^{2+}-dependence of histamine effect. The upper trace shows the channel activity in "Ca^{2+}-free" solution (0.1 μM Ca^{2+}). The middle trace shows the channel activity 6 min after addition of 10 μM histamine in "Ca^{2+}-free" solution. Note the absence of single-channel currents. The bottom trace shows the channel activity 30 sec after addition of 2 mM Ca. Cell-attached configuration. Data were filtered at 300 Hz.

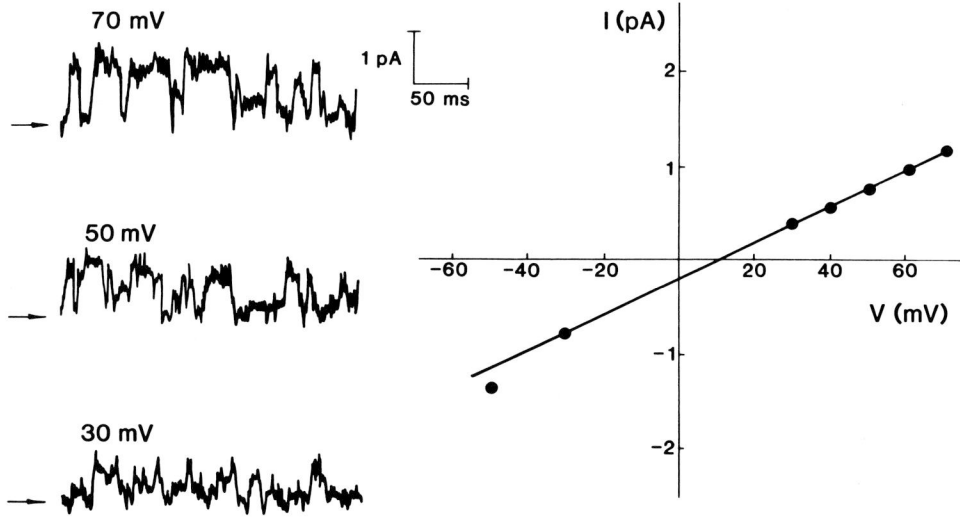

FIGURE 32. Single-channel I—V relationship and original records of single-channel currents, activated with calcium ionophore A23187. Cell-attached configuration. Left, single-channel currents at different membrane potential (70, 50, −30 mV). Arrow indicates the close state of channel. Right, I—V relationship of a single-channel, activated with Ca^{2+}.

20 min. The amount of thromboxane was increased tenfold, and that of prostacyclin by 20-fold compared to controls.

Experimental values obtained in kinetic studies for the synthesis of thromboxane/prostacyclin and GroInsP and InsP were normalized by maximal effect. Regression analysis revealed direct correlation dependences between the secretion of prostacyclin and the formation of GroInsP (r = 0.7492) and InsP (r = 0.8684), and also between the secretion of thromboxane and formation of GroInsP (4 = 0.8292) and InsP (r = 0.9493).

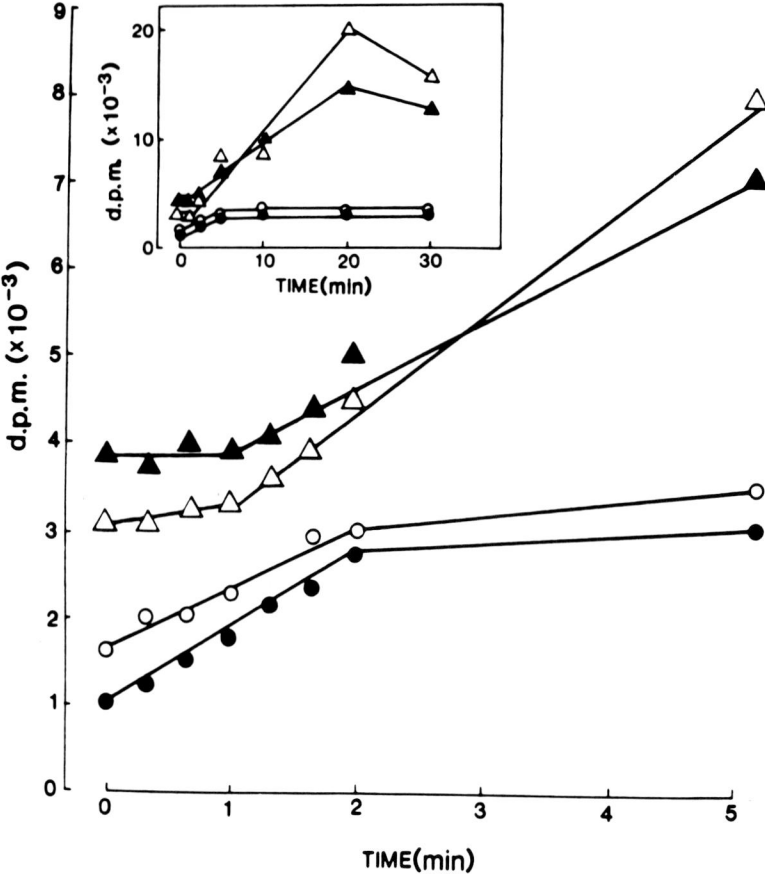

FIGURE 33. Kinetics of histamine effect on levels of inositol phosphates in human umbilical vein EC. Myo-[2-³H]inositol-prelabeled EC were exposed to 10^{-5} M histamine for the indicated time periods. After termination of incubations [³H]inositol phosphates in cell lysates were separated by anion exchange column chromatography. Radioactivity was measured in total eluate (7 mℓ) volumes. ($\triangle$), GroInsP eluted with 60 mM ammonium formiate; ($\blacktriangle$), InsP eluted with 200 mM ammonium formiate/0.1 M formic acid; ($\bigcirc$), InsP$_2$ eluted with 400 mM ammonium formiate/0.1 M formic acid; ($\bullet$), InsP$_3$ eluted with 1.0 M ammonium formiate/0.1 M formic acid. Values are means of duplicate incubations and are representative of a single typical experiment. Similar time courses were obtained from three other separate experiments.

These data were obtained from long-term culture of human umbilical vein endothelium. Qualitatively similar results were found in long-term cultures of human pulmonary artery endothelium.

Thus, it is suggested that the catabolism of phosphoinositides is part of the mechanism of the secretory response of EC to histamine. The involvement of phosphoinositide metabolism in secretion due to the interaction of agonist with receptor was earlier demonstrated for such secretory systems as cells of adrenal cortex,[70] neutrophils,[71] exocrine cells of pancreas,[72] parotid gland,[69] and platelets.[73]

Since the formation of InsP$_3$ is the earliest step in receptor-stimulated catabolism of phosphoinositides, the subsequent InsP$_3$-induced mobilization of Ca^{2+} from intracellular stores leads to the activation of the secondary calcium-dependent catabolism of phosphoinositides.[74-76] Thus, it is thought that histamine-induced immediate formation of InsP$_2$ and InsP$_3$ occurs at basal values of intracellular calcium concentration, whereas the formation

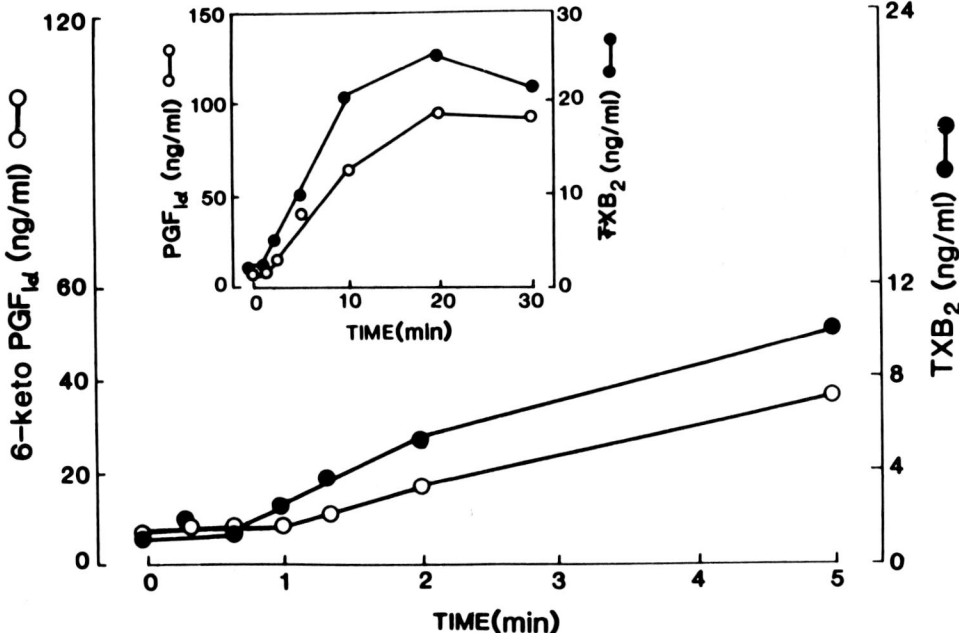

FIGURE 34. Tromboxane B$_2$ (TxB$_2$) and 6-keto prostaglandins F$_{1\alpha}$ (6-keto-PGF$_{1\alpha}$) release from umbilical vein EC in response to histamine kinetics. Results depicted in this figure were obtained from exactly the same experiment as that performed for Figure 33. Following incubation of EC with 10^{-5} histamine for the indicated time periods, the cell medium overlay was assayed serologically for TxB$_2$ (●) and 6-keto-PGF$_{1\alpha}$ (○) using ^{125}I-RIA kits. Values are means of duplicate incubations. Comparable time courses were obtained from the other separate experiments.

of InsP is initiated only after the elevation of intracellular calcium. An increased concentration of calcium in a cell is required to support the activity of A$_2$ and B phospholipases which catalyze the formation of GroInsP.[77]

Thus, our data and results of others[77-79] suggest the following sequence of the events in the interaction of histamine with H1 histaminergic receptor. The interaction of histamine with the receptor activates calcium flow into the cell (possibly, through the receptor-operated calcium channels). The increase in the concentration of calcium in the cytoplasm activates Ca^{2+}-dependent potassium or chloride channels, thus increasing integral conductance of the membrane (Figure 35). Histamine-receptor complex also activates phospholipase C, which is specific towards diphospho- and triphosphoinositides, and InsP$_2$ and InsP$_3$ are formed. InsP$_3$ induces additional elevation in the intracellular calcium concentration which, in turn, activates phospholipase A2 and phospholipase C, specific to phosphoinositides. Arachidonic acid, which was formed by the action of phospholipase A2, is used for the synthesis of prostanoids; InsP formed by activation of phospholipase C is used by cells to restore the pool of phosphoinositides.

Thus our data demonstrate that histamine increases the permeability and the secretory activity of EC membrane, which may lead to depolarization or hyperpolarization of a cell. The physiological role of this regulatory process is not yet clear. It is possible that the increased permeability of the membrane, the change in the resting potential of the cell, and the activation of phosphoinositide metabolism may eventually result in the production of the relaxation factor, which regulates the excitability of smooth muscle cells. This process may also lead to the changes in cell-cell interaction and the rearrangements of the morphology of EC.

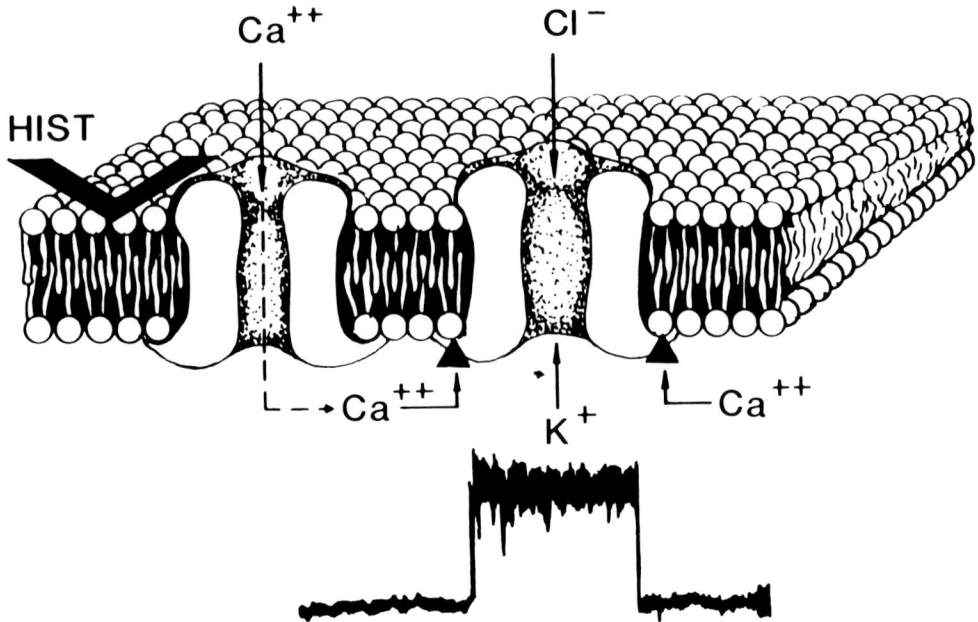

FIGURE 35.   Schematical representation of a possible mechanism of histamine effect on EC membrane.

VII. REGULATION OF ENDOTHELIAL CELL BEHAVIOR IN MONOLAYER: ROLE OF SECOND MESSENGERS

A. Morphology of Endothelium *In Situ*

It is known that the morphology of EC is affected by such factors as hemodynamic forces,[80,81] blood pressure,[82] vasoactive amines,[83,84] the products of lipid metabolism,[85] and other physiologically active substances.[86-88] Of special interest are the mechanisms which mediate the effects of these factors on endothelial morphology.

It is well known that the action of a majority of hormones, neuromediators, growth factors, and other endocrine and paracrine regulators is mediated by second messengers, which include cyclic nucleotides (cAMP and cGMP), Ca^{2+} ions, and the products of phosphoinositide metabolism, namely, triphosphoinositol and diacylglycerol (DAG).

Convenient tools to study second messengers are substances capable of selectively activating steps in each of these metabolic pathways. To study the effects of cAMP, we used forskolin, which is a known activator of adenylate cyclase, the enzyme which catalyzes cAMP formation. To study the role of the products of phosphoinositide metabolism, TPA was used which has a physiological analog, namely, diacylglycerol. Both forskolin and TPA were used in experiments on *in situ* perfusion of rabbit aorta to study their effects on the morphology of endothelial lining. We have described earlier[17,110,120] the manifestation of injury to the endothelial lining in the perfused rabbit aorta. In experiments on perfusion *in situ* in rabbit aorta, endothelial monolayer changes were determined by three criteria: (1) the EC shape index was determined as the ratio of cell length to cell width; (2) the EC injury index was determined as the sum of argyrophilic and esquamating cells per 1 mm²; (3) the intercellular junction injury index was determined as the sum of EC with wide intercellular junctions and EC with craters and stomata per 1 mm².

In experiments on perfusion *in situ* of rabbit aorta we found a dose-dependent change in the endothelial shape index in the presence of TPA (Table 4, Figure 36A and B). The shape index of endotheliocytes was 0.19 ± 0.006 in the control segments, and was elevated in

Table 4
DOSE-DEPENDENT EFFECTS OF FORBOL ESTER ON THE MORPHOLOGY OF RABBIT AORTIC EC

Concentration (μM)	EC injury per 1 mm²	Defect of intercellular junctions per 1 mm²	EC shape index
Control	2.3 ± 0.5	58 ± 10	0.09 ± 0.006
0.001	3.0 ± 0.9	44 ± 6	0.25 ± 0.007[a]
0.01	3.1 ± 0.7	48 ± 7	0.31 ± 0.008[a]
0.1	2.9 ± 0.6	50 ± 9	0.32 ± 0.008[a]

Note: Each M ± m value represents the results of at least three experiments. EC injury was determined using SEM as the sum of argyrophilic and esquamating EC per 1 mm². Defects of intercellular junctions were determined using SEM as the sum of EC with widening intercellular junctions and EC with craters and stomata per 1 mm². EC shape index was determined as the ratio of cell length to cell width. At least 1 to 2 cm² was analyzed in each experiment by SEM.

[a] P <0.05 compared to control according to Student's *t* test.

the presence of TPA to 0.25 ± 0.007, 0.31 ± 0.008 and 0.32 ± 0.008 at TPA concentrations 10^{-9}, 10^{-8} and 10^{-7} M, respectively. Neither injury index for endothelial cells, nor injury index for intercellular boundaries differed significantly from the control values.

Forskolin (10^{-5} M) did not cause the appearance of endothelial microdefects and did not change the shape index of EC. In the presence of forskolin (10^{-5} M), TPA (10^{-7} M) caused 40% denudation of luminal surface in perfused aortic segments. Other changes in the morphology of endothelial layer were not measured due to massive loss of endothelium. When TPA concentration was lowered from 10^{-8} to 10^{-9} M, denudation was less pronounced; however, the loss of separate cells was noticeable (Table 5, Figure 36 C and D). The injury index for endotheliocytes was 72 ± 8 and 70 ± 10 cells per square millimeter, respectively (2.3 ± 0.5 in the control). The injury index for intercellular contacts was equal to 156 ± 21 and 130 ± 18, respectively (58 ± 10 in the control). In addition, significant changes in endotheliocyte shape were observed. The cells normally elongated in the direction of blood flow became more round, with broadened intracellular junctions. The shape index elevated from 0.09 ± 0.006 in the control segments up to 0.52 ± 0.011 and 0.41 ± 0.01 at concentrations, 10^{-8} and 10^{-9} M of forskolin, respectively.

In 1982 to 1985 a number of authors investigated the effects of phorbol ester in culture and in vivo. Thus, Nawroth and co-authors[89] demonstrated that in bovine aortic endothelial cultures the addition of TPA caused a number of changes which led to cell death. When TPA was present in the incubation medium (10^{-7} to 10^{-8} M), cell shape was changed, which finally caused the disturbances in the integrity of the endothelial sheet. Intercellular contacts were destroyed, which was accompanied by increased synthesis of lactate dehydrogenase and prostacyclin. The elevation of prostacyclin synthesis is known to be a marker of injury for endothelium.[90,91] These changes finally led to the separation of endothelium from the substrate and cellular death. The authors concluded that the increase in the synthesis of prostacyclin and lactate dehydrogenase, and also the changes in cell shape in the presence of TPA, preceded cellular death. In 1985 Montesano and Orci studied the effect of TPA on cultured endothelium of bovine capillaries.[92] TPA (10^{-7} M) in 48 hr of incubation led to cell-shape changes in the endothelial monolayer grown on plastic. Endothelium growing on collagen substrate formed capillary-like structures. It is known that during angiogenesis proteases are produced by cells which cause the degradation of extracellular matrix.[93] In the presence of TPA the synthesis of collagenase and plasminogen activator by capillary EC

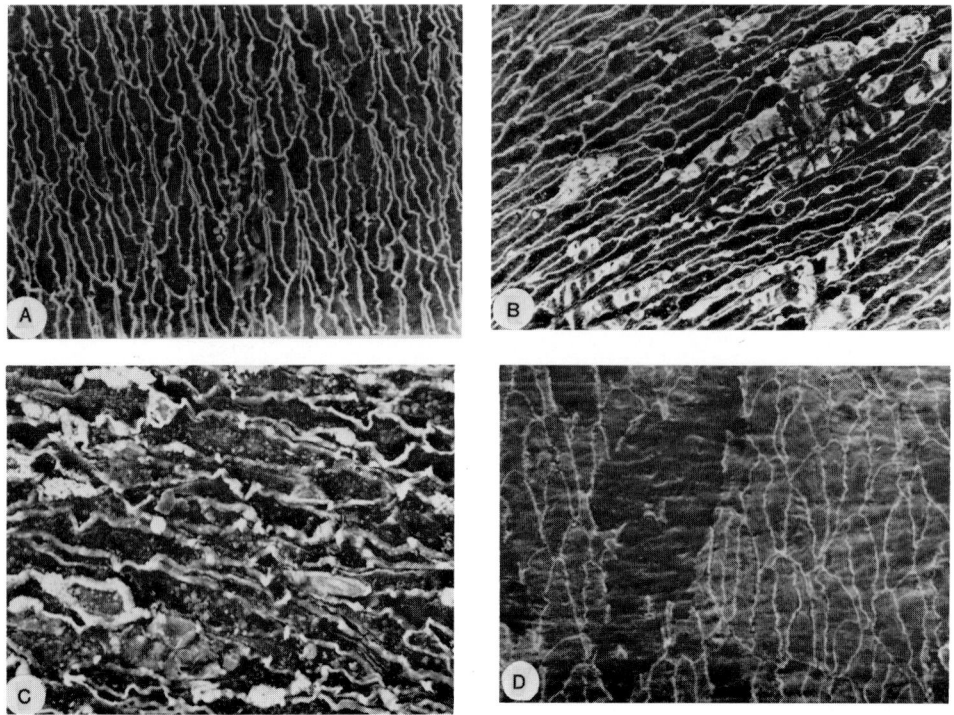

FIGURE 36.   (A) SEM of the control rabbit thoracic aorta segment perfused with medium 199 for 1 hr at 37°C under a pressure of 100 mmHg. Intercellular junctions are impregnated with silver nitrate. EC are oriented along vessel axis. There is no argyrophilic cells and denuded areas. (Magnification × 166.) (B) SEM of rabbit thoracic aorta perfused with 0.01 μ*M* TPA and 10 μ*M* forskolin. EC shape is changed. Argyrophilic and esquamating cells. Note single injuries of intercellular junctions. (Magnification × 166.) (C) SEM of rabbit thoracic aorta perfused with 0.01 μ*M* TPA and 10 μ*M* forskolin. Argyrophilic cells, cells of altered shape, and widening of intercellular junctions. (Magnification × 650.) (D) SEM of rabbit thoracic aorta perfused with 0.1 μ*M* TPA and 10 μ*M* forskolin. Extensive denuded areas. (Magnification × 330.)

Table 5
DOSE-DEPENDENT EFFECTS OF FORBOL ESTER IN THE PRESENCE OF 10 μ*M* FORSKOLIN ON THE MORPHOLOGY OF RABBIT AORTIC EC

Concentration (μ*M*)	EC injury per 1 mm²	Defects of intercellular junctions per 1 mm²	EC shape index
Control	2.3 ± 0.5	58 ± 10	0.09 ± 0.006
0.001	70 ± 10[a]	130 ± 18[a]	0.41 ± 0.01[a]
0.01	72 ± 18[a]	156 ± 21[a]	0.52 ± 0.011[a]
0.1[b]			

Note: Each M ± m value represents the results of, at least, three experiments.

[a]   P <0.05 compared to control according to Student's *t* test.
[b]   Data are now shown because 40% of the luminal surface was denuded.

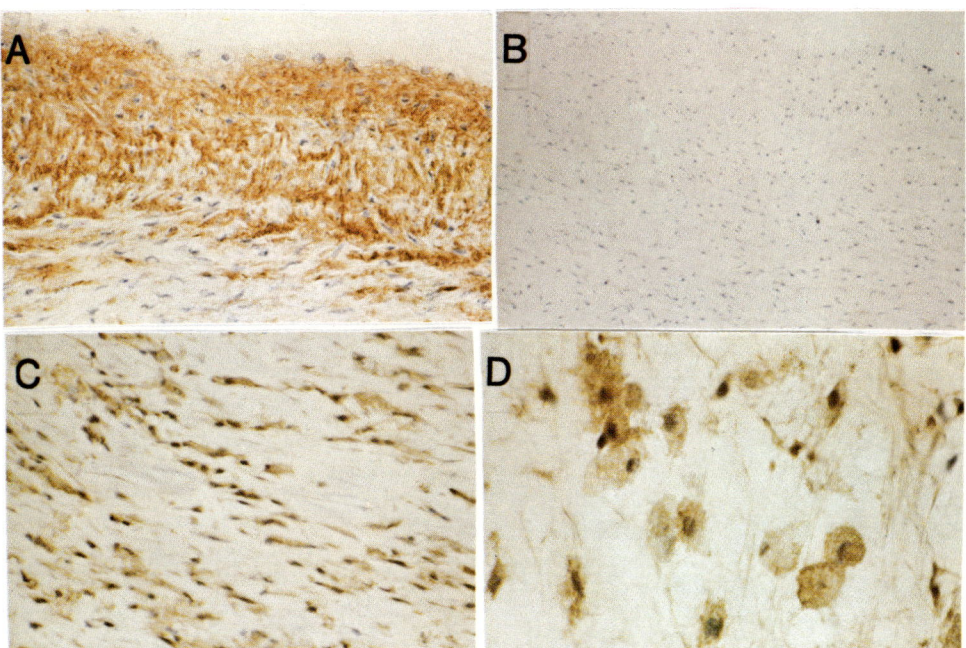

PLATE 2, Chapter 37.   Immunohistochemical stain of cryostate human carotid artery sections (5 μ thick). Carotid arteries were taken at autopsies of sudden death victims from 45 to 60 years of age within 3 to 4 hr after death. (A) Section of a grossly normal human carotid artery treated with monoclonal antibodies against native LDL (8G4), PAP-technique. (Intima, diffuse staining; magnification × 40.) (B) Section of a grossly normal human carotid artery treated with monoclonal antibodies against MDA-LDL (12E7), PAP-technique. (Negative staining; magnification × 26.) (C) Section of an atherosclerotic human carotid artery, intima layer, treated with monoclonal antibodies against MDA-LDL (12E7), PAP-technique. (Positive staining; magnification × 40.) (D) The same section, intima layer. (Positive staining of foam cells; magnification × 80.)

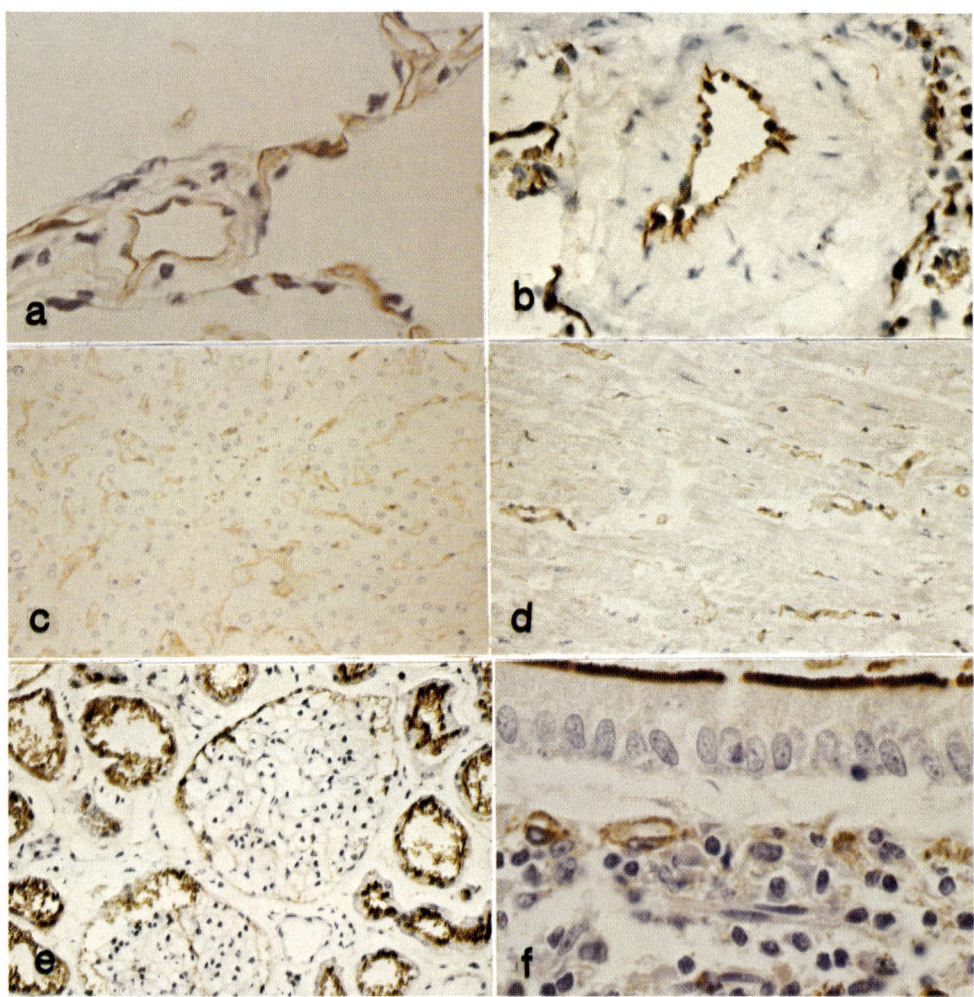

PLATE 3, Chapter 37.   ACE in different human organs. Immunohistochemical stain with monoclonal antibody
"9B9". (a, b) Human lung. EC of arterioles, capillaries (a, b), and blood vessels (b), are immunoreactive.
(c) Liver; (d) myocardium. EC of capillaries are immunoreactive. (e) Kidney. Proximal tubules are intensely
immunoreactive, EC in some capillaries are weakly stained. (f) Jejunum, the luminal surface of the epithelium
lining the villi and EC of capillaries are intensively stained. (Magnification: a, × 960; b, × 320; c, × 640; d,
× 160; e, × 160; f, × 960.)

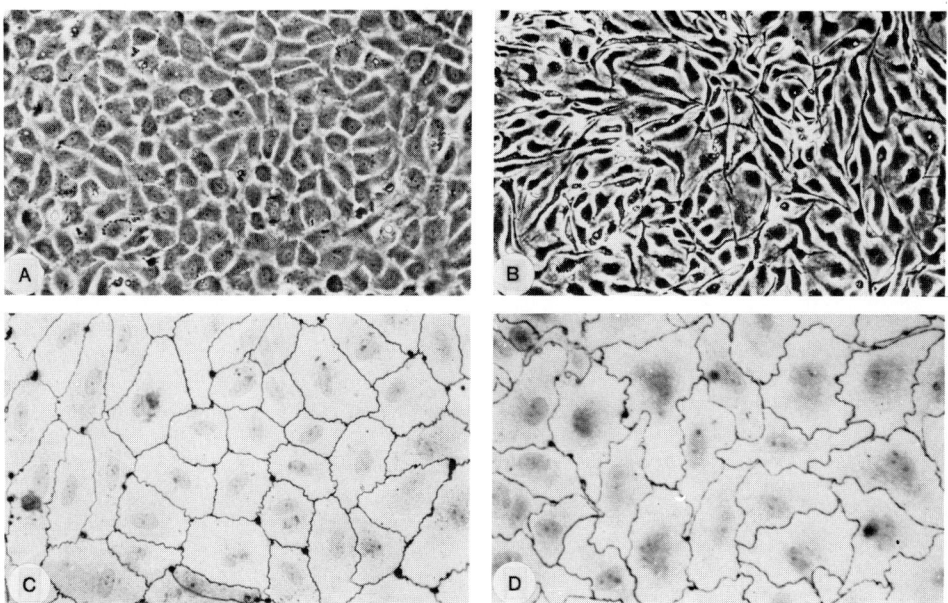

FIGURE 37. Effect of forskolin on morphological features of human umbilical vein EC in primary culture. (A, C) Control; cells were incubated in medium 199 with 0.5% BSA. (B, D) Cell culture after 1 hr incubation with forskolin (10^{-5} M); (A and C: phase contrast; magnification $\times$ 100; B and D: silver nitrate stain; magnification $\times$ 200.)

was significantly increased.[94] Angiogenesis in endothelial culture was accompanied by the invasion of cells into collagen substrate via degradation of collagen. The process of cell invasion into collagen substrate was accompanied by loss of the integrity of the monolayer and of intercellular contacts.[92]

Intravenous (i.v.) administration of TPA is often used to model acute respiratory distress syndrome. Shasby and co-authors[95] in 1982 found acute pulmonary edema caused by TPA. According to Mojarad and co-authors,[96] the basic pathological feature related to acute respiratory distress syndrome induced by TPA is endothelial damage. Morphologically endothelial injury is manifested as the vacuolization of cytoplasm, destruction of cytoplasmic membrane, and, finally, separation of cells from basal membrane.[97] Thus, in experiments on endothelial culture, *in situ* and in vivo TPA, by activation of kinase C, changes the shape of EC and destroys the integrity of intercellular contacts resulting in cell injury and death.[89,92]

In our experiments the most pronounced effects of TPA were seen in the presence of agents which increase the intracellular concentration of cAMP (forskolin, isobutylmethylxanthine, and dibutyril-cAMP). The involvement of cAMP- and DAG-dependent phosphorylation in the regulation of morphology and functional activity of endothelium was also found in the experiments using primary cultures of human vascular endothelium.

B. Morphology of Human Endothelium in Primary Culture

The addition of forskolin (10^{-5} M) to confluent primary culture of human umbilical vein endothelium caused rapid (40 to 80 min) and well expressed changes in the cell shape (Figure 37). These changes did not result in the loss of the integrity of the monolayer (Figure 37). Colchicine (10^{-6} M) completely abolished the effects of forskolin. Similar morphological changes were induced by dibutyril-cAMP (10^{-5} M) and by the inhibitor of cyclic nucleotide phosphodiesterase, isobutylmethylxanthine (10^{-4} M). Morphological changes caused by forskolin were maximal in 60 to 80 min, which was followed by restoration of the original shape (Figure 38).

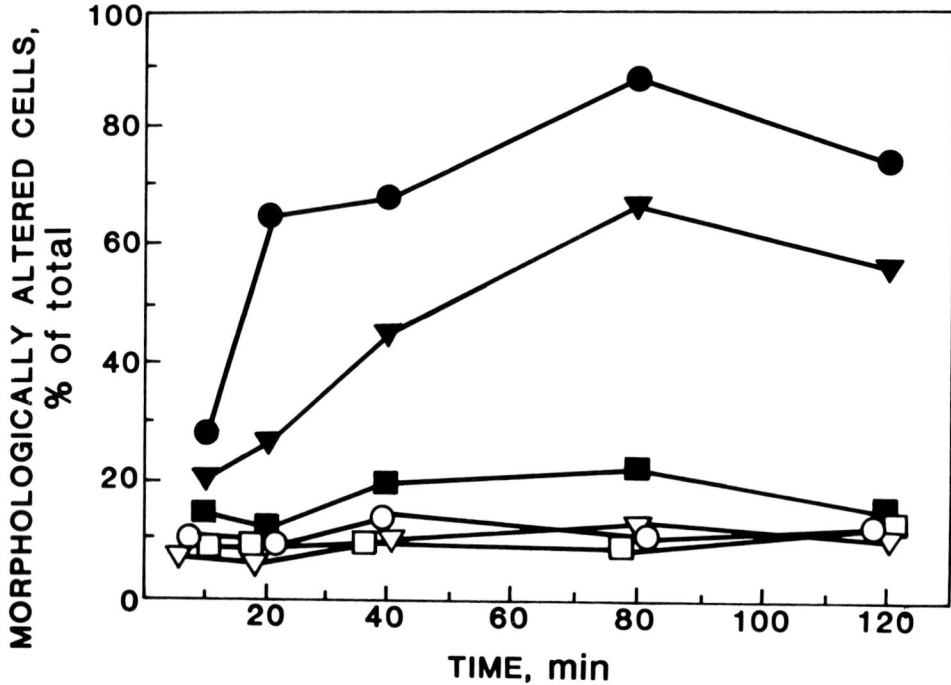

FIGURE 38.   Synergistic effects of TPA and forskolin on the morphology of human umbilical vein EC. Cultures were incubated in medium 199 with 0.5% BSA (□) or in medium 199 with 0.5% BSA supplemented with effectors: 10^{-5} *m* forskolin (▼); 10^{-7} *M* TPA (■); 10^{-6} *M* forskolin (▽); 10^{-8} TPA (○); 10^{-6} *M* forskolin with 10^{-8} M TPA (●). After the incubation, cultures were stained with silver nitrate and the shape index of EC was determined. EC of shape index $\leqslant 5$ were regarded as altered.

TPA (10^{-7} *M*) also caused changes in the morphology of human umbilical vein endothelium, but these changes were less pronounced than the effects of forskolin (Figure 38). Forskolin and TPA at concentrations of 10^{-6} and 10^{-8} *M*, respectively, did not affect cell morphology. At the same time, simultaneous addition to the cells of forskolin and TPA at these concentrations resulted in very rapid and well expressed changes in cell morphology (Figure 38). Morphological changes induced by the substances which elevate intracellular levels of cAMP and TPA were completely reversible in 30 to 60 min after removal of the agents from culture medium.

In human aortic EC, forskolin and TPA caused morphological changes which were similar to the changes in human umbilical endothelium (Figure 39). The most pronounced effect was registered with small- and medium-sized cells. The changes in large multinucleated cells were less evident.

In experiments with short-term tissue cultures of adult human aortic segments, shape changes in the presence of forskolin were also noticed (figure 40A and B). In these experiments, after addition of forskolin cells acquired the morphology characteristic of endothelium in regions of hemodynamic stress (Figure 40C). The most pronounced effect after addition of forskolin was found in homogeneous small-sized EC from child aorta (Figure 40 D and E).

All these findings demonstrate that the systems of cAMP and phosphoinositide metabolism participate in the regulation of endothelial morphology, and these systems are able to modulate each other.

Summarizing both our results and data in the literature, we may conclude that the activation of second messenger systems results in a broad spectrum of effects on endothelium in

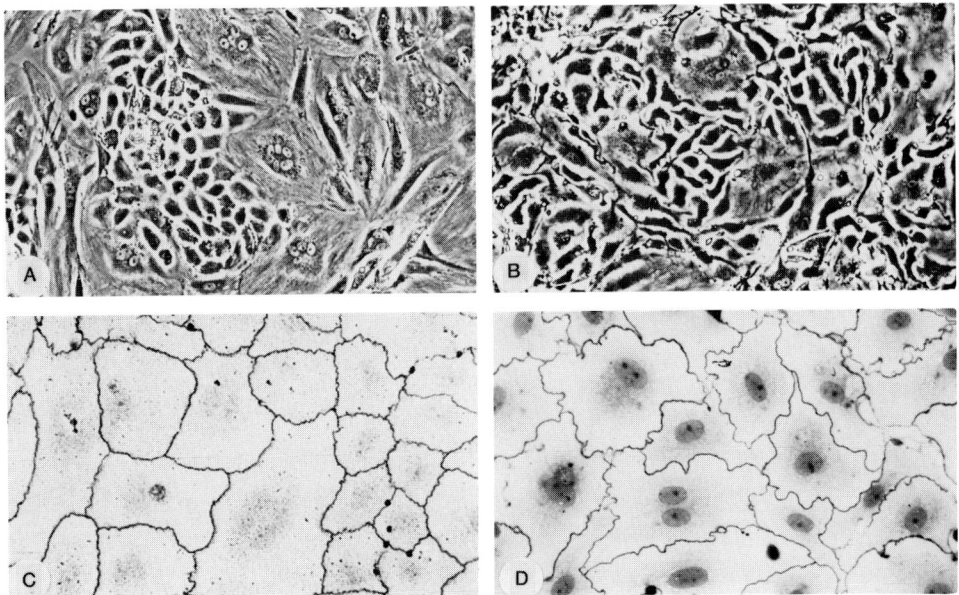

FIGURE 39.   Effect of forskolin on the morphological features of human aortic EC in primary culture. (A, C) Control; cells were incubated in medium 199 with 0.5% BSA. (B, D) Cell culture after 1 hr incubation with forskolin (10^{-5} *M*). (A and C: phase contrast; magnification × 100; B and D: silver nitrate stain; magnification × 200.)

experimental animal and human cells. These effects include changes in shape, loss of integrity of the endothelial monolayer, injury of intercellular junctions, death of separate endothelio-cytes, and/or large denudation of the aortic lumen. Some differences which were observed in the experiments on rabbit aortic segments *in situ* and in culture of human endothelium may be related to the specificity of the objects of the study and the species specificity of endothelium.

C. Proliferation and Formation of Cellular Contacts in Culture

We have discussed physiological processes which underlie the reorganization of the en-dothelial layer. It was shown that the morphology of EC in monolayer can be controlled by second messengers, which mediate the effect of a wide variety of endocrine, paracrine, and autocrine regulators. However, to a significant degree the morphological organization of endothelium can be predetermined at the stage of endothelial monolayer formation. Arbi-trarily, two steps can be distinguished in this process: the step of active cellular proliferation followed by the step of formation of intercellular contacts and the endothelial monolayer per se. The investigation of biochemical mechanisms of these events is required for under-standing the maintenance of morphological and functional integrity.

At present, quite a few growth factors are known which can stimulate proliferation of EC[98] and capillary growth.[98] Procedures of cloning endothelium and growing EC in serum-free synthetic media containing a number of hormones and growth factors have been de-veloped,[99] and a number of factors inhibiting proliferation of EC[100] and formation of capillaries[101] were described. Nevertheless, the mechanisms underlying the effects of growth factors, hormones, and the influence of extracellular matrix on proliferation of EC are still unknown. The least investigated are the factors which regulate the organization of EC in monolayer, their differentiation, and the formation of specific intercellular contacts.

Earlier we have shown that substances such as phorbol esters which elevate intracellular

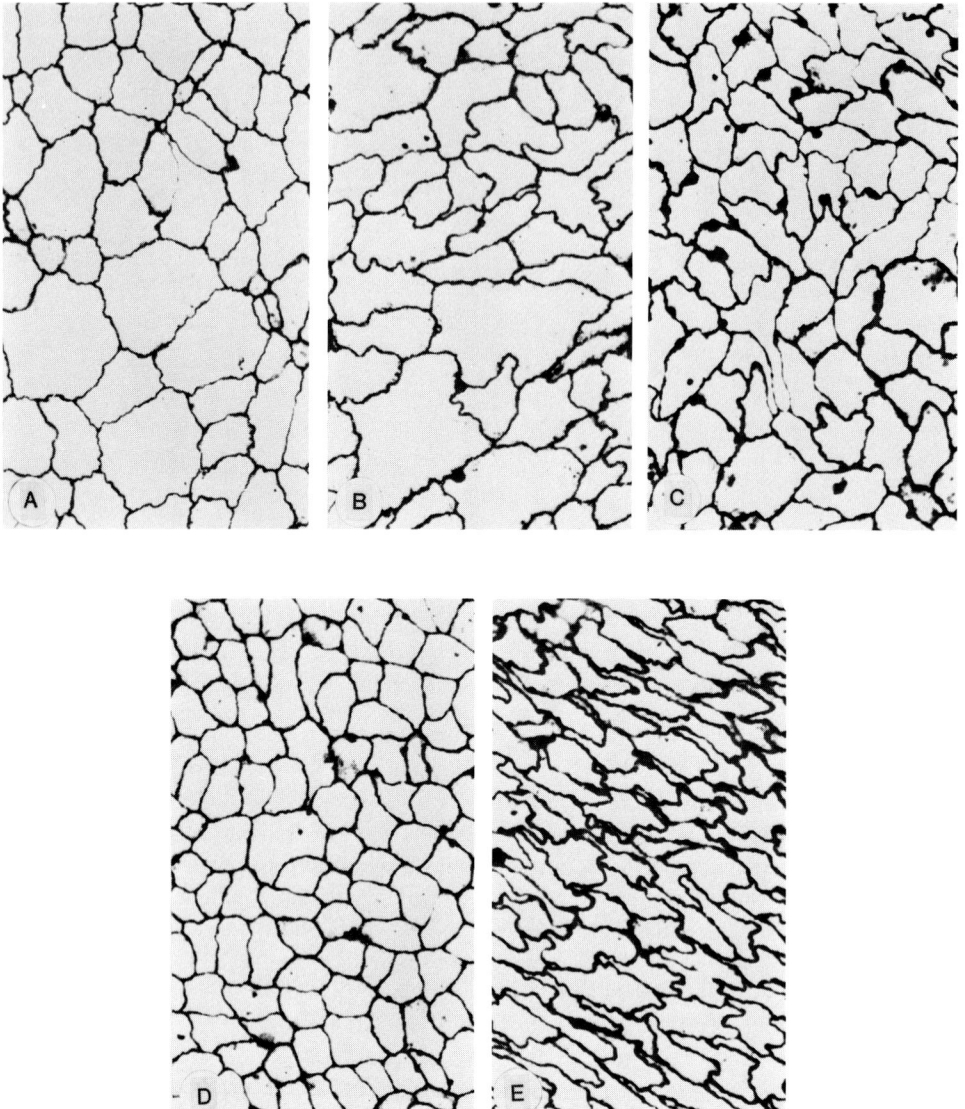

FIGURE 40. Effect of forskolin on EC from adult human (A, B) and child (D, E) aorta in short-term organ culture. Explants were incubated in medium 199 with 0.5% BSA (A, D) or in medium containing forskolin (10^{-5} M) (B, E) for 1 hr. (C) Endothelium in the zone of hemodynamic stress (ostia of arteria carotis). (Silver nitrate stain; magnification $\times$ 300.)

levels of cAMP and also kinase C activator, produce fast and reversible changes in the morphology of human umbilical vein endothelium and aortic endothelium in culture. By simultaneously affecting cAMP metabolism and kinase C, it was possible to make changes in the morphology occur much faster and be more pronounced in comparison with separate activation of kinase C or cAMP synthesis (see Figure 38).

Using selective activators of adenylate cyclase and kinase C, the role of second messengers systems in the regulation of proliferation of EC and the formation of intercellular contacts in culture were investigated.

It emerged that TPA in the range of concentrations 10^{-7} to 10^{-9} M exerted a powerful stimulating effect on both the proliferation of EC and the incorporation of ^{3}H-thymidine

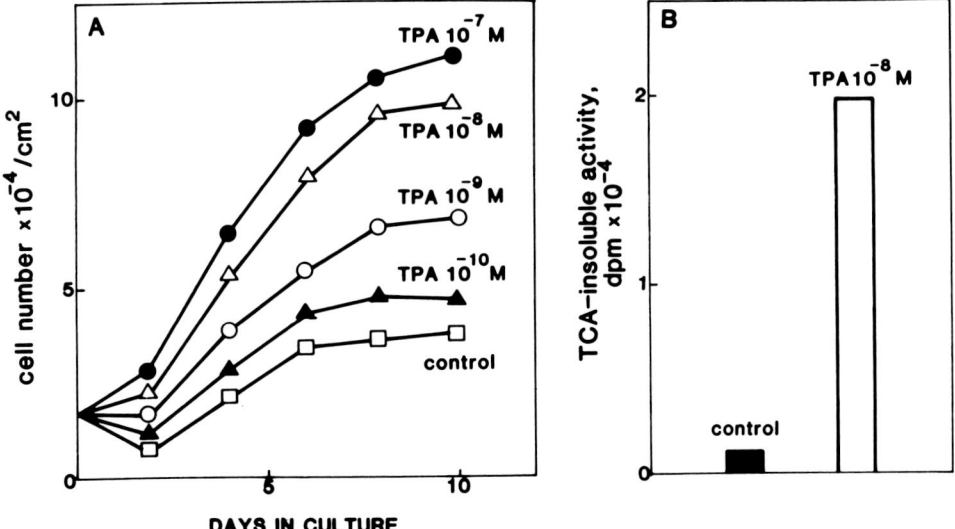

FIGURE 41. Effect of TPA on human umbilical vein EC. (A), dose-dependent effect of TPA on EC growth rate; (B), effect of TPA on [³H]thymidine incorporation in DNA in serum-free medium.

(Figure 41). If cells were grown in the presence of TPA (10^{-8} *M*), the average doubling time of the population shortened from 72 to 76 to 16 to 18 hr. The formation of confluent monolayers was achieved at seeding densities from 2×10^2 to 2×10^4 cells per square centimeter. The final density of the monolayer and the cell size in the monolayer correlated with initial seeding density. At low seeding densities and TPA-containing medium, monolayers with a density of 2- 3 $\times$ 10^4 cells per square centimeter were formed containing large and giant, often multinucleated, cells (Figure 42 A and B). High seeding density resulted in the formation of dense monolayers (8×10^4 to 10^5 cells per square centimeter,) were elongated and mutually oriented cells were predominant (Figure 42 C and D). These cultures cells reached contact inhibition, and the addition of TPA did not cause the proliferation of cells. We never observed the growth and migration of cells over each other, or the formation of multilayer cultures, even after prolonged treatment with TPA. This fact, in addition to the observation on the reversibility of TPA effects, led to the conclusion that mitogenic effects of TPA on EC may not be related to cell transformation. It is thought that TPA activates biochemical pathways which provide normal proliferative response of endotheliocytes.

The mitogenic effect of TPA depends on the concentration of serum in the medium. Nevertheless, even in serum-free medium a pronounced stimulation of proliferation and incorporation of ³H-thymidine was observed (Figure 41B). If cells were grown on plastic (seeding density 2×10^4 cells per square centimeter) without extracellular matrix (or other substrate) in a serum-free synthetic medium with no hormones or growth factors but in the presence of TPA (10^{-8} *M*), the formation of confluent monolayer was seen by the 7th to 11th day.

An adenylate cyclase activator, forskolin (10^{-5} *M*), inhibited the mitogenic effect of TPA by 30 to 50% and had no effect on the growth of endothelium in TPA-free medium with 10% of serum. The addition of forskolin (10^{-5} *M*) or dibutyril-cAMP (10^{-5} *M*) to preconfluent cultures where cell junctions were not yet stained with silver nitrate resulted in a fast (1 to 4 hr) reorganization of the monolayer with well defined cell boundaries easily stained with silver nitrate (Figure 43). These changes in the structure of the monolayer were accompanied by the accumulation of fibronectin in the regions of intercellular junctions. A

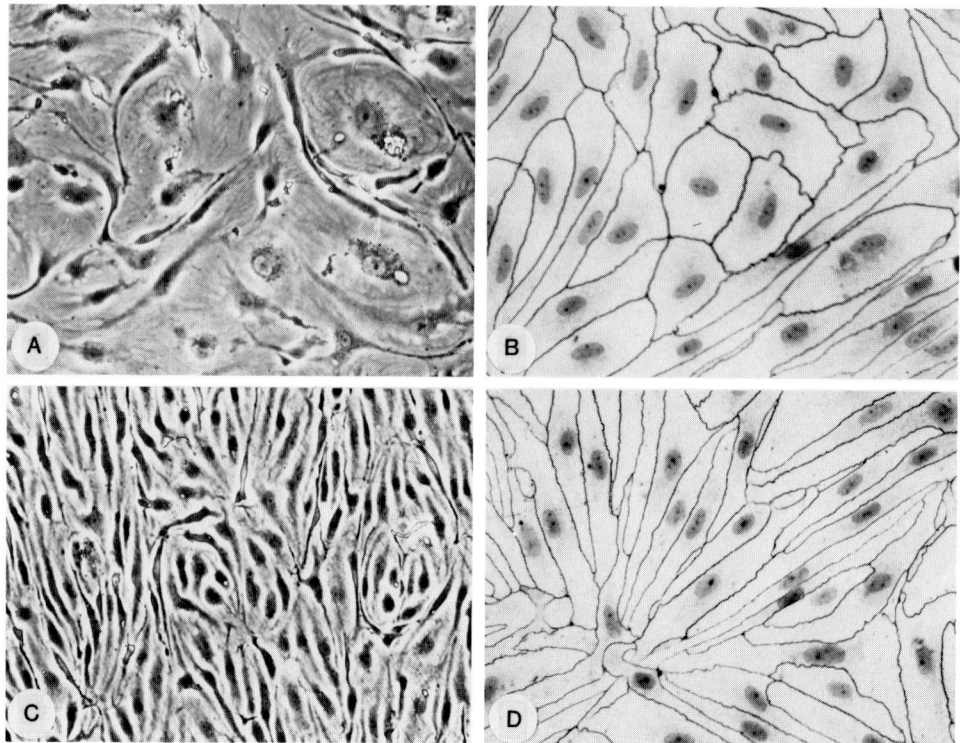

FIGURE 42.   The morphology of human umbilical vein EC, cultured in the presence of TPA at low (A, B) or high (C, D) seeding density. EC were seeded at a density of 10^3 cells per square centimeter (A, B) and 5×10^4 cells per square centimeter (C, D) and cultured in medium containing 10^{-8} M TPA until confluent monolayer was formed. (A and C): phase contrast; magnification $\times$ 120; B and D: silver nitrate stain; magnification $\times$ 240.)

similar effect of forskolin was observed in preconfluent cultures of endothelium grown in the presence of TPA. These findings show that second messengers are able to regulate the formation of endothelial monolayers and the products of phosphoinositide metabolism primarily stimulate proliferation of EC; the elevation of intracellular cAMP is important for the formation of intercellular contacts and the organization of the endothelial monolayer.

It should be noted that in EC TPA activates systems responsible for the initiation of proliferation, but not all the mechanisms which provide normal growth of endotheliocytes. This idea is supported by the fact that the mitogenic effect of TPA is potentiated by serum components. Growth factors in a number of cell types are known to activate both phosphoinositide metabolism and the synthesis of cGMP. At the same time, TPA is able to imitate the effect of only one of the products of phosphoinositide metabolism, namely, DAG. It is thought that for normal proliferation of EC it is necessary to switch on cGMP metabolism and, in parallel, to imitate the effect of the second product of phosphoinositide metabolism, triphosphoinositol. Selective activation of the systems which are affected by TPA is likely to result in the formation of certain morphotypes of EC (elongated, mutually oriented, and giant multinucleated cells).

We suggest that morphological heterogeneity of human endothelium may be functional. In other words, endothelial morphology can be reversibly adjusted to the changes in the intracellular cAMP and DAG concentration. These substances are formed by activation correspondingly of adenylate cyclase and the enzyme of phosphoinositide metabolism which are, in turn, affected by such endocrine factors as catecholamines, histamine, angiotensin, acetylcholine, prostaglandins, leukotrienes, etc. According to the data presented above, the

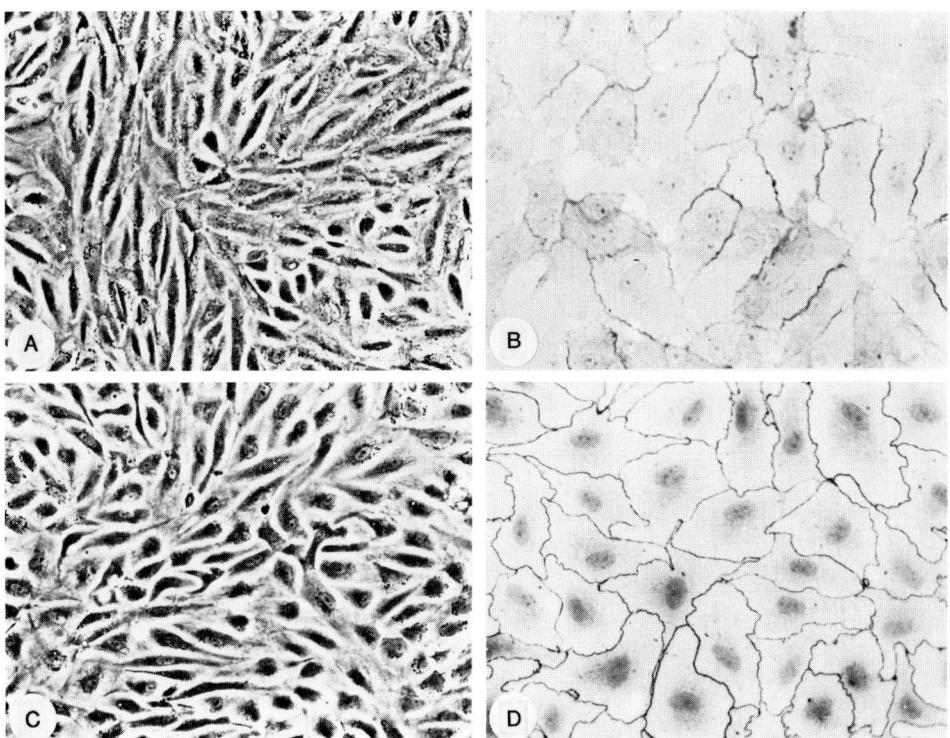

FIGURE 43.   Effect of forskolin of cell-cell contact formation in preconfluent human umbilical vein EC culture. (A, B) Control culture; (C, D) after 1 hr incubation in the presence of forskolin (10^{-5} M). (A and C: phase contrast; magnification $\times$ 120; B and D: silver nitrate stain; magnification $\times$ 180.)

hormones which act via adenylate cyclase largely affect cell morphology and the formation of intercellular contacts in the monolayer. The fact that all these effects are inhibited by antitubulin preparations shows that the cAMP elevation acts by changing cytoskeleton, rather than via the disturbance of osmotic processes. The hormones stimulating phosphoinositide metabolism may accelerate the proliferation of EC. In addition, the products of phosphoinositide metabolism (evidently, DAG) potentiate the morphological changes caused by adenylate cyclase activation.

EC and vascular wall may be rare, if not the only, examples of synergism between two important regulatory systems — cAMP- and DAG-dependent phosporylation of proteins. Molecular mechanisms of this synergism are yet to be understood. However, it is now evident that the targets for cAMP-dependent protein kinase and kinase C should be the proteins (or one and the same protein) which are involved in cell motility and the formation of intercellular contacts.

VIII. INJURY, REPAIR, AND PROTECTION OF ENDOTHELIUM

A. Receptor-Mediated Injury and Repair of Endothelium

Specific surface receptors for hormones, neutrotransmitters, and biologically active peptides mediate the response of endothelium to these agents in culture and *in situ*.[102] The receptors of EC are involved in the regulation of a number of functions of the cell, for example, transport, secretion, and the control of permeability.[102-107] Among vasoactive substances, the most important role belongs to catecholamines, which regulate vascular tone, and at high concentrations may damage vascular wall. The mechanism of the injurious effect

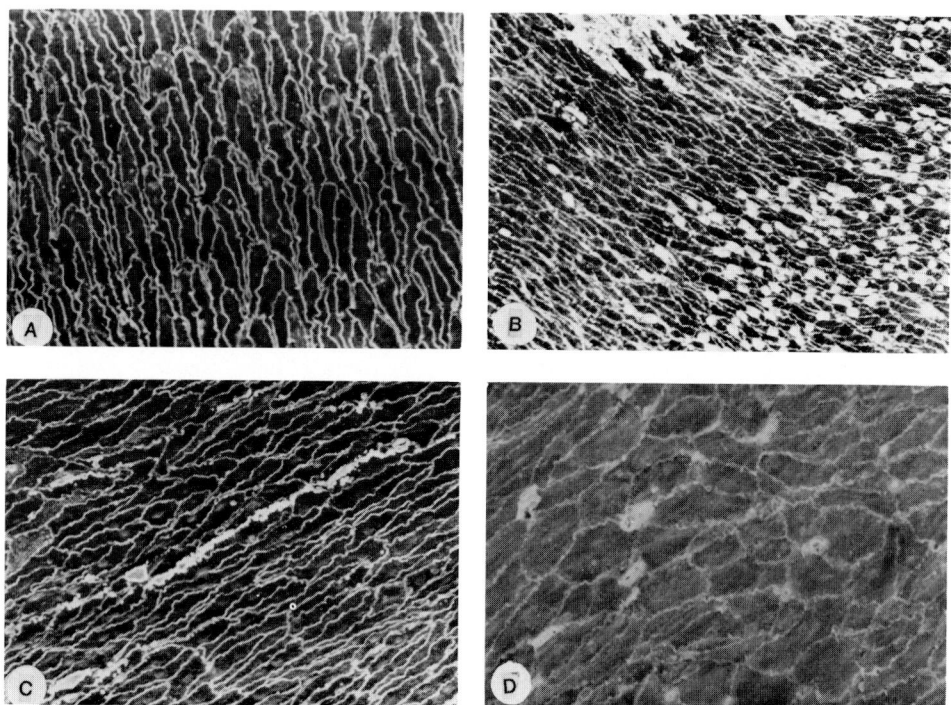

FIGURE 44. (A) SEM of the control sample shows an unaltered endothelial lining. Only intercellular junctions are silver stained. (Magnification × 166.) (B) SEM of thickened EC intercellular junctions. Perfusion with 0.1 μM epinephrine and 0.1 μM norepinephrine. (Magnification × 83.) (C) SEM of thickened intercellular borders of EC. Aorta perfused with 0.1 μM epinephrine and 0.1 μM norepinephrine. (Magnification × 166.) (D) SEM of endothelium from aorta perfused with 0.1 μM epinephrine and 0.1 μM norepinephrine. Single injuries of intercellular borders and shape change of EC. (Magnification × 333.) After perfusion segments were fixed with 2.5% glutaraldehyde, dehydrated in graded alcohols, stained with 0.1% silver nitrate, and observed with Phillips SEM 500 at 25 kV.

of catecholamines is not yet clear.[108,109] We have shown earlier[110] that epinephrine causes a specific combination of changes in endothelium and stimulates the incorporation of ^{125}I-labeled LDL in perfused rabbit aorta. Since both phenomena were completely inhibited by alpha- and beta-antagonists, it was suggested that the effect of epinephrine on the morphology of endothelium and the permeability of aorta for LDL is related to the simultaneous activation of both alpha- and beta-adrenoreceptors. To characterize receptor-dependent morphological and functional changes of rabbit endothelium, the effect of simultaneous stimulation of alpha- and beta-adrenoreceptors by selective agonists was investigated in perfused rabbit aorta. In addition, the effect of simultaneous stimulation of beta-adreno- and m-cholinoreceptors (acting like alpha$_1$-adreno-receptors, by the activation of phosphoinositide metabolism) on morphology of EC and permeability of aortic wall for LDL has been investigated.

The effects of catecholamines, agonists of adreno- and m-cholinoreceptors, were quantitated by monitoring the number and the type of morphological injuries per 1 mm^2 of the luminal surface and by the measurements of the incorporation of ^{125}I-LDL into aortic wall. It was found that catecholamines cause pronounced damage of endothelium (Figure 44) and stimulate the incorporation of ^{125}I-LDL into aortic wall. Epinephrine and norepinephrine exerted more pronounced injurious effects (Figure 45). Catecholamine-produced defects of endothelium and elevation of permeability of vessels for ^{125}I-LDL were completely prevented by the antagonist of beta-adrenoreceptors, propranolol, or the antagonist of alpha-adreno-receptors, phentolamine (Table 6). It is evident that the injuring effect of catecholamines under the conditions of our experiments is exerted through both alpha- and beta-adrenoreceptors.

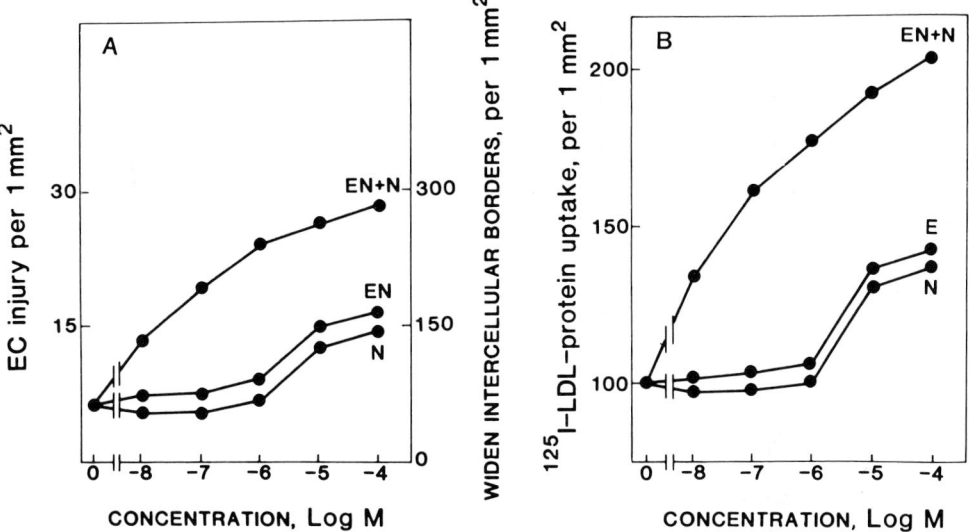

FIGURE 45.   Effects of epinephrine (EN) and norepinephrine (N) on the perfused rabbit aorta. (A) The dose-response relation of epinephrine, norepinephrine, and epinephrine with norepinephrine concentration on EC injury and widening of the intercellular borders was measured by SEM after perfusion of aorta segments for 1 hr with medium 199 containing the amounts of catecholamines. EC injury was assessed as a total number of esquamating and argyrophilic cells per 1 mm². (B) The dose-response relations of epinephrine, norepinephrine, and epinephrine with norepinephrine was also measured by ^{125}I-LDL incorporation into aorta. Perfusion was carried out with medium 199 containing 10% LDS and 5 μg/mℓ ^{125}I-LDL with the amounts of catecholamines for 1 hr. Then the segments were excised, adventitia was carefully removed, and the area of the luminal surface was measured. The content of the radioactive label was determined in a 1282 CompuGamma Y-counter (LKB, Sweden). ^{125}I-LDL uptake was expressed in picograms of LDL protein per square millimeter of luminal surface. Each point represents mean of at least five experiments.

It is well known that the stimulation of beta-adrenoreceptors is followed by an increase in intracellular cAMP concentration, whereas DAG, the second messenger of alpha$_1$-adreno- and m-cholinoreceptors, acts by activating protein kinase C.[111-113] To understand the mechanisms of the injuring effect of catecholamines, the effect of simultaneous stimulation of alpha- and beta-adrenoreceptors on the injury of EC and the role of simultaneous stimulation of m-cholinoreceptors and beta-adrenoreceptors were investigated. Adding isoproterenol (a selective stimulator of beta-adrenoreceptors) to the perfusion medium, we found no changes in the morphology of endothelium or in the incorporation of ^{125}I-LDL into rabbit aorta. Similarly, phenylephrine (selective stimulator of alpha-adrenoreceptors) produced no changes in the morphology of endothelium or in the incorporation of labeled LDL into aortic wall. Only simultaneous application of the agonists of alpha- and beta-adrenoreceptors (isoproterenol plus phenylephrin) produced significant elevation in the number of the injured EC, as well as the increase in the incorporation of ^{125}I-LDL (Table 6). The changes in endothelium induced by combined perfusion with phenylephrine and isoproterenol were reversible; this conclusion was confirmed by the experiments on perfusion with alpha$_1$ and beta-antagonists. Yohimbine, the antagonist of alpha$_2$-adrenoreceptors, was not effective in the prevention of endothelial damage (Table 6).

Thus, simultaneous stimulation of alpha$_1$ and beta-adrenoreceptors results in the injury and death of EC. Similar effects were observed with two agonists, which stimulate m-cholinergic- and beta-adrenoreceptors. Morphological and functional changes of endothelium were fully prevented by the corresponding antagonists, namely, atropine or propranolol (Table 6).

Thus, in the perfusion experiments on the segments of rabbit aorta, the phenomenon of

Table 6
RECEPTOR-MEDIATED INJURY OF EC AND ENHANCEMENT OF ^{125}I-LDL INCORPORATION INTO THE VESSEL WALL AND PROTECTIVE EFFECTS OF CORRESPONDING RECEPTOR ANTAGONISTS

Agent (μM)	EC injury per 1 mm^2	Defects of intercellular junctions per 1 mm^2	^{125}I-LDL incorporation (% from control)
	2.3 ± 0.5	58 ± 10	100 ± 12
Epinephrine (EN) + norepinephrine (N), 10 μM	21 ± 8[a]	160 ± 20[a]	200 ± 13[a]
EN + N + propropanolol, 10 μM	2.4 ± 0.4	49 ± 8	102 ± 14
E + NE + phentolamine, 10 μM	2.8 ± 0.6	61 ± 12	111 ± 14
Isoproterenol, 10 μM	2.2 ± 0.5	43 ± 10	92 ± 8
Phenylephrine, 10 μM	1.8 ± 0.4	45 ± 8	94 ± 7
Isoprotenerol + IP + phenylephrine PE, 10 μM	12 ± 3[a]	218 ± 21[a]	258 ± 22[a]
IP + PE + propranolol, 10 μM	2.10 ± 5	43 ± 12	97 ± 8
IP + PE + prazosin, 10 μM	2.1 ± 0.4	49 ± 6	101 ± 10
IP + PE + yohimbine, 10 μM	14 ± 4[a]	199 ± 18[a]	202 ± 15[a]
Carbamylcholine, 10 μM	1.6 ± 0.1	42 ± 8	108 ± 11
IP + carbamylcholine, 10 μM	16 ± 4[a]	149 ± 21[a]	208 ± 19[a]
IP + carbamylcholine + atropine, 10 μM	2.1 ± 0.6	58 ± 11	99 ± 7
IP + carbamylcholine + propranolol, 10 μM	2.4 ± 0.6	44 ± 8	108 ± 11

Note: Each M ± m value represents the results of at least five experiments.

[a] P <0.05 compared to the control, according to Student's *t* test.

receptor-dependent injury of the endothelium was demonstrated. This phenomenon is related to the synergism in the action of the activators of adenylate cyclase system and phosphoinositide metabolism.[114,115] The study of the mechanisms of these events at the cellular and molecular level may be important for better understanding of the cellular death in vessel pathology, including atherosclerosis.

It is commonly accepted that injury of endothelium plays an important role in atherogenesis.[116] Morphological criteria of endothelial injury were described by many authors and involve an increase in the staining of endotheliocytes by silver nitrate,[117] changes in the size and shape of cells,[3] and injury of intercellular junctions.[118] Using a number of experimental models, it was demonstrated that an increase in the permeability for plasma macromolecules may serve as a marker of endothelial injury.[118,119] The protection of endothelium from injury is the subject of many studies.[112,120] In experiments with cultured human and animal EC, it was found that an elevation of cAMP level accelerates lipid hydrolysis,[121-132] slows down its synthesis,[124] inhibits proliferation,[123,125] and lowers the rate of protein synthesis.[126] In other words, it normalizes the processes which are deteriorated in atherosclerosis. However, it is not known whether cAMP has any relation to endothelial injury. To answer this question, we investigated the effects of agents which increase the intracellular level of cAMP on endothelial morphology by SEM and by measuring the uptake of ^{125}I-LDL in perfused rabbit aorta (Figure 46).

In perfusing rabbit aorta under standard conditions,[120] it was found that the number of microdefects of endothelium (Figure 46) and permeability of aortic wall for ^{125}I-LDL grow linearly with the time of perfusion (Tables 7 and 8). The average cAMP content in rabbit

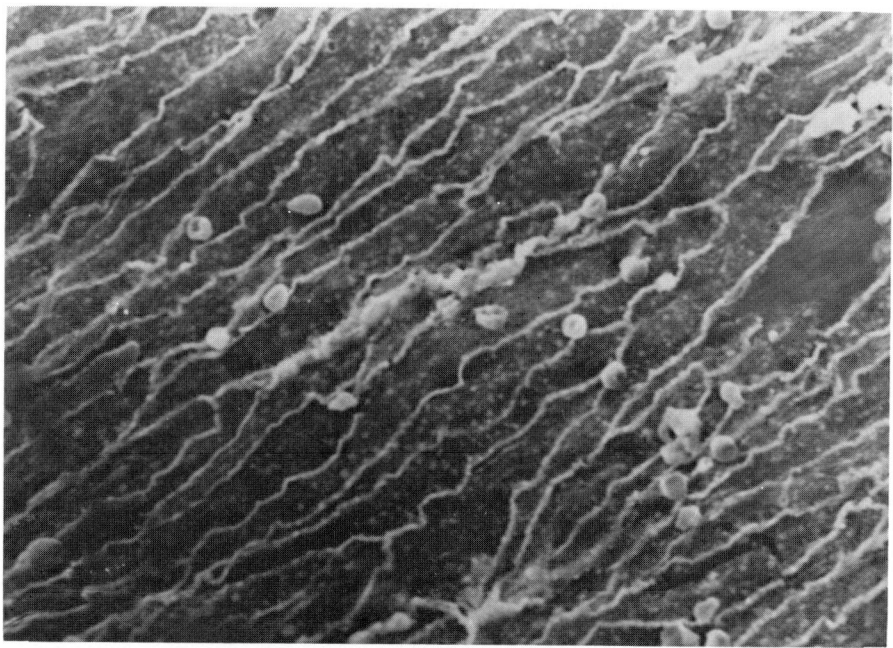

FIGURE 46.   SEM of endothelial monolayer from rabbit aorta after perfusion for 4 hr with medium 199 at 37°C under a pressure of 100 mmHg. Note widening of intercellular junctions and esquamating of the single cell. (Silver nitrate stain; magnification × 640.)

Table 7
EFFECT OF 1 mM IBMX AND 10 μM FORSKOLIN ON THE MORPHOLOGY OF EC IN PERFUSED RABBIT AORTA

Agent	Time of perfusion (hr)	EC injury per 1 mm^2	Defects of intercellular junctions per 1 mm^2
Control	1	4.4 ± 0.9	58 ± 10 (6)
IBMX + forskolin	1	5.9 ± 0.5	61 ± 8 (6)
Control	3	11.3 ± 0.8^a	44 ± 7 (4)
IBMX + forskolin	3	8.6 ± 2.5	52 ± 6 (4)

^{a} P <0.05 compared to control after 1 hr perfusion according to Student's *t* test.
 The number of experiments is given in parentheses.

aorta was equal to 222 ± 60 pmol/g of wet weight and did not significantly change within 3 hr of perfusion. To maximally increase cAMP level, the combination of forskolin and isobutylmethylxanthine (IBMX), an inhibitor of the cAMP phosphodiesterase, was applied. The maximal elevation in cAMP level (more than sixfold) was observed after 30 min of perfusion. Subsequently, cAMP concentration went down; in 60 min its value exceeds control by fivefold, and in 180 min by threefold (Table 9). In the aortic segments perfused with the agents elevating cAMP level, the number of injured endotheliocytes per 1 mm^2 of the luminal surface in 3 hr of perfusion was similar to the number of injured cells in control segments after 1 hr of perfusion (Table 7). The rise in cAMP level did not lead to changes in the shape of EC (Table 10). Longer perfusion times did not increase the incorporation of ^{125}I-LDL into segments perfused with forskolin and IBMX (Table 8). Thus, the elevation in intracellular cAMP concentration does prevent the development of morphological injuries of endothelium and stabilizes the level of LDL uptake by aortic wall.

Table 8
EFFECT OF 1 mM IBMX AND 10 μM FORSKOLIN ON ^{125}I-LDL INCORPORATION BY PERFUSED RABBIT AORTA

Agent	Perfusion time (hr)	^{125}I-LDL incorporation (pg/protein/mm²)
Control	1	95 ± 10 (6)
IBMX + forskolin	1	112 ± 14 (6)
Control	3	186 ± 19[a] (5)
IBMX + forskolin	3	130 ± 21 (5)

[a]   P <0.05 compared to control after 1 hr perfusion according to Student's *t* test. The number of experiments is given in parentheses.

Table 9
EFFECT OF ISOBUTYLMETHYLXANTHINE (IMPX) AND FORSKOLIN ON CYCLIC AMP CONTENT OF PERFUSED RABBIT AORTA

Agent	Time of perfusion (min)	cAMP content (% from control)
IBMX, 1 mM	15	278 ± 30[a] (5)
forskolin, 10 μM	15	357 ± 28[a] (5)
IBMX, 1 mM + forskolin, 10 μM	15	337 ± 62[a] (5)
IBMX, 1 mM + forskolin, 10 μM	30	654 ± 81[a] (6)
IBMX, 1 mM + forskolin, 10 μM	60	500 + 82[a] (6)
IBMX, 1 mM + forskolin, 10 μM	120	421 ± 108[a] (5)
IBMX, 1 mM + forskolin, 10 μM	180	322 ± 62[a] (5)

Note: Cyclic AMP content in control aortic segments was assumed for 100% (222 + 60 μg/g, wet weight).

[a]   P <0.05 compared to control according to Student's *t* test. The number of experiments is given in parentheses.

In this connection, of special interest is the understanding of the mechanisms involved in the regulation of homeostasis of EC, the maintenance of their viability and stability against external factors which produce cell damage. In the previous section it was shown that the behavior of EC in monolayer, the proliferation of endothelium, and the formation of contacts between cells may be controlled by the systems of the second messengers which mediate

Table 10
EFFECT OF 1 m IBMX AND 10 μ*M*
FORSKOLIN ON ENDOTHELIAL CELL
SHAPE INDEX

Agent	Perfusion time (hr)	Shape index
Control	1	0.19 ± 0.06 (6)
IBMX + forskolin	1	0.21 ± 0.06 (6)
Control	4	0.20 ± 0.05 (5)
IBMX + forskolin	4	0.22 ± 0.07 (5)

Note: Data shown as means ± SEM. The number of experiments is given in parentheses.

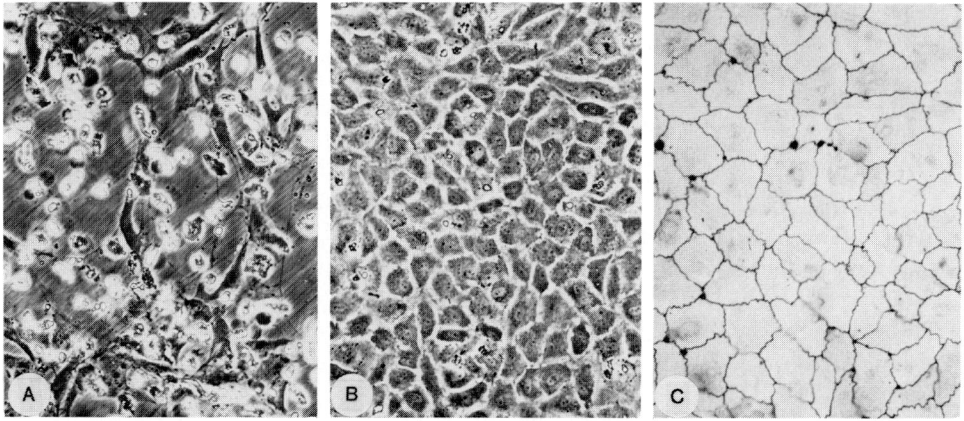

FIGURE 47.   Chemical alteration of cultured EC, protective effect of forskolin. (A) EC monolayer was incubated
for 24 hr in the presence of cholestantriol (5 μg/mℓ). (B, C) Cholestantriol with forskolin (10^{-5} *M*). (A and B:
phase contrast; magnification × 100; C: silver nitrate stain; magnification × 150.)

the external signals. The question arises whether these systems are also involved in the
control of endothelial homeostasis and its stability against external damaging effects.

An important role in atherogenesis is ascribed now to the products of cholesterol metabolism, particularly, to its oxidized derivatives which exert a powerful cytotoxic effect on
various cells. In Figure 47 a confluent monolayer of EC in primary culture is shown. The
addition to these cells of one of the product of cholesterol oxidation, 1,3,5-cholestantriole,
results in a death of endothelium within 35 hr (Figure 47A). The same derivative of cholesterol
has no toxic effect on similar cultures, when the level of cAMP is elevated by activation of
its synthesis or by the addition of cAMP analogs capable of penetrating into the cell (Figure
47B). Under these conditions the endothelium is completely preserved; even minimal de-
teriorations of cellular contacts are absent (Figure 47C). These data demonstrate that the
system of cAMP metabolism is closely involved in the regulation of the ability of endothelium
to sustain chemical damage.

However, this is not the only way by which the protective effect of cAMP is exerted.
The same system may play an important role in the reparation of the integrity of the endothelial
monolayer. We found that the elevation of cAMP in EC can significantly accelerate the
reparation of endothelium after mechanical injury. Figure 48A, shows the endothelial monolayer after mechanical injury. After addition of forskolin (10^{-5} *M*) the integrity of the

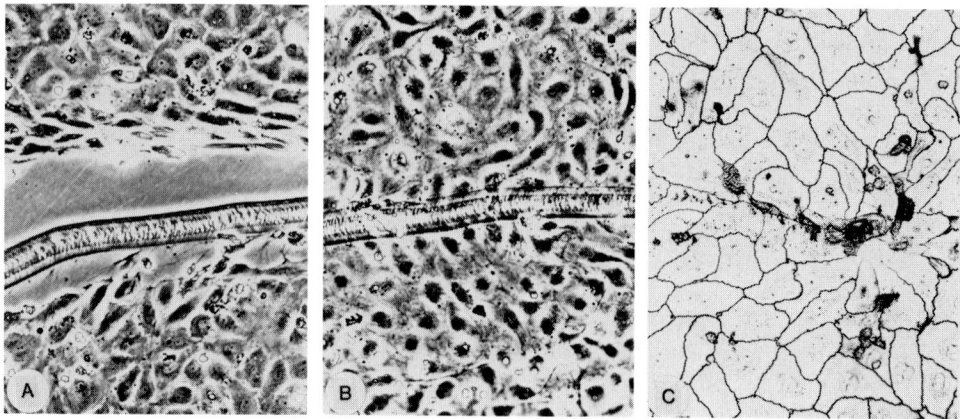

FIGURE 48. Effect of forskolin on EC monolayer reparation after mechanical damage. (A) EC culture after mechanical damage (a scrape). (B, C) Reparation of the monolayer after 2 hr incubation in the presence of forskolin (10^{-5} M) (A and B: phase contrast; magnification $\times$ 100; C: silver nitrate stain; magnification $\times$ 150.)

monolayer is completely restored within 1 to 2 hr (Figure 48B), whereas in its absence the reparation of endothelium after similar damage requires from 3 to 5 days. The effect of forskolin can not be explained by an increase in cell number: earlier in this review it was mentioned that cAMP does not stimulate, but rather inhibits, endothelial growth. The reparation on endothelial monolayer in this case is due to the stimulation of migration and spreading of endothelial cells. The cells migrate into the zone of injury and form a continuous monolayer with completely formed intercellular contacts, well seen after staining of the culture with silver nitrate (Figure 48C).

Thus, cAMP system may be involved in the regulation of not only stability and the protection of endothelium against chemical injury, but through this system the reparation of endothelial layer after mechanical denudation or atherosclerotic injury of vascular wall may occur.[121-125] The investigation along these lines may have an important impact on the development of the approaches to antiatherogenic therapy and for accelerating of endothelial reparation after transluminal angioplasty.

Studying the effects of the second messengers and hormones, the synergism between two important regulatory pathways, cAMP system and phosphoinositide metabolism, was found. These complementary effects can explain the injuring action of catecholamines on human vessels. The activation of cAMP system provides only "favorable" effect on endothelium. The activation of phosphoinositide metabolism does not lead to injuries either, but accelerates endothelial proliferation, not disturbing intercellular contacts. Thus, the binding of epinephrine and norepinephrine with either beta-receptors activating adenylate cyclase, or with alpha$_1$-receptors, which activate phosphoinositide-specific phospholipase C, should not per se lead to the disturbances in the endothelial layer. However, simultaneous activation of alpha$_1$- and beta-receptors does induce endothelial injury due to the synergism between cAMP- and DAG-dependent regulatory systems. In vascular wall cells, this synergism can be manifested in EC, as well as in subendothelial cellular population. Nevertheless, the potentiation of the effects of forskolin by phorbol ester found in EC culture suggests that the interaction of these two regulatory pathways takes place in one and the same EC.

Since epinephrine is more effectively bound to beta-receptors, while norepinephrine is to alpha-receptors, it is thought that the damaging effect on vessel wall is exerted by the combination of these two hormones. In our experiments the mixture of epinephrine with norepinephrine in the ratio 1:1 was more than tenfold more effective than each of these hormones individually. The synergism between cAMP- and DAG-dependent systems may also explain such an unusual phenomenon as the potentiation of the effects of adrenergic agents by cholinergic compounds. The effects of the second messengers found in our experiments on the morphology and function of EC allows us to predict possible changes in vascular wall due to various endocrine factors and to look for compounds capable of regulating the state of endothelium through changes in the activity of enzymes of cAMP system and phosphoinositide metabolism.

B. Targeting of Endothelium

The morphological polymorphism of EC described above even for relatively small areas of the human aortic lumen suggests that there may exist more profound differences between the populations of EC from various vascular territories. These differences may be related to the functions of these cells, as well as to the fine biochemical and immunological properties of the outer EC membranes. Attempts to find differences in the composition, structure, and immunological properties between the subpopulations of endothelium in human circulatory system have a fundamental significance and may find important applications for targeting of various physiological substances, including drugs, to the required zones of the circulation to produce specific local effects. Below, we present data collected in several laboratories of the U.S.S.R. Cardiology Research Center in a search for specific reagents or carriers for endothelium — the first attempts to prepare organ-specific carriers for endothelium and the results on model targeting to the parts of arteries with denuded endothelium.

The differential recognition of the cellular surface is, at present, most successfully made with monoclonal antibodies. In the literature, a number of attempts were described to prepare monoclonal antibodies against EC of animals and humans.[127,128] In our Center these studies are carried out to prepare monoclonal antibodies to surface antigens of human EC of unknown nature, as well as against components of the outer membrane of these cells. If successful, these antibodies could be used as carrier molecules to bring various effectors to certain areas of the circulatory system.

By immunizing mice with human endothelium from long-term cultures it was possible to prepare two monoclonal antibodies; the study of their properties is now being performed. Monoclonal antibody AME 78 (IgG) is bound by the surface of live and fixed EC isolated from human umbilical vein and cultured *in vitro*. The antigen is trypsin sensitive and is present on the surface of EC in primary cultures, as well as in long-term cultures. The binding of AME 78 to plasma proteins, platelets, fibroblasts, and human collagens of types I, III, IV, and V was not found in the range of concentrations of AME 78 of 0.4 to 100 µg/mℓ. Monoclonal antibody called AME 25 is bound to live and fixed EC isolated from human aorta, pulmonary artery, or umbilical vein cultured in vitro. This reaction is specific for EC, in contrast to plasma proteins, platelets, fibroblasts, human collagen types I, III, IV, and V in the range of concentrations of 0.05 to 100 µg/mℓ. Antigen is present on the surface of EC in primary cultures, as well as in long-term cultures. Unfortunately, in the range of concentrations of 0.05 to 10 µg/mℓ, these antibodies also interact with smooth muscle cells from human aorta. Nevertheless, this antibody is preferable as a carrier molecule for targeting to endothelium, since its binding to the cellular surface occurs at much lower concentrations than for AME 78, and the binding to blood components, collagens, and fibronectin is absent.

In an attempt to obtain a more specific reagent to recognize EC from various organs, we prepared monoclonal antibodies against angiotensin-converting enzyme (ACE) (EC.3.4.15.1). ACE is largely localized on the surface of vascular endothelium and catalyzes the transformation of angiotensin I into angiotensin II (a powerful vasoconstrictor agent) and the degradation of bradykinin which possesses vasodilator effects.[129] The enzyme plays a key role in the regulation of vascular tone. Using a highly purified preparation of ACE isolated from human lungs, a broad panel of monoclonal antibodies against this enzyme was prepared (25 primary populations). At present, 15 hybridoma clones producing monoclonal antibodies against, at least, 5 different epitopes of ACE molecules are available. These antibodies interact with the enzyme in solution and in an immobilized state, not inhibiting its enzymatic activity,[130] i.e., the active center is not in the binding site of ACE for these antibodies. Antibodies to ACE bind with high specificity to human endothelium. Plate 3* shows the results of interaction of the antibody 9B9 with cryostat sections of various human organs, detected by the PAP method.[131] It is evident that monoclonal antibodies against ACE react with antigenic determinants localized on the surface of vascular endothelium in various organs: lung, liver, heart, kidneys, and intestine. In the kidneys and intestine, these antibodies also recognize the epithelium of proximal part of tubules and villi, respectively. The application of these monoclonal antibodies may be useful for studying the pathogenesis and architecture of vascular tumors including hemangiomas. In particular, immunocytochemical study of the cases of nonspecific human aortoarteriitis using monoclonal antibody 9B9 demonstrated that this disease is accompanied by a dramatic propagation of vasa vasorum through adventitia.

Fortunately, antibody 9B9 interacts not only with human, but also with rat ACE, which allowed us to investigate the fate of this antibody in vivo. Intravenous administration of ^{125}I-labeled monoclonal antibody 9B9 into rats revealed unusually high accumulation of this antibody in the lungs (Figure 49A and B). It is important that the distribution of these antibodies between various organs does not depend on the route of administration (intravenously or intraarterially). From our point of view, there exist, at least, two explanations of this phenomenon: (1) 70% of the ACE of an organism is localized in lungs and, hence, the concentration of ACE in lungs is very high;[132] and (2) vascularization of lungs is unusually high.[133]

It is known that the administration of sufficiently high concentrations of polyclonal antibodies against ACE results in the death of animals due to lung edema caused by complement-dependent lysis of EC of lung capillary endothelim.[134] We found that the administration of large amounts of monoclonal antibody against ACE does not cause similar effects. This fact can be explained by the extremely low ability of monoclonal antibodies, compared to polyclonal, to fix complement. The activity of ACE in the lung, nevertheless, drops by five- to tenfold after monoclonal antibody administration, although 9B9 antibody does not inhibit enzyme activity (Figure 50). We believe that this immunological "enzyme ectomy" is the manifestation of the phenomenon of "antigenic modulation".[135]

A unique distribution of monoclonal antibody against ACE in the organism and the absence of side effects after its administration may open a number of possibilities for the research, diagnosis, and therapeutic application of these antibodies: (1) for targeting of drugs into lungs using direct conjugation of antibodies with a drug or a container which can carry the drug; (2) developing the animal models of specific injury of lung endothelium using antibodies conjugated with immunotoxins or, for example, with glucose oxidase, an enzyme, which catalyzes the formation of hydrogen peroxide, the best cytotoxic agent known in vivo; (3) for radioscintigraphy of the lung vessels using the antibodies conjugated with rapidly degraded radioactive label (for example, technicium or indium); (4) attempts to regulate ACE activity of the lung vessels by specific antibodies, so called "antigenic modulation"; and (5) to develop a simple and sensitive immunodiagnosticum to determine ACE in blood, which is important for diagnosis of diseases when the injury of pulmonary endothelium is the primary cause of the disease.

C. Targeting to Endothelium-Free Lumen

Of special interest is the development of approaches which allow recognition of sites

---

*   Plate 3 appears after page 180.

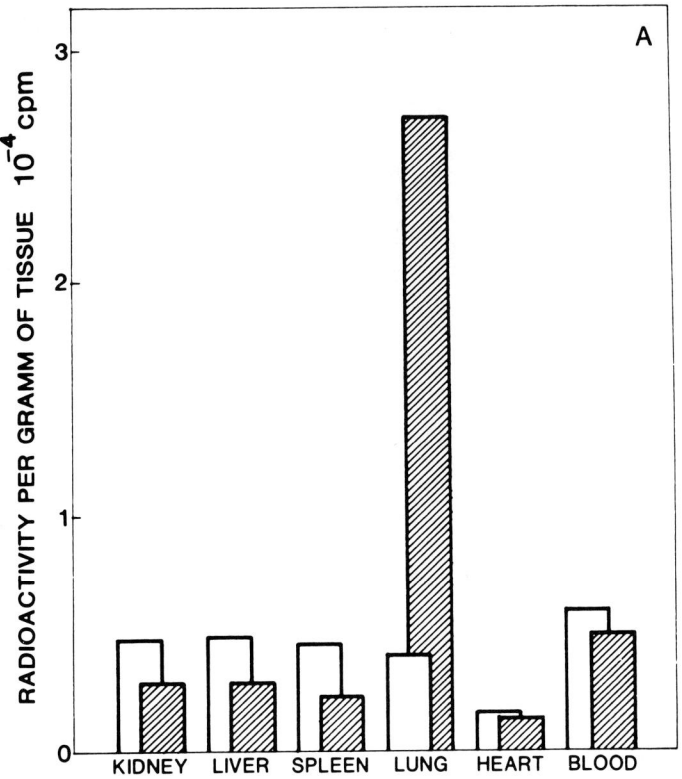

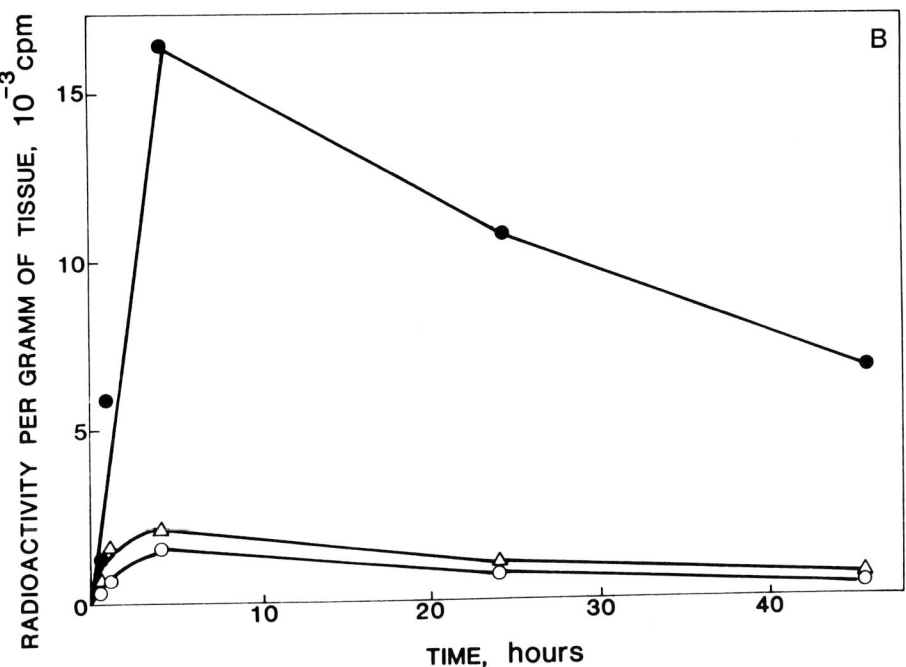

FIGURE 49.   The distribution of ^{125}I-monoclonal antibodies against ACE after injecting into blood of rats. (A) 20 hr after injecting 1 μg (1 μCi) of labeled and 100 μg of unlabeled antibodies. Antibodies against ACE, (▨); nonimmune mouse Ig, (☐). (B) Kinetics of accumulation of the antibodies against ACE in rat lung (●), kidney (△), and liver (○).

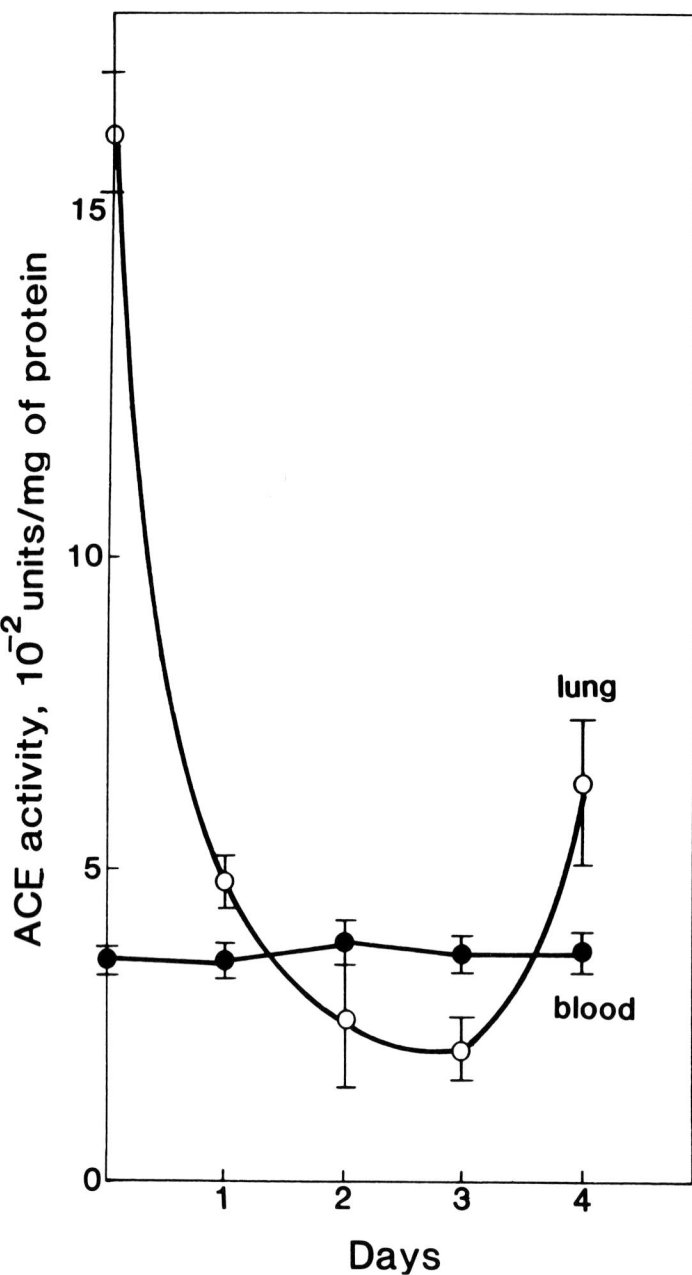

FIGURE 50.   ACE activity in rat lung and blood after injection 3 mg of antibodies against angiotensin converting enzyme (''9B9'').

where the loss of endothelium occurred. In a number of pathological conditions of the human cardiovascular system and various surgical interventions situations are seen when the morphological structures or chemical components of vascular subendothelium which are normally thrombogenic come into direct contact with blood. Loss of the integrity of the endothelial monolayer is observed in large human arteries in zones of high risk for atherogenesis and thrombosis including the regions of aortic ostia, bifurcations, and narrowing of the lumen of the vessel or changes in its curvature, which cause disturbances in laminar blood flow.[136]

Numerous investigations demonstrated that the formation of intracoronary thrombi which lead to myocardial infarction is always preceded by the rupture or fission of the atherosclerotic plaque cap.[137,138] The catheterization of vessels during angiography may also cause the loss of endothelium. Transluminal angioplasty of the stenotic vessels always results in vascular injury and exposure of intimal and often medial vascular structures to blood.[139] Routine bypass surgery on vessels performed without special precautions is often accompanied by deendothelialization of the shunted vessel, as well as of the venous graft.[140]

The necessity in all these circumstances to prevent thrombosis in the damage part of the vessel raises the following question: can this intervention be performed locally, by a targeted drug delivery into the site of injury to create high local concentration of a drug at a relatively low systemic concentration?

Since the atherosclerotic plaques contain large amounts of collagen type I,[141,142] it was thought that after cracking the surface of the plaque, the collagen structures underneath may become exposed to blood and used as the target to bring a drug to this thrombogenic surface. To check this suggestion, the partially denuded human carotid artery was perfused with radioactive antibodies to collagen type I, and the binding of these antibodies per unit area of the denuded and noninjured surface of the artery was measured. It was found that the binding of antibodies to the denuded area was 20-fold higher compared to the control areas (3.2×10^5 cpm/cm^2 and 1.5×10^4 cpm/cm^2, respectively). Preimmune labeled IgG was bound to the injured surface even somewhat less than to the intact surface (1.4×10^4 cpm^2/cm^2 and 1.9×10^4 cpm^2/cm^2, respectively). Thus, the antibodies to collagen type I can be, in principle, used as carriers which provide specific binding of containers filled with a drug to the injured vascular surface. As the second possible carrier fibronectin was suggested. The structure of this protein has a domain of high-affinity binding to collagen.[143]

At present, as potential containers for drugs, liposomes are actively studied.[144] Biodegradability, biocompatibility, and the possibilities of filling their internal space with various active substances make these structure attractive for drug targeting.[145] As seen from the data shown in Figure 51, liposomes, which contain on the outer surface the vector molecules, namely, the antibodies to collagen 1 or fibronectin, are selectively bound to the artificially denuded areas of bovine mesenteric artery (Figure 51A), rabbit aorta (Figure 51B), and human carotid artery (Figure 51C). In these cases, liposomes which carry preimmune rabbit immunoglobulin are bound to noninjured surface at the same level as to the injured area, or even better.

Other potential containers for carrying drugs in the organism are erythrocyte ghosts.[146,147] It was shown that syngeneic red blood cells filled with drugs circulate for a long time in the bloodstream;[148] these cytocontainers are also biocompatible and nonimmunogenic.[149] From the data presented in Figure 52A, it is seen that red blood cells, which carry on their surface the antibodies to collagen type I, are specifically bound to the injured surface of human vessels. The binding density for these immunoerythrocytes to the endothelium-free surface was equal to 5.4×10 cells per square centimeter, which is 20-fold higher than the binding to the noninjured surface. Immunoerythrocytes cover about 25% of the denuded surface. Figure 52B shows SEM microphotographs of the region of human carotid artery after perfusion with erythrocyte-antibody conjugates. It is seen that the significant part of the denuded surface is covered with immunoerythrocytes, whereas the intact surface binds only single erythrocytes. In all the experiments, erythrocytes which were carrying preimmune immunoglobulin did not provide selective binding to the endothelium-free surface.

Coming to in vivo models, which means targeting of the real drug to the real injured surface of the vessel, one encounters a number of difficulties related to the behavior of immunoliposomes and immunoerythrocytes in the organism. At the same time, recent studies demonstrate that by changing the size, the charge, the fluidity, and the composition of liposomes or modifying erythrocytes membranes, it is possible to affect significantly the stability and immunogeneicity of the containers and their distribution among tissues.[150-153]

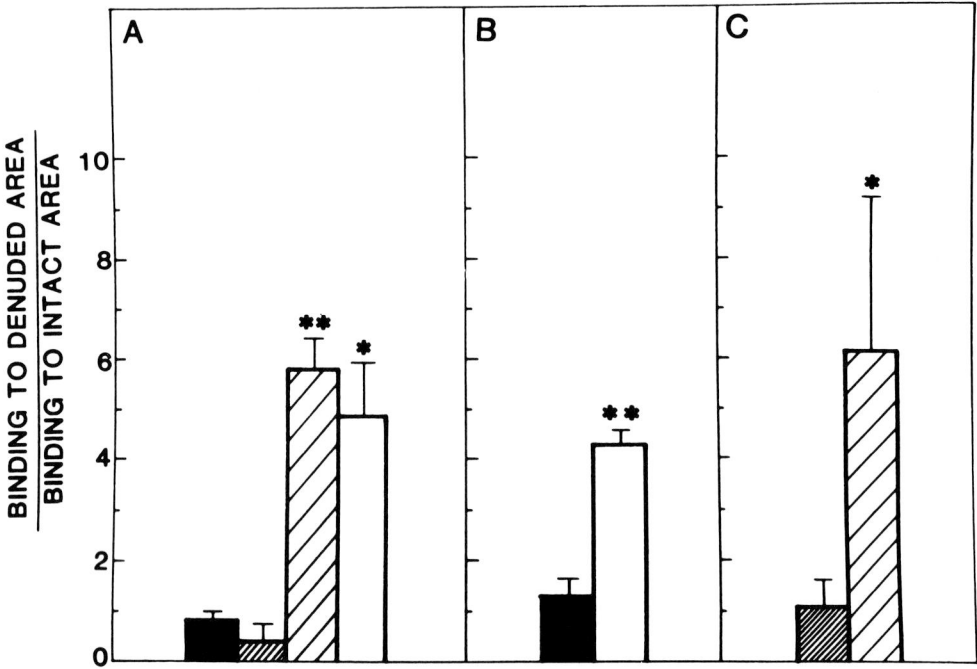

FIGURE 51.   Binding of ^{14}C-liposomes and ^{14}C-liposome-anticollagen antibody. ^{14}C-liposome-fibronectin and ^{14}C-liposome-IgG conjugates to the intact and denuded areas of bovine (A), rabbit (B), and human (C) arteries. Results are expressed as binding of ^{14}C-liposomes to 1 cm² of the denuded area relative to their binding to 1 cm² of the intact area (mean ± SEM for 15 (A), 7 (B), and 3 (C) experiments). *P <0.01; **P <0.001. (■), ^{14}C-liposomes; (▨), ^{14}C-liposomes-nonimmune IgG; (▧), ^{14}C-liposomes-antibody; (□), ^{14}C-liposomes-fibronectin.

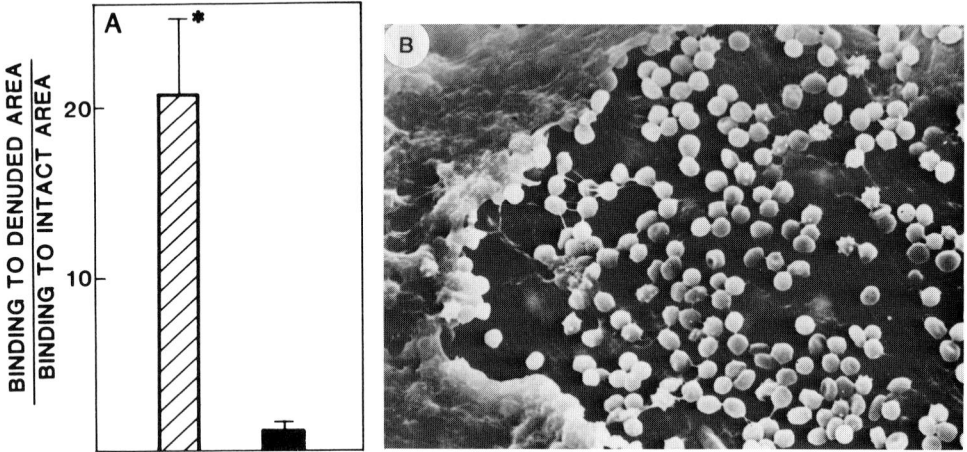

FIGURE 52.   Binding of erythrocyte-anticollagen antibody and erythrocyte-IgG conjugates to the intact and denuded areas of human carotid arteries. (A) Human carotid artery was perfused for 40 min at 37°C with the suspension of erythrocyte-antibody (□) or erythrocyte-IgG (■) conjugates. The amount of erythrocytes bound to 1 cm² of each area of the vessel was determined by direct counting using SEM. Results are expressed as the binding of erythrocytes to 1 cm² of denuded area relative to their binding to 1 cm² of the intact area (mean ± SEM) for five experiments, *P <0.001. (B) SEM of erythrocytes bound to the denuded and intact areas of human carotid artery. E, endothelium; DE, denuded area.

These results lead us to believe that the carrier systems which were tested in our experiments may be eventually used for local interventions in the effector systems of blood, which are activated in the areas with lost endothelium, for example, coagulation and complement systems,[154] platelets, and leukocytes.[155] Finally, it may be possible using immunoliposomes or immunoerythrocytes to create high local concentration of the stimulators of endothelial repair.

D. Protection of Endothelium from Hydrogen Peroxide-Induced Death

In inflammation vascular endothelium often becomes the target for the attack by activated macrophages.[156-160] One of the cytotoxic agents produced by macrophages is hydrogen peroxide,[161-163] which is also involved in the degradation of extracellular matrix,[164,165] regulates the activity of the serum proteases,[165] forms chemoattractants for leukocytes,[166] and modulates the functions of platelets.[167,168] The local increase in H_2O_2 concentration plays a key role in injury to ischemic myocardium,[169] in endothelial damage of the renal glomeruli in glomerulonephritis,[170] and in the injury of pulmonary endothelium caused by systemic activation of the complement, hyperoxygenation, and acute respiratory distress syndrome.[169,171,172]

Due to a high pathogenetic significance of the inflammatory injury, the ways to protect endothelium from death induced by oxygen metabolites were investigated. The use of catalase and superoxide dismutase,[169,170] PEG-derivatives of these enzymes,[172] and their liposomal form[171] is suggested. Since erythrocytes have very high levels of antioxidant enzymes,[173,174] ways are being developed to protect pulmonary endothelium from injury in hyperoxygenation by intratracheal administration of the suspension of intact erythrocytes.[175]

Remembering potential advantages of drug targeting compared to their systemic use,[176] we made an attempt to study local protection of EC from death induced by hydrogen peroxide by modeling the inflammatory process in a cell culture. First, the quantitative parameters of the cytotoxic effect of H_2O_2 on human EC in culture was investigated. It was found that the minimal cytotoxic dose of H_2O_2 depends strictly on the time-interval between the treatment with H_2O_2 and the detection of cell death: the cytotoxic effect of low (0.15 to 0.3 mM) doses is manifested only after a significant lag period (Figure 53). H_2O_2-treated cells after 1 day in culture acquire characteristic morphological changes (Figure 54): the swelling of nuclei is seen, cells lose their polyglonal shape, and the integrity of endothelial layer is disturbed. The dependence of the manifestation of a cytotoxic effect on the time of exposure to the toxic agent is similar to the findings recently obtained on the effect of ''nontoxic'' (i.e., not producing immediate cytotoxic effect) doses of H_2O_2 on some biochemical parameters of the endothelium: ATP level,[177] transendothelial transport,[178] synthesis and the release of prostacyclin,[179] and stability against injuring agents.[180] It is important that the effects of low doses of hydrogen peroxide may be realized in vivo in the zones of slowly developing chronic inflammation.

To demonstrate local protection of endothelium in culture, a vector system was used which included the pair of ''polyclonal mouse antiserum to human EC-polyclonal rabbit antibodies to mouse immunoglobulins''. On the surface of erythrocytes (which have been used as the natural catalase-rich containers), we immobilized rabbit antibodies to mouse immunoglobulins using chromium chloride;[181] catalase was conjugated to the same antibodies using m-maleimidobenzoic acid N-hydroxysuccinimide ester (MBA). The immunoerythrocytes and ''immunocatalase'' were specifically bound to the antigens adsorbed on plastic and effectively degraded exogenous H_2O_2. The local concentration of H_2O_2 in the surrounding of the immobilized immunoerythrocytes and immunocatalase was five- to tenfold lower compared to the average concentration of H_2O_2 in solution (Figure 55).

Immunoerythrocytes and immunocatalase were specifically bound to human EC pretreated with specific mouse antiserum: per square centimeter of the endothelial monolayer about 5

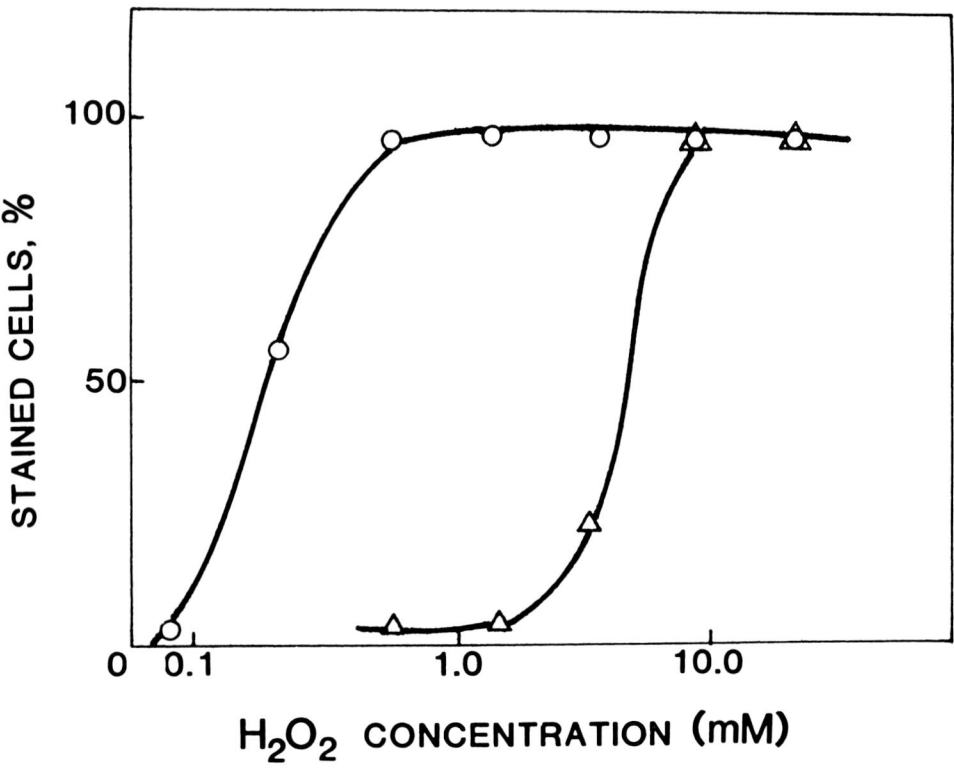

FIGURE 53.   Permeabilization of EC by exposure to hydrogen peroxide. The percentage of damaged cells was calculated by counting them after staining with trypan blue immediately after the treatment with hydrogen peroxide (△), or 24 hr later after its removal from the medium (○).

$\times$ 10^6 of immunoerythrocytes or four units of immunocatalase activity were bound. In a separate series of experiments it was shown that prolonged (minimum 24 hr) incubation of endothelial cultures with tightly bound immunoerythrocytes and immunocatalase does not affect the viability of targeted cells. Immunocatalase protects the cells in a dose-dependent manner from H_2O_2-induced injury (Figure 56); at the maximal targeting of the enzyme (about four units per square centimeter), the minimal H_2O_2 cytotoxic dose increases almost 100-fold. Immunoerythrocytes bound to the monolayer of endothelium in the range of H_2O_2 concentrations of 0.4 to 1.5 mM effectively and specifically protect this zone in the same well from death (compared to the control endothelium). Figure 57 shows the border which (during incubation with H_2O_2) was free of immunoerythrocytes (lower part) and the zone which had immunoerythrocytes on the cell surface (upper part). By morphological criteria and staining with trypan blue, the majority of dead cells were shown to be found in the unprotected region and the living cells were present largely in the protected area.

Inflammation of the vascular wall and phagocytic injury of endothelium may be also initiated by mechanical injury of vessel intima.[182] Since under these circumstances collagen of type I and III may be exposed to the lumen[183] and thus can serve as the target for drug delivery, immunoerythrocytes carrying anticollagen antibodies were prepared. In precon-fluent cultures, of EC these immunoerythrocytes were specifically bound to collagen gaps between endothelial cells, but not to the cell surface. It turned out that these immunoery-throcytes, bound to extracellular matrix, are nevertheless able to effectively protect EC from the H_2O_2 injury. All EC grown in the same well, but not on the collagen substrate (an thus not surrounded with immunoerythrocytes during incubation with H_2O_2) died.

The specificity of the targeting of immunoerythrocytes and immunocatalase, the absence

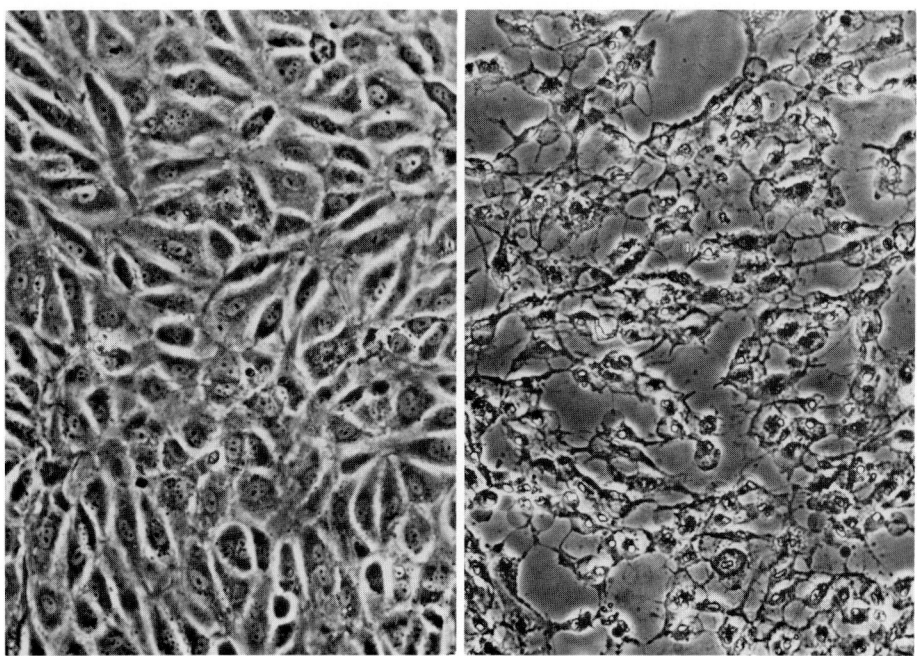

FIGURE 54.   Morphology of endothelial monolayer 24 hr after treatment with hydrogen peroxide at a cytotoxic concentration of 300 μ*M* (left) compared to intact cells (right). (Phase contrast; magnification × 100.)

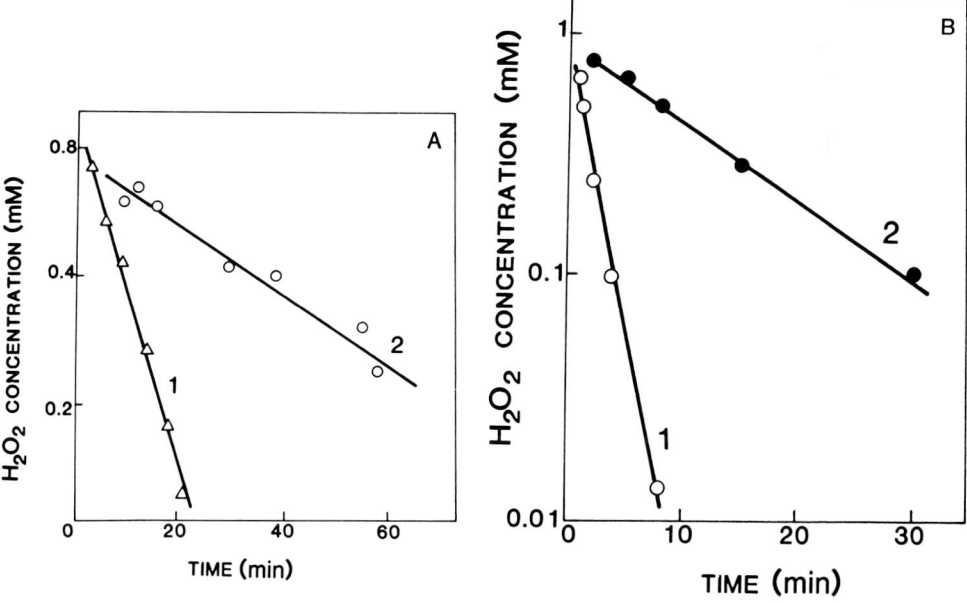

FIGURE 55.   Hydrogen peroxide degradation by immunoerythrocytes (A) and immunocatalase (B); (1), immunoerythrocytes (immunocatalase) antigen-bound; (2), immunoerythrocytes (immunocatalase) evenly suspended.

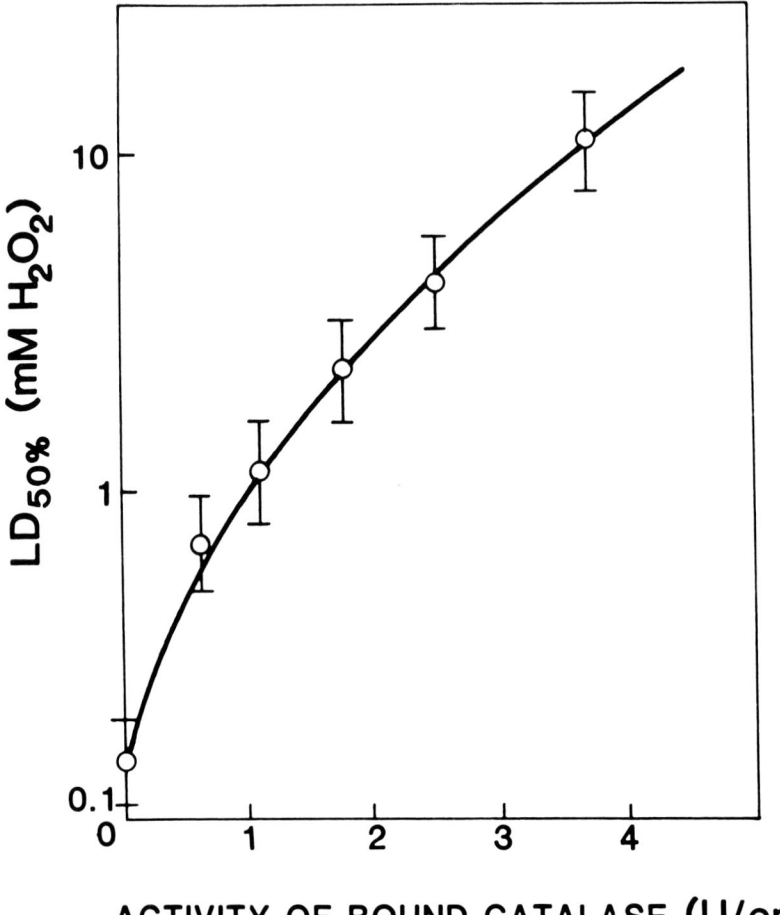

FIGURE 56.   Protection of endothelium against hydrogen peroxide with immunocatalase. Mouse antiserum raised against EC was added to EC culture and incubated with immunocatalase (catalase coupled with rabbit antibodies to mouse IgG). After washing, hydrogen peroxide was placed into each well. The percentage of damaged cells was calculated after trypan blue stain 24 hr after its removal from the medium.

of the negative effects on target cells in culture, their high effectiveness as protectors of target cells from H_2O_2 and, finally, the local nature of this protection suggest that these systems of targeted transport of antioxidant enzymes for local protection of cells in the inflammatory region are worth further exploration.

IX. CONCLUSION

For many years the viewpoint has been commonly shared that the decisive role in atherosclerotic transformation of the vessel wall belongs to the subendothelial layers of intima. In animal experiments it was demonstrated that the accumulation of cells, lipids, and extracellular matrix is the consequence of activation of proliferation, and metabolic and synthetic activity of smooth muscle cells. According to the ''response-to-injury'' hypothesis, the damage or partial loss of EC plays the role of only a trigger, which activates the cells located in deeper layers of vascular wall. However, the interrelations between the changes of endothelium and the activation of proliferation, metabolism, and function of smooth

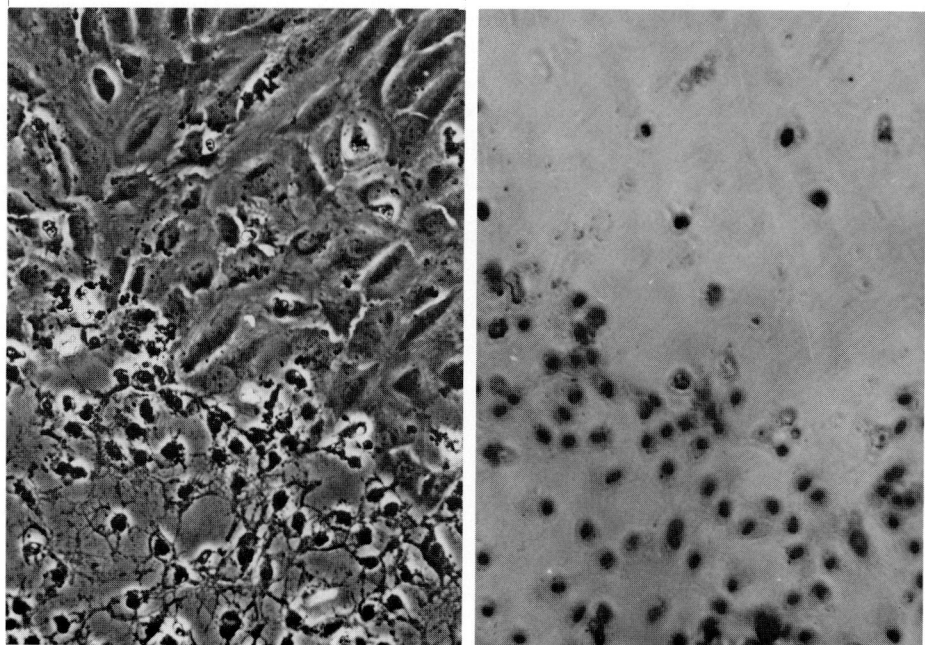

FIGURE 57. Protection of endothelium against hydrogen peroxide with immunoerythrocytes. All procedures were carried out as described for Figure 56. (Left: phase contrast; magnification × 100; right: trypan blue stain. Upper picture: protected area.)

muscle cells for a long time remained unexamined. Information on the involvement of endothelium in local accumulation of lipids, cholesterol, and components of extracellular matrix, is quite limited. Thus, it should be recognized that we know relatively little about the role of EC in atherogenesis.

The direct investigation of human vascular endothelium using quickly prepared autopsy material and cells in culture gradually leads to the accumulation of unique information on the peculiarities of morphology, behavior, and metabolism of EC from "atherosclerotic" vessels. Such almost-unknown phenomenon as the morphological, biochemical, and functional heterogeneity of EC becomes an experimental fact. This is a situation investigators rarely come across studying experimenta atherosclerosis on animal models.

The morphological and biochemical diversity of endothelial cells, from our point of view, should become the key line in the development of the methodology to study atherosclerosis. Below we will make an attempt to discuss morphological, biochemical, and functional characteristics which have been found in the study of human endothelium in our laboratories.

A. Human Vascular Endothelium *In Situ:* Polymorphism and Clusterization of Cells

The most evident feature of adult human vascular endothelium by which it differs from animal endothelial lining is its heterogeneity in size, shape, and mutual orientation of cells. In human adult vessels, two types of heterogeneous endothelium are found: (1) randomly distributed polymorphic cells and (2) clusters consisting of similar cells. In healthy adult vessels, and also in "atherosclerotic" vessels outside the atherosclerotic lesions, about 80% of all small EC are found as clusters, whereas clusters consisting of large and giant EC are only rarely observed. About 14.3% of plaques in human aorta are covered with heterogeneous endothelium with random distribution of polymorphic EC. The most diverse picture has been found on the periphery of the plaques in the transition zone to noninvolved intima.

According to our findings, about 95% of EC on the surface of the atheromatous plaques have primary cilia contacting the basal surface, whereas in the regions of aorta outside the lesions or over fibrous plaque the primary cilia is absent in 80% of cells. About 20% of cells in these areas have primary cilia located close to the luminal surface of EC. Polarization in the localization of centrioles is likely to reflect the direction of the secretory processes (abluminal or luminal) in different morphological types of human EC.

B. Human Endothelium in Culture: Polymorphism and Cellular Hypertrophy in Atherosclerosis

The substitution of dispase for collagenase significantly improved the effectiveness of harvesting of EC from normal and "atherosclerotic" human aorta. At the same time, the number of released contaminating cells of nonendothelial origin did not grow. This methodoligical innovation played an important role in the study of EC polymorphism in culture.

In primary culture within the first 10 days after seeding the morphological heterogeneity of EC found *in situ* is preserved. The ability of EC grown on artificial substrate to reproduce nonrandom shapes and organization of polymorphous cells into heterogeneous monolayer demonstrates that this reaction in culture is tightly controlled by cytoskeleton. In atherogenesis, the tendency toward elevation of morphological heterogeneity of EC in culture increases up to the appearance of "marker" subpopulations of hypertrophic multinucleated EC. These subpopulations are always absent in child aorta and rarely found in the healthy aorta.

According to flow cytometry data, in the lesioned human aorta the subpopulations of hypertrophied cells are significantly elevated. First, the percentages of tetraploid and hypertetraploid cells are higher. It is noteworthy that these cells are not labeled with thymidine.

Due to their low proliferation rate, EC from human aorta respond to physiological and pathological stimuli by hypertrophy of nuclear or cytoplasmic machinery, or by the simultaneous increase of DNA and protein in a cell. Such cells are not found in child aorta, with the exception of lipid streaks. Large hypertrophic EC rarely form clusters in human arteries, and the number of large cells in a cluster is limited to three to five. Unfortunately, data are not available on the dependence of the size of endotheliocytes on their volume.

The formation of hypertrophic smooth muscle cells in human aortic intima is rare, since these cells are actively proliferating. Earlier it was found that the cells of intimal layer only partially reproduce and preserve the polymorphism in culture that is characteristic *in situ*. These important differences in the responses of endothelial and subendothelial cells may be explained by the unique organization and unusual localization of EC in vascular wall. Earlier hypertrophic smooth muscle cells were found in the media of human and animal arteries in hypertension. It is not yet clear what are its triggering mechanisms which cause the formation of hypertrophic EC in atherosclerosis. Their formation may be related to the increased metabolic activity of these cells, or to the block in proliferation and cellular senescence.

Different methodological approaches described in this review all show significant degrees of morphological and biochemical diversity of EC in human aorta. With age, and especially in atherosclerosis, the variability of EC by morphological and biochemical criteria grows; as an ultimate state, the new subpopulations of EC appear which were absent in child vessels and in the healthy arteries of adults. Identical changes in disease and in old people are observed in the endothelium of human cornea: in the monolayer the total cell number decreases, and polymorphism and percentage of hypertrophic cells grow. Thus, in different tissues the endothelial monolayer responds by stereotypic cellular reaction. It becomes apparent that the important trend in the study of human endothelium in atherosclerosis should be oriented to the development of the techniques for cloning of different EC and multiparametric analysis of individual cells.

C. Metabolism of Modified LDL is Changed in the Atherosclerotic Intima: the Role of Endothelium

In perfused human arteries with intact endothelium, the rate of the uptake of native and modified LDL is increased. In the atherosclerotic lesions of intima, the accumulation of modified LDL was found. In primary culture of EC from "atherosclerotic aorta", an unproportional (with respect to uptake) decrease in the degradation rate of MDA-LDL is observed. This defect of degradation may explain the excessive accumulation of lipids and cholesterol and its esters in the EC over lesions. It is likely that some modified LDL reaches the subendothelial spaces by retroendocytosis, and some native LDL may be transformed into modified LDL within the vessel wall. Thus, an important role in the development of lipoidosis in intima belongs to endothelium and modified lipoproteins. EC are involved in the formation, as well as in the catabolism of modified LDL. The study of heterogeneous subpopulations of EC should finally permit elucidation of the nature of resistance and sensitivity of human vascular wall to lipoidosis.

It is possible that in the vessels of different people, nonidentical defects of lipid and lipoprotein metabolism within the cells of vascular wall will finally be found.

D. Receptors and Second Messengers in the Regulation of EC Behavior

The study of morphological variability of human EC, especially from "atherosclerotic" vessels, allowed us to model morphological rearrangements of endothelium in culture using certain chemical interventions. It was shown that the changes in the morphology of the endothelial monolayer may be reproduced using hormones and some vasoactive substances. Nevertheless, it turned out that most fast and reversible changes in the morphology of EC are effectively controlled by changes in the concentrations of the second messengers cAMP and DAG. These substances are formed by activation of adenylate cyclase and via the cascade of the enzymes of phosphoinositide metabolism caused by such compounds as catecholamines, angiotensin, histamine, prostaglandins, leukotrienes, etc.

According to our findings, hormones and vasoactive compounds which act by increasing the concentration of cAMP affect mostly the morphology of cells and the state of intercellular contacts. Since all their effects are blocked by antitubulin drugs, it is possible that the elevators of cAMP act by reorganization of the cytoskeleton, and not through ionic mechanisms and/or changes in osmotic processes. Hormones, which stimulate the phosphoinositide metabolism, can trigger or accelerate the proliferation of EC. Both systems of the second messengers are complementary. The activators of cAMP basically act on the population of nonproliferating EC, increasing their viability and resistance to a number of cytotoxins and unfavorable factors. In zones of injury, these agents may accelerate repair processes due to faster cell migration and intensive spreading of EC. Simultaneously, the formation of dense intercellular contacts is stimulated by cAMP.

The phosphoinositide metabolism activators restore and support morphological integrity in a different way, acting by increasing the number of cells in the monolayer. Both mechanisms are essential for normal morphological integrity of the endothelial lining. At the same time, the products of phosphoinositide metabolism (apparently DAG) induce morphological injuries and the death of EC. Given separately, the stimulators of DAG and cAMP formation are not able to produce this effect. Ths synergism found between cAMP- and DAG-dependent regulatory processes explains a new phenomenon: the potentiation of the injuring effects of epinephrine by norepinephrine and by other cholinergic agents. The role of second messengers in the control or morphology and the injury of endothelium allow predictions of the likely changes in the vascular wall at different combinations of endocrine factors and stress mediators and the suggestion of a way to look for potential drugs which protect endothelium from damage by selective effects on receptors or on the systems of second messengers.

E. Targeting to Arterial Lumen

Due to their unique location in the circulatory system, EC are the permenent targets for such risk factors as atherogenic lipoproteins, stress mediators, viruses, bacterial toxins, antibodies, etc. It is known that regionally the death of EC may be the event which triggers or accelerates the development of atherosclerotic lesions. In lungs, for example, the death and desquamation of vascular endothelium results in the development of acute respiratory distress syndrome.

Some injuring agents can act by binding to the specific sites on plasma membranes (toxins, hormones, antibodies, viruses). Lipid peroxides and oxiderivatives of cholesterol, simularly to ionophores, open ionic channels. The active forms of oxygen and H_2O_2 simultaneously affect many components of plasma membrane and intracellular enzymes.

It is known that lesions and partial loss of endothelium are the "hot spots" of thrombi formation and other related events, which often lead to the fatal outcome of cardiovascular disease. The delivery of a potential drug to the zones of vascular damage as a preventive measure against thrombosis can be based on the use of carrier molecules able to recognize and be bound to the surface of exposed subendothelium or to the surface of injured EC. It was demonstrated in in vitro experiments on perfused arteries that monoclonal and polyclonal antibodies to interstitial collagen may be used for local delivery and accumulation of liposomes and cytocontainers in the zones where collagen type I and III are exposed. Monoclonal antibodies to the surface antigens and enzymes of EC provided in vitro effective delivery of liposomes and erythrocyte ghosts to the surface of EC. In future, these systems may be a useful starting point to develop methods for protection of EC from damage and also for the construction of noninvasive techniques to study the state of vascular endothelium in patients.

Potentially of high interest, at present, are the attempts to prepare monoclonal antibodies to various subpopulations and clones of human EC. In future, these tools would allow the study of topography and behavior of EC in embryogenesis at the various stages of postnatal growth and differentiation of the vessels, and also the distribution of heterogeneous subpopulations of EC in atherosclerotic vessels.

The investigation of endothelium is at the stage now when it is possible to analyze these cells by the methods of cell biology, biochemistry, physiology, immunology, and molecular biology. It becomes plausible to look for the relations between earlier unrelated morphological, biochemical, and physiological findings concerning these cells. These approaches would lead to deeper understanding of the regularities in the organization and functioning of endothelium of healthy and diseased vessels.

ACKNOWLEDGMENTS

We gratefully thank Drs. A. Antonov, V. Babaev, P. Brezhestovsky, V. Bystrevksaya, T. Voyno-Yasenetskaya, G. Grigoryan, S. Danilov, S. Domogatsky, T. Klueva, M. Lukashev, V. Muzykantov, M. Nikolaeva, S. Preobrazhensky, N. Perov, O. Prinseva, Yu. Romanov, T. Romanova, A. Rudin, D. Sakharov, D. Sviridov, M. Smirnov, I. Trakht, researchers of the Institute of Experimental Cardiology of the U.S.S.R. Cardiology Research Center, and T. Resink (Basel Cantonal Hospital, Switzerland), who were involved in the experiments and actively participated in the preparation of this review. The authors are also very grateful to Dr. R. Latsis for his valuable assistance in the preparation of the manuscript.

REFERENCES

1. **Kamenskaja, N. L.,** Morphology of endothelium in human thoracic aorta, *Dokl. Akad. Nauk SSSR,* 83, 741, 1952.
2. **Vikhert, A. M. and Rosinova, V. N.,** On the endothelial lining of human arteries at genesis of the atherosclerotic plaque, in *Vessel Wall in Athero- and Thrombogenesis — Studies in the USSR,* Chazov, E. I. and Smirnov, V. N., Eds., Springer-Verlag, Berlin, 1982, chap. 1.
3. **Repin, V. S., Dolgov, V. V., Zaikina, O. E., Novikov, I. D., Antonov, A. S., Nikolaeva, M. A., and Smirnov, V. N.,** Heterogeneity of endothelium in human aorta — a quantitative analysis by scanning electron microscopy, *Atherosclerosis,* 50, 35, 1984.
4. **Davica, P. F., Reidy, M. A., Goode, T. B., and Bowyer, D. E.,** SEM in the evaluation of endothelium integrity of the fatty lesion atherosclerosis, *Atherosclerosis,* 25, 125, 1976.
5. **Schwartz, M. S. and Benditt, E. P.,** Clustering of replicating cells in aortic endothelium, *Proc. Natl. Acad. Sci. U.S.A.,* 73, 651, 1976.
6. **Tucker, R. W., Pardee, A. B., and Fujiwara, K.,** Centriole ciliation in related to quiescence and DNA synthesis in 3T3 cells, *Cell,* 17, 527, 1979.
7. **Albrecht-Buechler, G. and Bushnell, A.,** The ultrastructure of primary cilia in quiescent 3T3 cells, *Exp. Cell Res.,* 126, 427, 1980.
8. **Trelstad, R. L., Hay, E. D., and Revel, J.-P.,** Cell contact during early morphogenesis in the chick embryo, *Dev. Biol.,* 16, 78, 1967.
9. **Ito, S. and Winchester, R. J.,** The fine structure of the gastric mucosa in the bat, *J. Cell Biol.,* 16, 541, 1969.
10. **Zeligs, J. D.,** Association of centrioles with clusters of apical vesicls in mitotic thyroid epithelial cells. Are centrioles involved in directing secretion?, *Cell Tissue Res.,* 201, 11, 1979.
11. **Latta, H., Maunsbach, A. B., and Madden, S. C.,** Cilia in different segments of the rat nefron, *J. Biophys. Biochem. Cytol.,* 11, 248, 1961.
12. **Anderson, R. and Brenner, R.,** The formation of basal bodies (centrioles) in the rhesus monkey oviduct, *J. Cell Biol.,* 50, 10, 1971.
13. **Chambant-Guerin, A., Muller, P., and Rossinghol, B.,** Microtubules and protein secretion in rat lacrimal glands, *J. Biol. Chem.,* 253, 3870, 1978.
14. **Howell, S. L. and Tyhurst, M.,** Role of microtubules in the intracellular transport of growth hormone, *Cell Tissue Res.,* 190, 163, 1978.
15. **Sandoval, J. V., Bonifacio, J. S., Klausner, R. D., Hankart, M., and Wehland, J.,** Role of microtubules in the organization and localization of the Golgi apparatus, *J. Cell Biol.,* 99, 113, 1984.
16. **Shepro, D. and Amore, P. A.,** Endothelial cell metabolism, in *Advances in Microcirculation,* Vol. 9, Altura, B. M., Ed., S. Karger, Basel, 1980, 161.
17. **Repin, V. S., Dolgov, V. V., Zaikina, O. E., and Pozdnykov, O. M.,** Endothelium injury and polymorphism — a quantitative analysis by scanning electron microscopy, in *Vessel Wall in Athero- and Thrombogenesis — Studies in USSR,* Chazov, E. I. and Smirnov, V. N., Eds., Springer-Verlag, Berlin, 1982, 13.
18. **Antonov, A. S., Nikolaeva, M. A., Klueva, T. S., Romanov, Yu.A., Babaev, V. R., Bystrevskaya, V. B., Repin, V. S., and Smirnov, V. N.,** Primary culture of endothelial cells from athtrosclerotic human aorta. I. Identification, morphological and ultrastructural characteristics of two endothelial cell subpolations, *Atherosclerosis,* 59, 1, 1986.
19. **Babaev, V. R., Antonov, A. S., Romanov, Yu.A., Rukosuev, V. S. and Repin, V. S.,** Distribution of myosin in primary cell culture of human aortic intima, *Bull. Exp. Biol. Med. (USSR),* 3, 350, 1984.
20. **Haudenschild, C. C., Cotran, R. S., Gimbrone, M. A., and Folkman, J.,** Fine structure of vascular endothelium in culture, *J. Ultrastruct. Res.,* 50, 22, 1975.
21. **Stohr, M. and Goerttler, K.,** The Heidelberg flow analyser and sorter (HEIFAS) approach on the pre-screening of uterine cancer, *J. Histchem. Cytochem.,* 27, 564, 1979.
22. **Brodsky, V. V. and Uryvaeva, I. V.,** Cell polyploidy: its relation to tissue growth and function, *Int. Rev. Cytol.,* 4, 139, 1977.
23. **Higgins, P. J., Kessler, G. K., Nisselbaum, J. S., and Melamed, M. R.,** Characterisation and cell cycle kinetics of hepatocyte populations isolated from adult liver tissue by a nonenzymatic procedure, *J. Histchem. Cytochem.,* 33, 672, 1985.
24. **Fowlkes, B. J., Herman, C. J., and Cassidy, M.,** Flow microfluorometric system for screening gynecologic cytology specimens using propidium odide-fluorescein isothiocyanate, *J. Histochem. Cytochem.,* 24, 322, 1976.
25. **Ho, D. D., Rota, T. R., and Hirsch, M. S.,** Infection of human endothelial cells by human T-lymphotropic virus type 1, *Proc. Natl. Acad. Sci. U.S.A.,* 81, 7588, 1984.
26. **Tulloss, J. H. and Booyse, F. M.,** Effect of various agents and physical damage on giant cell formation in bovine aortic endothelial cultures, *Microvasc. Res.,* 16, 51, 1979.

27. **Gerrity, R. G., Richardson, M., Somer, J. B., Bell, F. P., and Schwartz, C. J.,** Endothelial cell morphology in areas of *in vivo* evans blue uptake in the aorta of young pigs, *Am. J. Pathol.,* 89, 313, 1977.

28. **Meyrick, B. and Reid, L.,** Endothelial and subintimal changes in rat hilar pulmonary artery during recovery from hypoxia. A quantitative ultrastructural study, *Lab. Invest.,* 42, 603, 1980.

29. **Booyse, F. M., Osikowicz, C., and Radek, J.,** Effect of nicotine on cultured bovine aortic endothelial cells, *Thromb. Res.,* 23, 169, 1981.

30. **Hansson, G. K., Bondjers, G., and Nilsson, L-A.,** Plasma protein accumulation in injured endothelial cells. Immunofluorescent localisation of IgG and fibrinogen in the rabbit aortic endothelium, *Exp. Mol. Pathol.,* 30, 12, 1979.

31. **Owens, G. K. and Schwartz, S. M.,** Alterations in vascular smooth muscle mass in the spontaneously hypertensive rat: role of cellular hypertrophy, hyperploidy and hyperplasia, *Circ. Res.,* 51, 280, 1982.

32. **Vlodavsky, I., Fielding, P. E., Johnson, L. K., and Gospodarovwicz, D.,** Inhibition of low density lipoprotein uptake in confluent endothelial cell monolayers correlates with a restricted surface receptor redistribution, *J. Cell Physiol.,* 100, 481, 1979.

33. **Preobrazhensky, S. N., Dolgov, V. V., Flegel, H. G., Repin, V. S., and Smirnov, V. N.,** ^{125}I-LDL uptake in rabbit arteries perfused *in situ*. Effect of HDL on intact and deendothelialized vessels, *Atherosclerosis,* 48, 147, 1983.

34. **Vasile, E., Simonescu, M., and Simonesku, N.,** Vizualization of the binding endocytosis and transcytosis of low density lipoprotein in the arterial endothelium *in situ, J. Cell Biol.,* 96, 1677, 1983.

35. **Santillan, G. G., Schun, J., Chan, S. I., and Bing, R. J.,** Binding and internalization of low density lipoproteins by perfused arteries, *Biochem. Biophys. Res. Commun.,* 95, 1410, 1980.

36. **Wiklund, O., Carew, T. E., and Steinberg, D.,** Role of the low density lipoprotein receptor in penetration of low density lipoprotein into rabbit aortic wall, *Arteriosclerosis,* 5, 135, 1985.

37. **Nicoll, A., Duffield, R., and Lewis, B.,** Flux of plasma lipoproteins into human arterial intima. Comparison between grossly normal and atheromatous intima, *Atherosclerosis,* 39, 229, 1981.

38. **Roberts, A. B., Lees, A. M., Lees, R. S., Strauss, H. W., Fallon, J. T., Taveras, J., and Kopiwoda, S.,** Selective accumulation of low density lipoproteins in damaged arterial wall, *J. Lipid Res.,* 24, 1160, 1983.

39. **Falcone, D. J., Hajjar, D. P., and Minick, C. R.,** Lipoprotein and albumin accumulation in reendothelialized and deendothelialized aorta, *Am. J. Pathol.,* 114, 112, 1984.

40. **Smith, E. B. and Ashall, Ch.,** Low-density lipoprotein concentration in interstitial fluid from human atherosclerotic lesions. Relation to theories of endothelial damage and lipoprotein binding, *Biochim. Biophys. Acta,* 784, 249, 1983.

41. **Mahley, R. W., Innerarity, T. L., Weisgraber, K. H., and Oh, S. Y.,** Altered metabolism (*in vivo* and *in vitro*) of plasma lipoproteins after selective chemical modification of lysine residues of the apoproteins, *J. Clin. Invest.,* 64, 743, 1980.

42. **Haberland, M. E., Fogelman, A. M., and Edwards, P. A.,** Specificity of receptor-mediated recognition of malondialdehyde-modified low density lipoproteins, *Proc. Natl. Acad. Sci. U.S.A.,* 79, 1712, 1982.

43. **Nagelkerke, J. F., Barto, K. P., and van Berkel, T. J. C.,** *In vivo* and *in vitro* uptake and degradation of acetylated low density lipoprotein by rat liver endothelial, kupffer, and parenchymal cells, *J. Biol. Chem.,* 254, 12224, 1983.

44. **Henriksen, T., Mahoney, E. M., and Steinberg, D.,** Enhanced macrophage degradation of low density lipoprotein previously incubated with cultured endothelial cells: recognition by receptors for acetylated low density lipoproteins, *Proc. Natl. Acad. Sci. U.S.A.,* 78, 6499, 1981.

45. **Gerrity, R. J.,** The role of the monocyte in atherogenesis, *Am. J. Pathol.,* 103, 181, 1981.

46. **Vlodavsky, I., Fielding, P. E., Fielding, C. J., and Gospodarowicz, D.,** Role of contact inhibition in the regulation of receptor-mediated uptake of low density lipoprotein in cultured vascular endothelial cells, *Proc. Natl. Acad. Sci. U.S.A.,* 75, 353, 1978.

47. **Stein, O. and Stein, Y.,** Bovine aortic endothelial cells display macrophage-like properties towards acetylated ^{125}I-labeled low density lipoprotein, *Biochem. Biophys. Acta,* 620, 631, 1980.

48. **Fogelman, A. M., Haberland, M. E., Seager, J., Hokom, M., and Edwards, P. A.,** Factors regulating the activities of the low density lipoprotein receptor and the scavenger receptor on human monocyte-macrophages, *J. Lipid. Res.,* 22, 1131, 1981.

49. **Goldstein, J. L., Ho, Y. K., Basu, S. K., and Brown, M. S.,** Binding site on macrophage that mediates uptake and degradation of acetylated low density lipoprotein, producing massive cholosterol deposition, *Proc. Natl. Acad. Sci. U.S.A.,* 76, 333, 1979.

50. **Anderson, R. G. W., Goldstein, J. L., and Brown, M. S.,** Fluorescence visualization of receptorbound low density lipoprotein in human fibroblasts, *J. Receptor Res.,* 1, 17, 1980.

51. **Pitas, R. E., Innerarity, T. L., Weinstein, J. N., and Mahley, R. W.,** Acetoacetylated lipoproteins used to distinguish fibroblasts from macrophages *in vitro* by fluorescence microscopy, *Arteriosclerosis,* 1, 177, 1981.

52. **Pitas, P. E., Boyles, J., Mahley, R. W., and Bissell, D. M.,** Uptake of chemically modified low density lipoproteins *in vivo* is mediated by specific endothelial cells, *J. Cell Biol.,* 100, 103, 1985.

53. **Anderson, R. G. W., Brown, M. S., Beisiegel, U., and Goldstein, J. L.,** Surface distribution and recycling of the low density lipoprotein receptor as visualized with antireceptor antibodies, *J. Cell Biol.,* 93, 523, 1982.

54. **Van Der Schroeff, J. G., Havekes, L., Emeis, J. J., Wijsman, M., van Der Meer, H., and Vermeer, B. J.,** Morphological studies on the binding of low-density lipoproteins and acetylated low-density lipoproteins to the plasma membrane of cultured monocytes, *Exp. Cell Res.,* 145, 95, 1983.

55. **Traber, M. G., Kallman, B., and Kayden, H. J.,** Localization of the binding sites of native and acetylated low-density lipoprotein (LDL) in human monocyte-derived macrophages, *Exp. Cell Res.,* 148, 281, 1983.

56. **Patelsky, J., Bower, D. E., and Howard, A. N., Jonnings, J. W., Thorne, C. J. R., and Gresham, G. A.,** Modification of enzyme activities in experimental atherosclerosis in the rabbit, *Atherosclerosis,* 12, 41, 1970.

57. **Aulinskas, T. H., Oram, J. F., Bierman, G. A., Coetzee, G. A., Gevers, W., and Westhuyzen, D. R.,** Retroendocytosis of low density lipoprotein by cultured human skin fibroblasts, *Arteriosclerosis,* 5, 45, 1985.

58. **Hamill, O. P., Marty, A., Neher, E., Sakmann, B., and Sigworth, F. J.,** Improved patch-clamp techniques for high-resolution current recording from cells and cell-free membrane patches, *Pfluegers Arch.,* 391, 85, 1981.

59. **Freudenberg, N., Riese, K.-H. and Freudenberg, M. A.,** *The Vascular Endothelial System,* Gustav Fischer Verlag, Stuttgart, 1983, 21.

60. **Northover, B. J.,** The membrane potential of vascular endothelial cells, *Adv. Microcirc,* 9, 135, 1980.

61. **Richter, R. and Halle, W.,** Membrane potential of vascular mono- and multinuclear endothelial cells cultured *in vitro, Experientia,* 39, 55, 1983.

62. **Sinback, C. N. and Shain, W.,** Chemosensitivity of single smooth muscle cells to acetylcholine, noradrenaline, and histamine *in vitro, J. Cell. Physiol.,* 102, 99, 1980.

63. **D'Amore, P. and Shepro, D.,** Stimulation of growth and calcium influx in cultured bovine aortic endothelial cells by platelets and vasoactive substances, *J. Cell. Physiol.,* 92, 177, 1977.

64. **Baenziger, N. L., Fogerty, F. J., Mertz, L. F., and Chernuta, L. F.,** Regulation of histamine-mediated prostacyclin synthesis in cultured human vascular endothelial cells, *Cell,* 24, 915, 1981.

65. **McIntyre, T. M., Zimmerman, G. A., Satoh, K., and Prescott, S. M.,** Cultured endothelial cells synthesize both platelet-activating factor and prostacyclin in response to histamine, bradykinin and ATP, *J. Clin. Invest.,* 76, 271, 1985.

66. **Ingerman-Wojensky, C., Silver, M. J., Smith, J. B., and Macarak, E.,** Bovine endothelial cells in culture produce thromboxane as well as prostacyclin, *J. Clin. Invest.,* 67, 1292, 1981.

67. **Crutchley, D. J., Ryan, J. W., Ryan, U. S., and Fisher, G. H.,** Bradykinin-induced release of prostacyclin and thromboxanes from bovine pulmonary artery endothelial cells: studies with lower homologs and calcium antagonists, *Biochim. Biophys. Acta,* 751, 99, 1983.

68. **Bounassisi, V. and Venter, J. C.,** Hormone and neurotransmitter receptors in an established vascular endothelial cell line, *Proc. Natl. Acad. Sci. U.S.A.,* 73, 1612, 1976.

69. **Berridge, M. J., Dawson, R. M. C., Downes, C. P., Heslop, G. P., and Irvine, R. F.,** Changes in the levels of inositol phosphates after agonist-dependent hydrolysis of membrane phosphoinositides, *Biochem. J.,* 212, 473, 1983.

70. **Schrey, M. P. and Rubin, R. P.,** Characterization of a calcium-mediated activation of arachidonic acid turnover in adrenal phospholipids by corticotropin, *J. Biol. Chem.,* 254, 11234, 1979.

71. **Rubin, R. P., Sink, L. G., and Freer, R. J.,** On the relationship between formylmethionyl-leucylphenylalanine stimulation of arachidonyl phosphatidylinositol turnover and lysosomal enzyme secretion by rabbit neutrophils, *Mol. Pharmacol.,* 19, 31, 1981.

72. **Rubin, R. P., Kelly, K. L., Halenda, S. P., and Laychock, S. G.,** Arachidonic acid metabolism in rat pancreatic acinar cells: calcium-mediated stimulation of the lipoxigenase system, *Prostaglandins,* 24, 179, 1982.

73. **Rittenhouse-Simmons, S.,** Differential activation of the platelet phospholipases by thrombin and ionophore A23187, *J. Biol. Chem.,* 256, 4153, 1981.

74. **Putney, J. W., Jr., Burgess, G. M., Godfrey, P. P., and Aub, D. L.,** Messages of the phosphoinositide effect, in *Inositol and Phosphoinositides: Metabolism and Regulation,* Bleasdale, J. E., Eichberg, J., and Hauser, G., Eds., Humana Press, Clifton, N. J., 1985, 337.

75. **Berridge, M. J. and Irvine, R. F.,** Inositol triphosphate and calcium mobilization, in *Inositol and Phosphoinositides: Metabolism and Regulation,* Bleasdale, J. E., Eichberg, J., and Hauser, G., Eds., Humana Press, Clifton, N.J., 1985, 351.

76. **Burgess, G. M., Godfrey, P. P., McKinney, J. S., Berridge, M. J., Irvine, R. F., and Putney, J. W., Jr.,** The second messenger linking receptor activation to internal Ca release in liver, *Nature (London),* 209, 63, 1984.

77. **Van den Bosch, H.,** Intracellular phospholipases A, *Biochim. Biophys. Acta,* 604, 191, 1980.
78. **Brotherton, A. F. and Hoak, J. C.,** Role of Ca^{2+} and cyclic AMP in the regulation of the production of prostacyclin by the vascular endothelium, *Proc. Natl. Acad. Sci. U.S.A.,* 79, 495, 1982.
79. **Irvine, R. F.,** How is the level of free arachidonic acid controlled in mammalian cells?, *Biochem. J.,* 204, 3, 1982.
80. **Langille, B. L. and Adamson, S. L.,** Relationship between blood flow direction and endothelial cell orientation at arterial branch sites in rabbit and mice, *Circ. Res.,* 48, 481, 1981.
81. **Levesque, M. J., Liepsch, D., Moravec, S., and Nerem, R. M.,** Correlation in endothelial cell shape and wall sheer stress in a stenosed dog aorta, *Arteriosclerosis,* 6, 220, 1986.
82. **De Chastonay, C., Gabbiani, G., Elemer, G., and Huttner, I.,** Remodeling of the rat aortic endothelial layer during experimental hypertensia: changes in replication rate, cell density, and surface morphology, *Lab. Invest.,* 48, 45, 1983.
83. **Shepro, D., Welles, S. L., and Hecjtman, H. B.,** Vasoactive agonists prevent erythrocyte extravasation in thrombocytopenic hamsters, *Thromb. Res.,* 53, 421, 1984.
84. **Welles, S. L., Shepro, D., and Heichtman, H. B.,** Vasoactive amines modulate actin cables (stress fibres) and surface area in cultured bovine endothelium, *J. Cell. Physiol.,* 123, 337, 1985.
85. **Sandra, A., Bar, R. S., Dolash, S., Marshall, S. J., Kaduce, T. L., and Spector, A. A.,** Morphological alterations in cultured endothelial cells induced by arachidonic acid, *Exp. Cell. Res.,* 158, 484, 1985.
86. **Galdal, K. S., Evensen, S. A., and Brosstad, F.,** Effects of divalent cations and various vasoactive and haemostatically active agents on the integrity of monolayers of cultured human endothelial cells, *Thromb. Res.,* 21, 273, 1981.
87. **Galdal, K. S., Evensen, S. A., and Nilsen, E.,** Thrombin-induced shape changes of cultured endothelial cells: metabolic and functional observations, *Thromb. Res.,* 32, 57, 1983.
88. **Galdal, K. S., Evensen, S. A., and Brosstad, F.,** Effects of divalent cations and various vasoactive and haemostatically active agents on the integrity of monolayers of cultured human endothelial cell. *Thromb. Res.,* 21, 273, 1981.
89. **Nawroth, P. P., Stern, D. M., Kaplan, K. L., and Nossel, H. L.,** Prostaglandin production by perturbed bovine aortic endothelial cells in cultre, *Blood,* 64, 801, 1984.
90. **Boeynaems, J. M., Galand, N., and Ketelbant, P.,** Prostacyclin production by the deendothelialized rabbit aorta, *J. Clin. Invest.,* 76, 7, 1985.
91. **Goldsmith, J. C.,** Contribution of the subendothelium to prostacyclin release after vascular injury, *J. Lab. Clin. Med.,* 100, 574, 1982.
92. **Montesano, R. and Orci, L.,** Tumor-promoter phorbol esters induce angiogenesis *In vitro, Cell,* 42, 469, 1985.
93. **Gross, J. L., Moscatelli, D., and Rifkin, D. B.,** Increased capillary endothelial cell protease activity in response to angiogenic stimuli *in vitro, Proc. Natl. Acad. Sci. U.S.A.,* 80, 2623, 1983.
94. **Gross, J. L., Moscatelli, D., Jaffe, E. A., and Rifkin, D. B.,** Plasminogen activator and collagenase production by cultured capillary endothelial cells, *J. Cell Biol.,* 95, 974, 1982.
95. **Shasby, D. M., Van Benthysen, K. M., and Tabe, R. M.,** Granulocytes mediate acute edematous lung injury in rabbits and isolated rabbit lungs perfused with phorbol myristate acetate: role of oxygen radicals, *Am. Rev. Respir. Dis.,* 125, 443, 1982.
96. **Mojarad, M., Cox, C. P., and Said, S. I.,** The platelet activating factor (PAF) and acute lung injury, in *Pulmonary Circulation and Acute Lung Injury,* Said, S. I., Ed., New York, 1982, 375.
97. **Johnson, K. J. and Ward, P. A.,** Acute and progressive lung injury after contact with phorbol myristate acetate, *Am. J. Pathol.,* 107, 29, 1982.
98. **Gospodarovicz, D., Brown, K. K., Birdwell, C. R., and Zetter, B. R.,** Control of proliferation of human vascular endothelial cells. Characterization of the response of human umbilical vein endothelial cells to fibroblast growth factor, epidermal growth factor, and thrombin, *J. Cell Biol.,* 77, 774, 1978.
99. **Gospodarowicz, D., Moran, J., Braun, D., and Birdwell, C.,** Clonal growth of bovine vascular endothelial cells: fibroblast growth factor as a survival agent, *J. Cell Biol.,* 73, 4120, 1976.
100. **Frater-Schroder, M., Muller, G., Birchmeier, W., and Bohlen, P.,** Transforming factor-beta inhibits proliferation, *Biochem. Biophys. Res. Commun.,* 137, 295, 1986.
101. **Sorgente, N., Bullard, D. L., Jakovljevic, L., and Dorey, K.,** Endogenous regulation of endothelial cell proliferation, *Cell Tissue Kinet.,* 17, 573, 1984.
102. **Schafer, A. I., Gimbrone, M. A., and Handin, R. I.,** Regulation of endothelial cell function by cyclic nucleotide, in *Biology of Endothelial Cells,* Jaffe, E. A., Ed., Martinus Nijhoff, Boston, 1984, 248.
103. **Fielding, P. E. and Fieldins, C. J.,** Role of lipoproteins in the regulation of cultured endothelial cell cholesterol metabolism, in *Biology of Endothelial Cells,* Jaffe, E. A., Ed., Martinus Nijhoff, Boston, 1984, 356.
104. **Johnson, A. R.,** The metabolism of vasoactive peptide by human endothelial cells, in *Biology of Endothelial Cells,* Jaffe, E. A., Ed., Martinus Nijhoff, Boston, 1984, 302.

105. **Ryan, U. S. and Ryan, J. W.,** Cell biology of pulmonary endothelium, *Circulation,* 70, 46, 1984.
106. **Gardal, K. S., Lyberg, T., Evensen, S. A., Nilsen, E., and Prydz, H.,** Thrombin induces thromboplastin synthesis in cultured vascular endothelial cells. *Thromb. Haemost.,* 54, 373, 1985.
107. **Shimada, K. and Ozawa, T.,** Release of heparan sulfate proteoglycans from cultured aortic endothelial cells by thrombin, *Thromb. Res.,* 39, 387, 1985.
108. **Lee, J. C. and Sponenberg, D. P.,** Role of alpha$_1$-adrenoceptors in norepinephrine-induced cardiomyopathy, *Am. J. Pathol.,* 121, 316, 1985.
109. **Waldenstrom, P., Hjalmarson, C., and Thornell, L.,** A possible role of norepinephrine in the development of myocardial infarction, *JAMA,* 95, 43, 1978.
110. **Makhmudov, R. M., Dolgov, V. V., Voyno-Yasenetskaya, T. A., Ivanov, V. O., Preobrazhensky, S. N., Mamedov, Y. D., and Repin, V. S.,** Incorporation of ^{125}I-low density lipoproteins into aorta under the impact of adrenaline, *Cardiologia (Moscow),* 24, 96, 1984.
111. **Nadler, E., Nijjar, M. S., and Oron, Y.,** Phosphoinositide breakdown in isolated rat parotid membranes. Stimulation by cholinergic and alpha$_1$-adrenergic agonists, *FEBS Lett.,* 178, 278, 1984.
112. **Makarski, J. S.,** Stimulation of cyclic AMP production by vasoactive agents in cultured bovine aortic and pulmonary artery endothelial cells, *In Vitro,* 17, 450, 1981.
113. **Berridge, M. J. and Irvine, R. F.,** Inositol trisphosphate, a novel second messenger in cellular signal transduction, *Nature (London),* 321, 315, 1984.
114. **Archer, C. B., Frohlich, W., Page, C. P., Paul, W., Morley, J., and McDonald, D. M.,** Synergistic interaction between prostaglandins and PAF-acether in experimental animals and man, *Prostaglandins,* 27, 495, 1984.
115. **Yoshimura, K., Nezu, E., and Yoneyama, T.,** Augmentation of isoproterenol-stimulated tissue cyclic AMP level by cholinergic agonists in rat parotid gland, *Jpn. J. Physiol.,* 35, 765, 1985.
116. **Reidy, M. A.,** Biology of disease. A reassessment of endothelial injury and arterial lesion formation, *Lab. Invest.,* 53, 513, 1985.
117. **Christensen, B. C.,** Repair of arterial tissue. A scanning electron microscopic (SEM) and light microscopic study on the endothelium of rabbit thoracic aorta following noradrenaline in toxic doses, *Virchows Arch. A,* 363, 33, 1974.
118. **Zimmerman, M. and McGeachie, J.,** Quantitation of the relationship between aortic endothelial intercellular morphology and permeability to albumin, *Atherosclerosis,* 59, 277, 1986.
119. **Robertson, A. L. and Khairallah, P. A.,** Arterial endothelial permeability and vascular disease. The "trap door" effect, *Exp. Mol. Pathol.,* 18, 241, 1973.
120. **Dolgov, V. V., Voyno-Yasenetskaya, T. A., and Repin, V. S.,** Modeling and quantitative analysis of risk-factors of atherosclerosis using perfused rabbit arteries, in *Human Atherosclerosis,* Vol. 2, (Soviet Union Reviews, Section Cardiology), Chazov, E. I. and Smirnov, V. N., Eds., Gordon and Brich, in press.
121. **Hajjar, D. P. and Weksler, B. B.,** Metabolic activity of cholesteryl esters in aortic smooth muscle cells is altered by prostaglandin I$_1$ and E$_2$, *J. Lipid Res.,* 24, 1176, 1983.
122. **Hajjar, D. P.,** Prostaglandins and cyclic nucleotides. Modulators of arterial cholesterol metabolism, *Biochem. Pharmacol.,* 34, 295, 1985.
123. **Tertov, V. V., Orekhov, A. N., Repin, V. S., and Smirnov, V. N.,** Dibutyryl cyclic AMP decreases proliferative activity and the cholesteryl ester content in cultured cells of atherosclerotic human aorta, *Biochem. Biophys. Res. Commun.,* 109, 1228, 1982.
124. **Stout, R. W.,** Relative intensity of sterol synthesis in cultured rat aortic smooth muscle cells. Effect of dibutyryl cyclic AMP, *Diabetologia,* 15, 323, 1978.
125. **Stout, R. W.,** Cyclic AMP: a potent inhibitor of DNA synthesis in cultured arterial endothelial and smooth muscle cells, *Diabetologia,* 22, 51, 1982.
126. **Lundholm, L., Jacobsson, L., and Andersson, R.,** Relationship between cyclic AMP and protein synthesis in atherosclerotic pig aorta, in *Vascular Neuroeffector Mechanisms,* Bevan, J. A., Ed., Raven Press, New York, 1980, 257.
127. **Auerbach, R., Alby, L., Grieves, J., Joseph, J., Lindgren, C., Morrissey, L., Sidky, Y., Tu, M., and Watt, S.,** Monoclonal antibodies against angiotensin-converting enzyme: its use as a marker for murine, bovine and human endothelial cells, *Proc. Natl. Acad. Sci. U.S.A.,* 79, 7891, 1982.
128. **Danilov, S., Allikmets, E., Sakharov, I., Dukhanina, E., and Trakht, I.,** Monoclonal antibodies to human lung angiotensin converting enzyme, *Bull. Exp. Biol. Med. (USSR),* in press.
129. **Cushman, D. W. and Ondetti, M. A.,** Inhibitors of angiotensin-converting enzyme, in *Progress in Medicinal Chemistry,* Ellis, G. and West, G., Eds., Elsevier, Amsterdam, 1980, 42.
130. **Danilov, S., Allikmets, E., Sakharov, I., Dukhanina, E., and Trakht, I.,** Monoclonal antibodies to human lung angiotensin converting enzyme, *Bull. Exp. Biol. Med. (USSR),* in press.
131. **Straus, W.,** Imidasole increases the sensitivity of cytochemical reaction for peroxidase with diaminoenzidine at neutral pH, *J. Histochem. Cytochem.,* 30, 491, 1982.

132. **Ryan, J. and Ryan, U.,** Biochemical and morphological aspects of the actions and inactivations of kinase and angiotensins, in *Enzymatic Release of Vasoactive Peptides,* Gross, H. and Vogel, G., Eds., Raven Press, New York, 1980, 259.

133. **Fishman, A.,** Dynamics of the pulmonary circulations, in *Handbook of Physiology,* Vol. 2, Hamilton, W. F. and Dow, P., Eds., American Physiological Society, Washington, D.C., 1963, 1667.

134. **Barba, L., Caldwell, P., Downie, G., Camussi, G., Brentjens, J., and Andres, G.,** Lung injury mediated by antibodies to endothelium, *J. Exp. Med.,* 158, 2141, 1983.

135. **Danilov, S., Muzykantov, V., martynov, A., Sakharov, I., Trakht, I., Medvedev, O., and Smirnov, V.,** unpublished data 1986.

136. **Cornhill, S. F., Barret, W. A., Herderick, E. E.,Mahley, R. W., and Fry, D. L.,** Topographic study of sudanophilic lesions in cholesterol-fed minipigs by image analysis, *Arteriosclerosis,* 5, 415, 1985.

137. **Ridolfi, R. L. and Hutchiks, G. M.,** The relationship between coronary artery lessons and myocardial infacts: ulceration of atherosclerotic plaque precipitating coronary thrombosis, *Am. Heart J.,* 93, 568, 1977.

138. **Horie, T., Sekigicki, M., and Hirosawa, K.,** Coronary thrombosis in pathogenesis of acute myocardial infarction. Histopathological study of coronary necropsied cases using serial section, *Br. Heart J.,* 40, 153, 1978.

139. **Hollman, J., Gruentzig, A. R., Douglas, J. S., King, S. B., Ischinger, T., and Meier, B.,** Acute occlusion after precutaneous transluminal coronary angioplasty — a new approach, *Circulation,* 68, 725, 1983.

140. **Walts, A. E., Fishbein, M. C., Sustaita, H., and Matloff, S. M.,** Ruptured atheromatous plaques in saphenous vein coronary artery bypass grafts: a mechanism of acute, thrombolitic, late graft occlusion, *Circulation,* 65, 197, 1982.

141. **Morton, L. F. and Barnes, M. J.,** Collagen polymorphism in the normal and diseased blood vessel wall, *Atherosclerosis,* 42, 41, 1982.

142. **McCullagh, K. G., Duance, V. C., and Bishop, K.,** The distribution of collagen types 1, 111, and V (AB) in normal and atherosclerotic human aorta, *J. Pathol.,* 130, 45, 1980.

143. **Koteliansky, V. E., Glukhova, M. A., Benjanan, M. A., Smirnov, V. N., Filimonov, V. V., Zalite, O. M., and Benyaminov, S. Y.,** A study of the structure of fibronectin, *Eur. J. Biochem.,* 119, 619, 1981.

144. **Torchilin, V. P.,** Immobilized enzymes and the immobilization principles for drug targeting, in *Targeted Drugs,* Goldberg, E., Ed., John Wiley & Sons, New York, 1983, 127.

145. **Papahadjopoulos, D., Wilson, T., and Taber, R.,** Liposomes as vesicles for cellular incorporation of biological active macromolecules, *In Vitro,* 16, 45, 1980.

146. **Ihler, G. and Glew, R.,** Enzyme-loaded erythrocytes, in *Biomedical Applications of Immobilized Enzymes,* Chang, T. M. S., Ed., Plenum Press, New York, 1977, 219.

147. **Samokhin, G. P., Smirnov, M. D., Muzykantov, V. R., Domogatsky, S. P., and Smirnov, V. N.,** Targeting of red blood cells to collagen-coated surface, *FEBS Lett.,* 154, 257, 1983.

148. **Updike, S. S., Wakamiya, R. T., and Lightfoot, E. N.,** Asparaginase entrapped in red blood cells, *Science,* 193, 681, 1976.

149. **Kinoshita, K., Jr. and Tsong, T. Y.,** Survival of sucrose-loaded erythrocytes in the circulation, *Nature (London),* 272, 258, 1978.

150. **Poste, G.,** Liposomes targeting *in vivo:* problems and opportunities, *Biol. Cell.,* 47, 19, 1983.

151. **Poznansky, M. J. and Juliano, R. L.,** Biological approaches to the controlled delivery of drugs: a critical review, *Pharmacol. Rev.,* 36, 277, 1984.

152. **Torchilin, V. P., Berdichevsky, V. R., Barsukov, A. A., and Smirnov, V. N.,** Coating liposomes with protein decreases their capture by macrophages, *FEBS Lett.,* 111, 184, 1980.

153. **Hunt, C. A.,** Liposome disposition *in vivo, Biochim. Biophys. Acta,* 719, 450, 1982.

154. **Colman, R. W.,** Surface-mediated defense reactions. The plasma contact and activation system, *J. Clin. Invest.,* 73, 1249, 1984.

155. **Shaub, R. G., Simmons, C. A., Koets, M. H., Romano, P. J., and Stewart, G. J.,** Early events in the formation of a venous thrombus following local trauma and stasis, *Lab. Invest.,* 51, 218, 1984.

156. **Fantone, J. C. and Ward, P. A.,** Role of oxygen-derived free radicals and metabolites in leukocytes-dependent inflammatory reactions, *Am. J. Pathol.,* 107, 397, 1982.

157. **Weiss, S. J. and LoBuglio, A. F.,** Phagocyte-generated oxygen metabolites and cellular injury, *Lab. Invest.,* 47, 5, 1982.

158. **Weiss, S. J., Young, J., LoBuglio, A., and Slivka, A.,** Role of hydrogen peroxide in neutrophil-mediated destruction of cultured endothelial cells, *J. Clin. Invest.,* 68, 714, 1981.

159. **Harlan, J. M., Killen, P. D., Harker, L. A., and Striker, G. E.,** Neutrophil-mediated endothelial injury, *in vitro, J. Clin. Invest.,* 68, 1394, 1981.

160. **Varani, J., Fligiel, S. E. G., Till, G. O., Kunkel, R. G., Ryan, U. S., and Ward, P. A.,** Pulmonary endothelial cell killing by human neutrophils. Possible involvement of hydroxyl radical, *Lab. Invest.,* 53, 656, 1985.

161. **Badwey, J. A. and Karnovsky, M. L.,** Active oxygen species and the functions of phagocytic leukocytes, *Annu. Rev. Biochem.,* 49, 695, 1980.
162. **Nathan, C. and Cohn, Z.,** Role of oxygen-dependent mechanisms in antibody-induced lysis of tumor cells by activated macrophages, *J. Exp. Med.,* 152, 198, 1980.
163. **Nathan, C. F., Silverstein, S. C., Brukner, L. H., and Cohn, Z. A.,** Extracellular cytolysis by activated macrophages and granulocytes, *J. Exp. Med.,* 149, 100, 1979.
164. **Fligiel, S. E. G., Lee, E. C., McCoy, L. P., Johnson, K. J., and Varani, J.,** Protein degradation following treatment with hydrogen peroxide, *Am. J. Pathol.,* 115, 418, 1984.
165. **Weiss, S. J., Peppin, G., Ortiz, X., Ragsdale, C., and Test, S. T.,** Oxidative autoactivation of latent collagenase by human neutrophils, *Science,* 227, 747, 1985.
166. **Shingu, M. and Nobunaga, M.,** Chemotactic activity generated in human serum from the fifth component of complement by hydrogen peroxide, *Am. J. Pathol.,* 117, 201, 1984.
167. **Canoso, R. T., Rodvein, R., Scoon, K., and Levine, P. H.,** Hydrogen peroxide and platelet function, *Blood,* 43, 645, 1974.
168. **Del Principe, D., Menichelli, A., De Matteis, W., Di Corpo, M. L., Di Giulio, S., and Finazzi-Agro, A.,** Hydrogen peroxide has a role in the aggregation of human platelets, *FEBS Lett.,* 185, 142, 1985.
169. **Hammond, B., Kontos, H. A., and Hess, M. L.,** Oxygen radicals in the adult respiratory distress syndrome, in myocardial ishemia and reperfusion injury, and in cerebral vascular damage, *Can. J. Physiol. Pharmacol.,* 63, 173, 1984.
170. **Renah, A., Johnson, K. J., Wiggins, R. C., Kunkel, R. G., and Ward, P. A.,** Evidence for the role of oxygen radicals in acute nephrotoxic nephritis, *Lab. Invest.,* 51, 396, 1984.
171. **Freeman, B. A., Turrens, J. F., Mirza, Z., Crapo, J. D., and Young, S. L.,** Modulation of oxidant lung injury by using liposome-entrapped superoxide dismutase and catalase, *Fed. Proc.,* 44, 2591, 1985.
172. **Ward, P. A., Till, G. O., Hatherill, G. R., Annesley, T. M., and Kunkel, R. G.,** Systemic complement activation, lung injury, and products of lipid peroxidation, *J. Clin. Invest.,* 76, 517, 1985.
173. **Agar, N. S., Sadizadeh, S. M. N., Hallaway, P. E., and Eaton, G. W.,** Erythrocyte catalase: a somatic oxidant defense?, *J. Clin. Invest.,* 77, 319, 1986.
174. **Toth, K. M., Gilford, D. P., Berger, E. M., White, C. W., and Repine, J. E.,** Intact human erythrocytes prevent hydrogen peroxide-induced damage to isolated perfused rat lungs and cultured bovine pulmonary artery endothelial cells, *J. Clin. Invest.,* 74, 292, 1984.
175. **Van Asbeck, B. S., Hoidal, J., Vercellotti, C. M., Schwartz, B. A., Moldow, C. F., and Jacob, H. S.,** Protection against lethal hyperoxia by insufflation of erythrocytes: role of red cell glutathione, *Science,* 227, 756, 1985.
176. **Gregoriadis, G.,** Targeting of drugs, *Nature (London),* 265, 131, 1977.
177. **Spragg, R. G., Hinshaw, D. B., Hyslop, P. A., Schraufstatter, I. U., and Cochrane, C. G.,** Alterations in adenosine triphosphate and energy charge in cultured endothelial and P388D1 cells after oxidant injury, *J. Clin. Invest.,* 76, 1471, 1985.
178. **Shasby, D. M., Lind, S. E., Shasby, S. S., Goldsmith, J. C., and Hunninghake, G. W.,** Reversible oxidant-induced increases in albumin transfer across cultured endothelium: alterations in cell shape and calcium homeostasis, *Blood,* 65, 605, 1985.
179. **Whorton, A. R., Montgomery, M. E., and Kent, R. S.,** Effect of hydrogen peroxide on prostaglandin production and cellular integrity in cultured porcine aortic endothelial cells, *J. Clin. Invest.,* 76, 295, 1985.
180. **Ager, A. and Gordon, J. L.,** Differential effects of hydrogen peroxide on indices of endothelial cell function, *J. Exp. Med.,* 159, 592, 1984.
181. **Romagnani, S., Amerigogna, F. A., Giudzu, G. M., and Ricci, M.,** Rosette formation with protein A-coated erythrocytes: a method for detecting both IgG-bearing cells and another subset of human peripheral blood B lymphocytes, *J. Immunol. Meth.,* 33, 11, 1980.
182. **Shaub, R. G., Simmons, C. A., Koets, M. H., Romano, P. J., and Stewart, G. J.,** Early events in the formation of a venous thrombus following local trauma and stasis, *Lab. Invest.,* 51, 218, 1984.
183. **Chazov, E. I., Samokhin, G. P., Smirnov, M. D., Muzykantov, V. R., Domogatsky, S. P., and Smirnov, V. N.,** Targeted transport of red blood cells to denuded vascular areas, *Circulation,* 68(II), 83, 1983.

Chapter 38

ENDOTHELIAL SEEDING OF SYNTHETIC VASCULAR GRAFTS

Linda M. Graham, James C. Stanley, and William E. Burkel

TABLE OF CONTENTS

I. INTRODUCTION

The thrombogenicity of synthetic vascular grafts contributes to the failure of many clinical arterial reconstructive procedures. Evidence of this is the 2-year patency rate of approximately 50% with ePTFE grafts in the femoro-popliteal position.[1] Early failures are often due to the thrombogenicity of the graft surface, and late failures are usually secondary to anastomotic intimal hyperplasia or progression of downstream atherosclerosis. The development of a less thrombogenic synthetic conduit would be advantageous in decreasing thrombosis, especially under conditions of low flow, and perhaps in decreasing late failures by minimizing platelet activation and other systemic perturbations caused by vascular prostheses. Considerable research efforts in the past have been directed toward the development of less thrombogenic grafts by altering the chemical and physical character of the prostheses. In addition, attempts have been made to stimulate spontaneous endothelial cell coverage of grafts by fabricating prostheses which facilitate endothelial cell ingrowth and surfacing. Recently, experimental studies have utilized endothelial cell seeding of grafts at the time of implantation, or in vitro seeding of prostheses prior to implantation, to promote the early evolution of an endothelial cell surface.

II. DEVELOPMENT OF ENDOTHELIAL CELL SEEDING

In 1978 Herring and colleagues from Indiana University described an endothelial cell seeding technique in which canine endothelium was derived by passing a steel-wool pledget through a segment of excised vein.[2] Endothelial cells obtained in this manner were then washed from the pledget, collected, and mixed with blood used to preclot a 6-mm I.D. knitted Dacron® prosthesis which was implanted into the infrarenal aorta in a canine model. When the study was concluded 4 weeks after graft implantation, thrombus-free luminal surface coverage was approximately 76% in the seeded grafts, compared to 22% in unseeded control grafts. The lining of the autologous cell-seeded grafts was shown to be endothelium by identification of Weibel-Palade bodies and by immunofluorescent microscopy utilizing antibodies to Factor VIII-related antigen.[3,4]

Experiments at the University of Michigan demonstrated the feasibility of harvesting adult canine endothelial cells using 0.1% trypsin and 0.5% collagenase solutions.[5] In early studies cells harvested by this method were cultivated in vitro for 2 weeks, then used to seed 6-mm I.D., 25- to 30-cm long, knitted Dacron® grafts which were implanted as throacoabdominal bypasses in a canine model.[6] Seeding density was approximately 8000 to 10,000 cells per square centimeter of graft surface. Two weeks after implantation the surface of unseeded grafts consisted of fibrin and platelets, but endothelium covered over 60% of the luminal surface of grafts seeded with autologous cells. Four weeks postimplantation the fibrin-platelet layer in unseeded grafts had increased in thickness, and endothelium was present only as pannus ingrowth which extended to a maximum of 10 mm from each anastomosis (Figure 1A). Seeded conduits at the same time had developed endothelial cell coverage over at least 80% of the luminal surface (Figure 1B). Immunofluorescent microscopy using antibodies to Factor VIII-related antigen confirmed that the cellular surface of seeded grafts was endothelium. In further studies at the University of Michigan, seeding of endothelial cells into Dacron® grafts immediately following enzymatic harvesting produced essentially identical results to those obtained using cultured cells.[7]

Sequential studies undertaken to follow the development of the endothelial surface in seeded grafts demonstrated no difference between seeded and unseeded knitted Dacron® grafts by gross inspection, light microscopy, or scanning electron microscopy (SEM) at 1 or 2 days after implantation.[8] However, by 4 days isolated patches of endothelium could be identified on seeded grafts using SEM (Figure 2). By 7 days the surface coagulum was

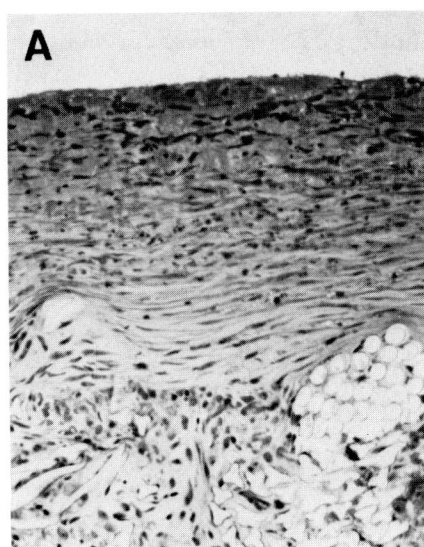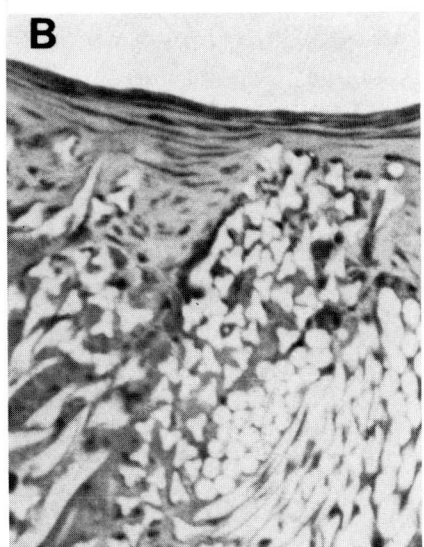

FIGURE 1. Light micrographs of Dacron® grafts removed 4 weeks postimplantation. (A) The luminal surface is covered by a thick fibrin-platelet coagulum with underlying smooth muscle cells. The tissue between graft fibrils consists of fibroblasts, collagenous fibers, macrophages, and giant cells. (Methylene blue-basic fuchsin; magnification × 150.) (B) The surface of the seeded graft is covered by typical endothelial cells. The subendothelial tissue is composed of collagenous fibers and smooth muscle cells, while the tissue in the interstices of the graft is similar to that of the nonseeded graft. (Methylene blue-basic fuchsin; magnification × 150.)

thicker in control grafts, but a cellular luminal lining could be identified by light microscopy in isolated areas of seeded grafts. By 14 days postimplantation unseeded grafts demonstrated progressive accumulation of fibrin and platelets with endothelial cells present only in the 1- to 2-mm of pannus ingrowth adjacent to anastomoses. However, further thinning of the fibrin-platelet layer occurred in seeded grafts, and endothelial cells covered 60 to 70% of the luminal surface. By 28 days endothelium was present on unseeded grafts only as limited anastomotic pannus ingrowth, but on seeded grafts endothelium covered over 80 to 90% of the luminal surface. Beneath this endothelium was an organized subendothelial tissue composed of collagen and smooth muscle cells (Figure 3).

The stability of the endothelial surface on seeded Dacron® grafts was verified in studies extending to 12 months.[9] The mean thickness of the inner capsule of seeded grafts was 10, 33, and 58 μm at 4, 6, and 12 months, respectively. This contrasted with the thicker inner capsule of unseeded grafts which averaged 46, 155, and 141 μm at the same time periods. The continuing increase in thickness of the inner capsule in seeded grafts raised concern as to whether this thickening was self limited, or would continue to progress with eventual luminal compromise. Progressive inner capsule thickening may be related to release of smooth muscle cell mitogens from surface cells or it may simply be a reflection of transinterstitial tissue ingrowth through a porous graft as part of the healing process. This phenomenon and means to control it remain to be better defined.

Endothelial cell seeding of ePTFE thoracoabdominal grafts in the canine model was shown to produce a luminal monolayer of endothelium.[10] Coverage averaged 64% of the inner surface at 2 weeks, and over 90% at 4 weeks in seeded grafts (Figure 4). At 4 weeks postimplantation, unseeded ePTFE grafts exhibited a very thin coagulum lining the luminal surface (Figure 5A). The inner capsule of endothelial cell seeded grafts was only one or two cells thick (Figure 5B). In contrast to Dacron® grafts, subendothelial tissue in seeded

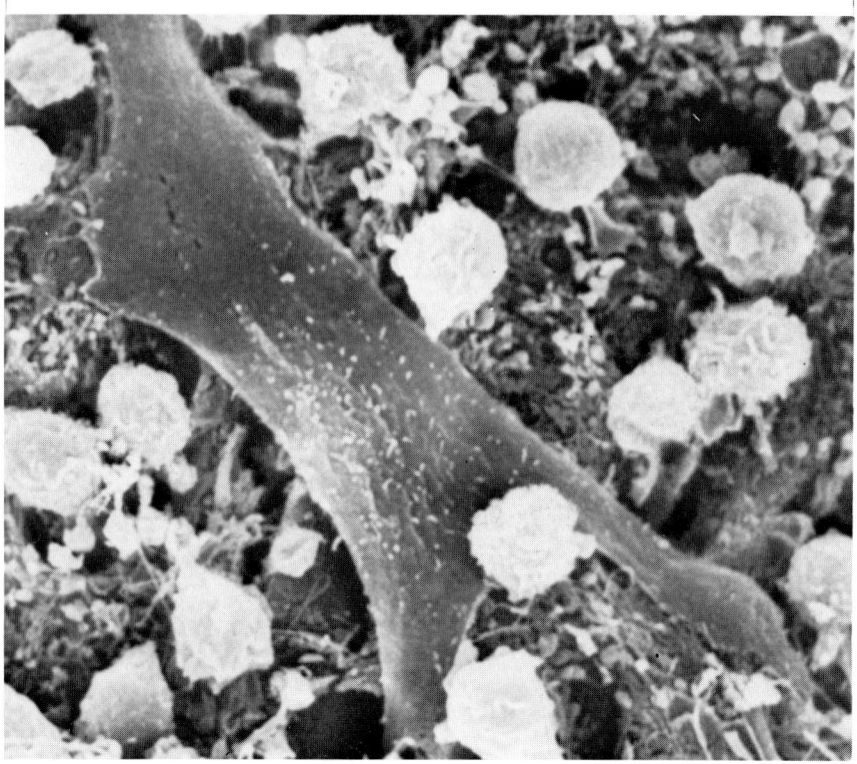

FIGURE 2.   Scanning electron micrograph of a single endothelial cell on the surface of a seeded Dacron®
graft 4 days after implantation. The cell lies on the bed of fibrin and platelets and is surrounded by white
blood cells. (Magnification × 3200.)

ePTFE grafts was scant at 4 weeks. However, in other studies carried out to 16 weeks,[11]
subendothelial mesenchymal tissue was well developed (Figure 6). Transmission electron
microscopy demonstrated that the lining of control grafts was composed of whole and
degranulated platelets embedded in fibrin, with no cellular tissue as part of the lining (Figure
7A). Seeded grafts were surfaced by endothelial cells with attenuated cytoplasmic extensions
and interdigitating tight junctions (Figure 7B).

III. FUNCTIONAL STUDIES

After the anatomic success of endothelial seeding in the canine model, experiments were
undertaken to determine the functional integrity of the endothelial lining and, perhaps more
importantly, the effects of its antithrombogenic properties on vascular graft patency. The
lining of endothelial cell-seeded grafts was shown to produce prostacyclin. Clagett demonstrated that 12 weeks after implantation of seeded thoracoabdominal grafts in dogs, the
luminal production of prostacyclin, as measured by its stable metabolite 6-keto-PGF$_{1\alpha}$, was
0.90 ± 0.72 ng/cm^2, which was not significantly different from normal aortic production.[12]
However, this level was significantly greater than the 0.15 ± 0.11 ng/cm^2 produced in
unseeded grafts.

A. Platelet Studies

In canine studies, implantation of knitted Dacron® or ePTFE thoracoabdominal grafts
depressed platelet survival.[13] However, when such grafts were seeded with endothelium,

FIGURE 3. Transmission electron micrograph of a seeded Dacron® graft 4 weeks postimplantation. The endothelial cells (EC) of the surface display typical features of mature endothelium. In the subendothelial tissue are smooth muscle cells (SMC) in their synthetic phase surrounded by collagenous fibers. (Magnification × 5500.)

platelet survival time returned to normal or near normal levels within 8 weeks after implantation, while depression of platelet survival continued in animals with unseeded grafts.[12]

Platelet serotonin content was used as a measure of in vivo platelet activation and release phenomenon associated with graft implantation. Twelve weeks after placement of a thoracoabdominal graft in the canine model, platelet serotonin content was found to be significantly reduced in animals with unseeded grafts (1.28 ± 0.16 ng/10^9 platelets) when compared to animals with endothelial cell seeded grafts (1.83 ± 0.10 ng/10^9 platelets).[12]

Gamma camera imaging of endothelial cell seeded and unseeded grafts in dogs using ^{111}In-labeled platelets provided another means of assessing platelet kinetics and graft-platelet interactions. In a study of Dacron® thoracoabdominal grafts, at 1 week postimplantation 88% of unseeded prosthesis and 80% of seeded grafts were imaged using indium-labeled platelets.[14] By 8 weeks postimplantation graft-platelet interaction was decreasing, with images evident in 50% of unseeded grafts, but absent in seeded grafts. Animals with a negative scan 8 weeks after graft implantation were found to have over 99% of the graft surface covered by endothelium, whereas a positive scan was associated with endothelial coverage of only 72%. Gamma activity of excised grafts correlated inversely with both endothelial surface coverage and platelet survival times. Gamma camera imaging of ePTFE thoracoabdominal grafts was also studied in the canine model.[15] Again, graft visualization decreased as graft-platelet interactions diminished with the development of an endothelial cell surface. In addition to documenting diminished graft-platelet interactions, these radionuclide studies offer a potential noninvasive means of following the development of an endothelial cell surface on grafts.

Investigators at Washington University School of Medicine studied indium-labeled platelet deposition on 4-mm I.D., external velour Dacron® carotid artery and femoral artery inter-

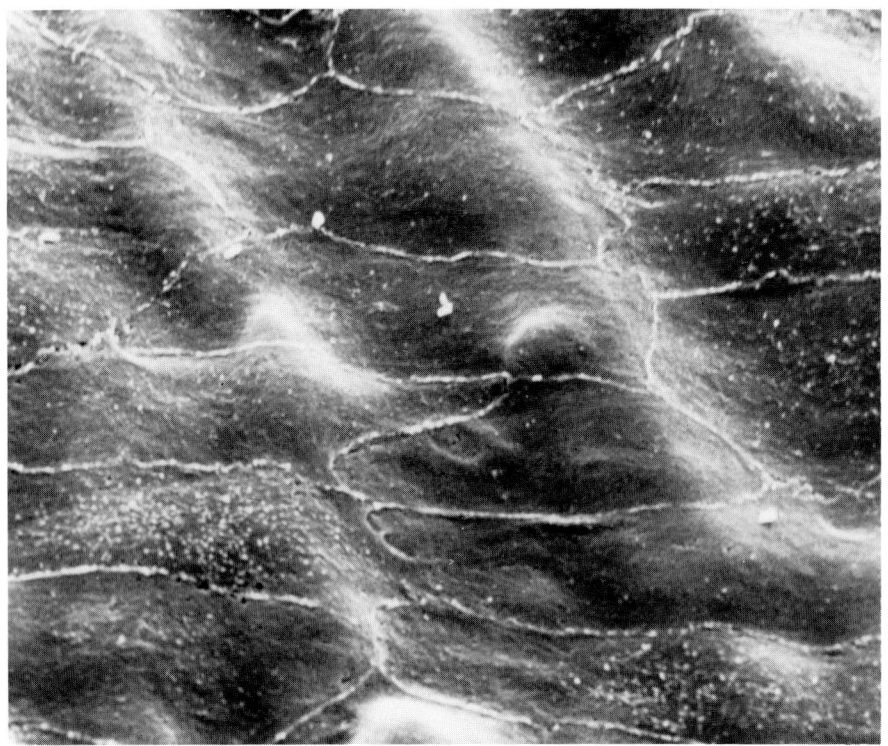

FIGURE 4. Scanning electron micrograph of the luminal surface of a seeded ePTFE graft 4 weeks postimplantation. The surface cells are elongated in the direction of blood flow and display intercellular junctions and microvilli typical of endothelium. There is virtually no subendothelial tissue at this time and, therefore, the nodes of the underlying graft are clearly visible as vertical ridges beneath the endothelial cells. (Magnification × 2500.)

position grafts.[16] Although endothelial cell-seeded grafts accumulated significantly more indium-labeled platelets 24 hr after implantation than did nonseeded grafts, by 2 weeks postimplantation platelet deposition on seeded grafts was significantly less than on control grafts. Platelet deposition did not change significantly upon aspirin withdrawal in animals with seeded grafts, but increased by a factor of five in control dogs when aspirin was stopped.

Investigators at Tufts University studied platelet deposition using ^{111}In-labeled platelets in 4-mm I.D., 5-cm-long, knitted Dacron® carotid interposition grafts in baboons.[17] Platelet accumulation as measured by a gamma camera was not significantly different between seeded and control grafts 24 hr postimplantation. However, platelet accumulation at 2 and 4 weeks on grafts seeded with endothelium was significantly less than on unseeded grafts. Although patency rates at 5 weeks were similar in seeded and control grafts utilizing these short prostheses, the luminal surface of seeded grafts had less thrombus, a thinner inner capsule with a lining of cells which stained for Factor VIII-related antigen, and the SEM appearance of endothelium. This study provided evidence of endothelialization in primates, supporting the potential efficacy of endothelial seeding in humans.

B. Patency Studies

The importance of a functionally intact endothelial cell lining in improving small-caliber graft patency has been demonstrated in a number of experiments.[11,16,18-20] In a University of Michigan study, the 16-week cumulative patency of 4-mm I.D., 10-cm-long, knitted

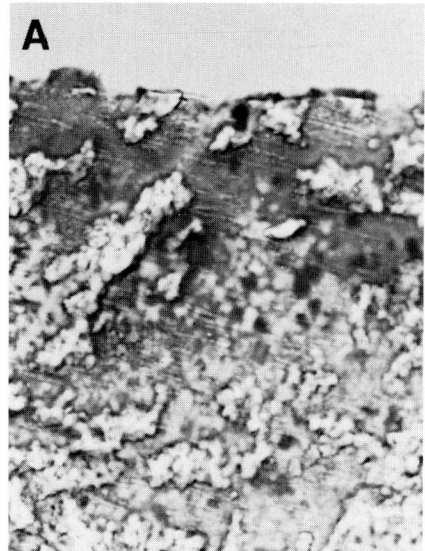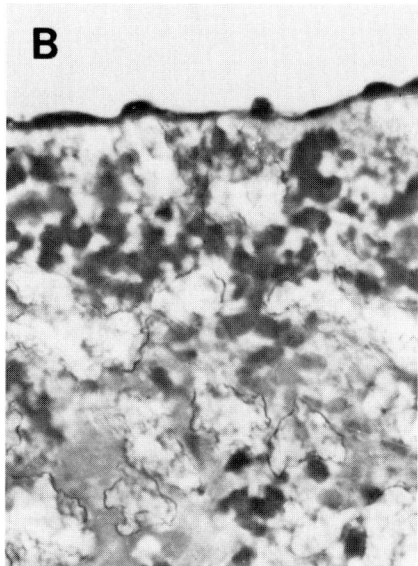

FIGURE 5. Light micrograph of ePTFE grafts 4 weeks postimplantation. (A) The surface of the nonseeded grafts is covered by fibrin and platelets with interspersed bare areas as shown here. (Methylene blue-basic fuchsin; magnification × 300.) (B) The seeded grafts are surfaced by a monolayer of endothelium. Macrophages are readily apparent in the interstices of the graft. (Methylene blue-basic fuchsin; magnification × 650.)

Dacron® iliofemoral grafts in dogs receiving antiplatelet drugs preoperatively and for 14 days postoperatively, was 73% for endothelial cell seeded grafts and 27% for unseeded grafts.[18] Investigators at the University of Akron and Akron City Hospital found that successfully seeded 4-mm I.D., 6-cm-long, double velour Dacron® carotid interposition grafts exhibited 100% patency when followed up to 4 weeks as compared to 75% patency in control grafts.[19] In a study from Washington University School of Medicine, the cumulative 7-month patency rate of 96% among seeded 4-mm I.D. Dacron® carotid and femoral artery interposition grafts was significantly greater than the 29% patency rate of unseeded controls.[16] In yet another study, 16-week patency rates of seeded and unseeded 4-mm I.D., 12- to 16-cm-long, ePTFE aortoiliac grafts were 50 and 0%, respectively, in dogs receiving acetylsalicylic acid preoperatively and for 14 days postoperatively.[11] At the University of Cincinnati Medical Center endothelial seeding of more porous (45 vs. the standard 30-μm internodal distance), unreinforced, 4-mm I.D., ePTFE carotid interposition grafts was also associated with improved patency rates.[20] In contrast to the standard ePTFE grafts, these more porous grafts demonstrated a thicker cellular inner capsule.

Investigators from Akron City Hospital and Northeastern Ohio Universities College of Medicine documented improved patency of endothelial cell-seeded grafts during periods of reduced blood flow produced by constricting devices.[21] Three weeks after implantation all seeded grafts remained patent during a 4-hr period of 70% reduction in flow, but 50% of nonseeded grafts occluded. Five weeks after implantation, all seeded grafts again remained patent during the induced low flow state, but 25% of nonseeded grafts occluded. Furthermore, final flow rates through the conduits after removal of constricting devices were only 20% of the initial flow rates in unseeded grafts, compared to 93% of the baseline flow rates in seeded grafts. Seeded grafts retained thrombus-free surfaces during the periods of low flow, while control grafts accumulated heavy thrombus deposition often leading to occlusion.

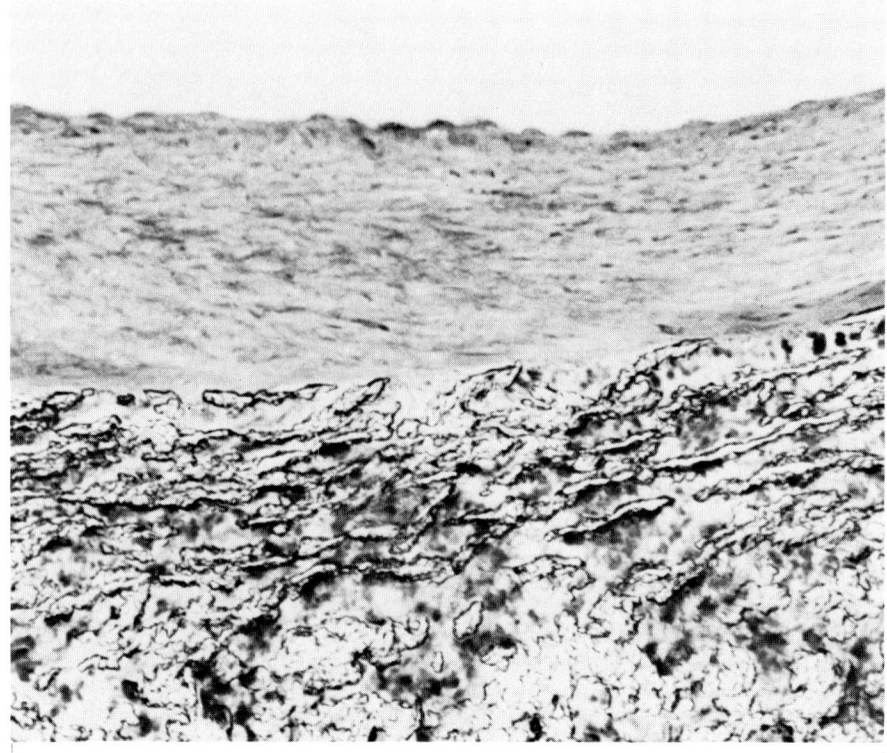

FIGURE 6.   Light micrograph a seeded ePTFE graft 16 weeks postimplantation. In some regions of these grafts, the inner lining, while covered by endothelium, is considerably thickened by layers of smooth muscle cells and collagenous tissue. (Methylene blue-basic fuchsin; magnification × 300.)

IV. CELL ADHERENCE

Endothelial cell adherence to various graft materials was measured utilizing ^{111}In-oxide-labeled endothelial cells. Reported cell adherence to knitted Dacron® grafts ranged from 46 to 74%,[22,23] but to ePTFE grafts was markedly less, in the range of 1.5 to 13.4%.[24] This prompted research into ways of improving cell attachment. In more porous (45-μm internodal distance) ePTFE grafts initial endothelial cell adherence was 19.8%.[25] However, when these same grafts were coated with fibronectin, initial endothelial cell adherence increased to 46.7%. When implanted in the arterial system for 24 hr, fibronectin-coated grafts retained 21.3% of initially adherent cells, compared to retention of only 3.4% of cells on uncoated grafts. Seeger, in studies using standard ePTFE, demonstrated a seeding efficiency of 67% on fibronectin-coated grafts compared to 5% on uncoated grafts, with retention of 25% of initially adherent cells after 1 hr of exposure to blood flow through an arteriovenous shunt.[26] Although fibronectin increased attachment of endothelial cells, it also increased adherence of platelets to grafts, thus potentially increasing early thrombogenicity. This effect may be counteracted by use of antiplatelet agents.

Several studies demonstrated that endothelial cell surfaces increased resistance to bacteremia-induced graft infections. In one study grafts exhibiting greater than 80% endothelial surface coverage had a 16.7% infection rate following intravenous administration of 10^8 staphylococcus aureus, compared to a 57% infection rate when there was less than 80% endothelial cell coverage.[27] In another study fewer radiolabeled bacteria adhered to seeded grafts than to control grafts, suggesting that early endothelialization was associated with diminished bacterial adherence to grafts following intravenous introduction of organisms.[28]

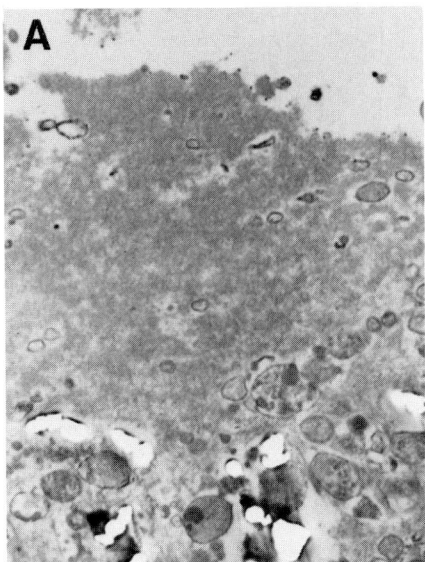 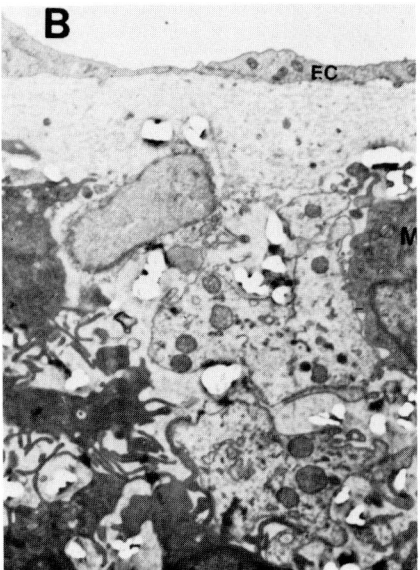

FIGURE 7. Transmission electron micrographs of ePTFE grafts 4 weeks postimplantation. (A) Fibrin and platelets cover the surface of nonseeded grafts and fill the interstices. (Magnification × 1000.) (B) The seeded graft is surfaced by endothelium cells (EC) lying on collagenous connective tissue. The interstices contain large numbers of macrophages (M). (Magnification × 9000.)

V. CURRENT AREAS OF INVESTIGATION

Methods to increase cell harvesting efficiency while maintaining cell viability are major areas of research. Collagenase, with or without trypsin, has been utilized for deriving endothelium in most reported studies. Several concerns with these enzymes exist. First, trypsin decreases cell viability and alters surface enzymes.[29] Secondly, most commercial collagenase preparations are impure, containing enzyme contaminants including nucleases. The latter carry the risk of damage to the cell or its DNA with attendant potential for cellular transformation. Neutral proteases have also been utilized in the harvesting of endothelial cells and may eliminate some of these concerns.[30]

Because of the limited number of veins available for clinical cell harvesting, recent investigations have been directed towards other potential sources of endothelium. Omentum has been utilized to derive both mesothelial and endothelial cells for graft seeding.[31,32] However, use of cells obtained from omentum has been complicated by contamination with fibroblasts and smooth muscle cells, resulting in substantial subendothelial tissue in seeded grafts.[32] Endothelium may also be harvested from periaortic, perinephric, or subcutaneous adipose tissue.[33,34] Endothelial cells are derived from adipose tissue by incubation in collagenase, with subsequent serial filtration through a 250-μm nylon mesh to remove the larger tissue fragments, then through a 30-μm nylon mesh to collect clumps of endothelial cells for further purification.[33] The time-consuming nature of this process may limit its use in the clinical setting. Investigators at Jefferson Medical College have utilized collagenase followed by centrifugation in a 45% Percoll® density gradient to separate endothelium from perinephric fat.[34] Approximately 10^6 endothelial cells are derived per gram of fat by this method.

Modification of graft surfaces to improve endothelial cell adherence would be advantageous. Ideally, substances or combinations of substances utilized to increase endothelial cell attachment should also promote proliferation and migration of the endothelium. In this regard investigators at Loyola University are studying attachment of endothelial cell growth factor (ECGF) to vascular grafts in an effort to promote spontaneous endothelialization, but attachment of ECGF may also prove beneficial as an adjunct to endothelial seeding.[35] The importance of limiting platelet and leukocyte adherence to seeded grafts has not been defined, although leukocytes have been shown to be detrimental to endothelial cells after implantation of prostheses seeded in vitro.[36]

Endothelial seeding of vascular grafts in vitro for later implantation has the theoretic benefit of providing a nonthrombogenic surface immediately following graft placement. However, it has several potential drawbacks. Operative handling of the graft may damage the cellular surface, producing greater graft thrombogenicity if subendothelial tissue is exposed. Furthermore, the time required for autologous endothelial cell graft coverage in vitro may be prohibitively long in the clinical setting. Lastly, the cultivation step required for in vitro seeding introduces possible microbial contamination of the prosthesis, as well as potential alteration of cells leading to their eventual transformation.

Some investigators have suggested that grafts seeded with nonautologous (homologous or heterologous) endothelium develop an endothelial surface more rapidly than nonseeded control grafts.[37-39] These studies are controversial in that the source of the eventual cellular lining has not been documented to be the inoculum of seeded cells. Given the known antigenicity of endothelium, it seems unlikely that the cells surfacing the grafts in these studies are of a nonautologous source. However, release of growth factors by endothelial cells or leukocytes may serve to hasten the development of an endothelial surface by host cells, but this hypothesis remains unproven.

Endothelial cells in culture can express procoagulant activity and produce platelet-derived growth factor.[29,40] The degree or duration of this activity by seeded endothelial cells remains undefined. However, mitogens produced by seeded endothelium may provide a continuing stimulus to smooth muscle cell proliferation and migration. This may lead to progressive inner capsule thickening, or may be a temporary phenomenon subsiding when an intact endothelial lining is established. Methods to limit these potentially adverse functions of endothelium must be studied.

Clinical studies have been initiated at Indiana University and several European centers. In an initial study on seeding mechanically harvested endothelial cells, Herring and colleagues found 18-month patency rates for femoropopliteal grafts in nonsmokers of 100% with seeded grafts compared to 60% with unseeded grafts.[41] However, cigarette smokers exhibited a patency rate of 45% in seeded grafts compared to 80% in nonseeded grafts. This untoward response to seeding in smokers may have been due to contamination of the endothelial cells with smooth muscle cells during the mechanical harvesting process. These cells subjected to increased mitogens from platelets or endothelial cells in smokers may have produced the observed detrimental effect. In another report from this group, endothelium was demonstrated 10 cm from the distal anastomosis of a patent, endothelial cell-seeded ePTFE graft which required exploration 90 days after implantation.[42] This unique finding, although anecdotal, is encouraging evidence that seeding of endothelial cells into vascular grafts in humans may lead to eventual endothelialization of the prostheses.

VI. CONCLUSION

It has been well documented in experimental models that endothelial seeding of prosthetic vascular grafts promotes the early development of an endothelial lining. This surface has functional properties similar to that of endothelium of normal vessels, producing prostacyclin and promoting early normalization of platelet survival times. Patency rates of small-caliber grafts in the canine model clearly are improved with endothelial cell seeding.

Although many areas of endothelial cell seeding require further research and definition, this technology carries promise for improving the function of vascular prostheses in the clinical setting by directly decreasing graft thrombogenicity. Furthermore, the early development of an endothelial cell surface may diminish prosthesis-related platelet activation and other systemic perturbations which may contribute to accelerated atherosclerosis in regions far distant from the implanted graft.

REFERENCES

1. **Graham, L. M. and Bergan, J. J.,** Expanded polytetrafluoroethylene vascular grafts: clinical and experimental observations, in *Biologic and Synthetic Vascular Prostheses,* Stanley, J. C., Burkel, W. E., Lindenauer, S. M., Bartlett, R. H., and Turcotte, J. G., Eds., Grune & Stratton, New York, 1982, 563.

2. **Herring, M., Gardner, A., and Glover, J.,** A single-staged technique for seeding vascular grafts with autogenous endothelium, *Surgery,* 84, 498, 1978.

3. **Herring, M. B., Dilley, R., Jersild, R. A., Jr., Boxer, L., Gardner, A., and Glover, J.,** Seeding arterial prostheses with vascular endothelium. The nature of the lining, *Ann. Surg.,* 190, 84, 1979.

4. **Dilley, R., Herring, M., Boxer, L., Gardner, A., and Glover, J.,** Immunofluorescent staining for Factor VIII-related antigen: a tool for study of healing in vascular prostheses, *J. Surg. Res.,* 27, 149, 1979.

5. **Ford, J. W., Burkel, W. E., and Kahn, R. H.,** Isolation of adult canine venous endothelium for tissue culture, *In Vitro,* 17, 44, 1981.

6. **Graham, L. M., Vinter, D. W., Ford, J. W., Kahn, R. H., Burkel, W. E., and Stanley, J. C.,** Endothelial cell seeding of prosthestic vascular grafts. Early experimental studies with cultured autologous canine endothelium. *Arch. Surg.,* 115, 929, 1980.

7. **Graham, L. M., Burkel, W. E., Ford, J. W., Vinter, D. W., Kahn, R. H., and Stanley, J. C.,** Immediate seeding of enzymatically derived endothelium in Dacron vascular grafts, *Arch. Surg.,* 115, 1289, 1980.

8. **Burkel, W. E., Vinter, D. W., Ford, J. W., Kahn, R. H., Graham, L. M., and Stanley, J. C.,** Sequential studies of healing in endothelial seeded vascular prostheses. Histologic and ultrastructure characteristics of graft incorporation, *J. Surg. Res.,* 30, 305, 1981.

9. **Burkel, W. E., Ford, J. W., Vinter, J. W., Kahn, R. H., Graham, L. M., and Stanley, J. C.,** Fate of knitted Dacron velour vascular grafts seeded with enzymatically derived autologous canine endothelium, *Trans. Am. Soc. Artif. Intern. Organs,* 28, 178, 1982.

10. **Graham, L. M., Burkel, W. E., Ford, J. W., Vinter, D. W., Kahn, R. H., and Stanley, J. C.,** Expanded polytetrafluoroethylene vascular prostheses seeded with enzymatically derived and cultured canine endothelial cells, *surgery,* 91, 550, 1982.

11. **Graham, L. M., Stanley, J. C., and Burkel, W. E.,** Improved patency of endothelial-cell-seeded, long, knitted Dacron and ePTFE vascular prostheses, *Asaio J.,* 8, 65, 1985.

12. **Clagett, G. P., Burkel, W. E., Sharefkin, J. B., Ford, J. W., Hufnagel, H., Vinter, D. W., Kahn, R. H., Graham, L. M., Stanley, J. C., and Ramwell, P. W.,** Platelet reactivity in vivo in dogs with arterial prostheses seeded with endothelial cells, *Circulation,* 69, 632, 1984.

13. **Clagett, G. P., Graeber, G. M., Robinowitz, M., Langloss, J. M., and Ramwell, P. W.,** Differentiation of vascular prostheses in dogs with serial tests of in vivo platelet reactivity, *Surgery,* 95, 331, 1984.

14. **Whitehouse, W. M., Jr., Wakefield, T. W., Vinter, D. W., Ford, J. W., Swanson, D. P., Thrall, J. H., Forehlich, J. W., Burkel, W. E., Graham, L. M., and Stanley, J. C.,** Indium-111-oxide labeled platelet imaging of endothelial seeded Dacron thoracoabdominal vascular prostheses in a canine experimental model, *Trans. Am. Soc. Artif. Intern. Organs,* 29, 183, 1983.

15. **Wakefield, T. W., Lindblad, B., Graham, L. M., Whitehouse, W. M., Jr., Ripley, S. D., Petry, N. A., Spaulding, S. A., Burkel, W. E., and Stanley, J. C.,** Nuclide imaging of vascular graft-platelet interactions: comparison of indium excess and technetium subtraction techniques, *J. Surg. Res.,* 40, 388, 1986.

16. **Allen, B. T., Long, J. A., Welch, M. J., Hopkins, K. T., Sicard, G. A., and Clark, R. E.,** Influence of endothelial cell seeding on platelet deposition and patency in small-diameter Dacron arterial grafts, *J. Vasc. Surg.,* 1, 224, 1984.

17. **Shepard, A. D., Eldrup-Jorgensen, J., Keough, E. M., Foxall, T. F., Ramberg, K., Connolly, R. J., Mackey, W. C., Gavris, V., Auger, K. R., Libby, P., O'Donnell, T. F., and Callow, A. D.,** Endothelial cell seeding of small-caliber synthetic grafts in the baboon, *Surgery,* 99, 318, 1986.

18. **Stanley, J. C., Burkel, W. E., Ford, J. W., Vinter, D. W., Kahn, R. H., Whitehouse, W. M., Jr., and Graham, L. M.,** Enhanced patency of small-diameter, externally supported Dacron iliofemoral grafts seeded with endothelial cells, *Surgery,* 92, 994, 1982.

19. **Belden, T. A., Schmidt, S. P., Falkow, L. J., and Sharp, W. V.,** Endothelial cell seeding of small diameter vascular grafts, *Trans. Am. Soc. Artif. Intern. Organs,* 28, 173, 1982.

20. **Kempczinski, R. F., Rosenman, J. E., Pearce, W. H., Roedersheimer, L. R., Berlatzky, Y., and Ramalanjaona, G.,** Endothelial cell seeding of a new PTFE vascular prosthesis, *J. Vasc. Surg.,* 2, 424, 1985.

21. **Hunter, T. J., Schmidt, S. P., Sharp, W. V., and Malindzak, G. S.,** Controlled flow studies in 4 mm endothelialized Dacron grafts, *Trans. Am. Soc. Artif. Intern. Organs,* 29, 177, 1983.

22. **Graham, L. M., Ford, J. W., Vinter, D. W., Burkel, W. E., and Stanley, J. C.,** Endothelial cell adherence to Dacron vascular prostheses: optimization of seeding technique, *Am. Soc. Artif. Intern. Organs,* 13(Abstr.), 26, 1986.

23. **Sharefkin, J. B., Latker, C., Smith, M., and Rich, N. M.,** Endothelial cell labeling with indium-111-oxine as a marker of cell attachment to bioprosthetic surfaces, *J. Biomed. Mater. Res.,* 17, 345, 1983.

24. **Lindblad, B., Wright, S., Sell, R., Graham, L. M., Burkel, W. E., and Stanley, J. C.,** Alternate techniques of seeding cultured endothelial cells to ePTFE grafts of different diameters, porosities and surfaces, *J. Biomed. Mater. Res.,* 21, 1013, 1987.

25. **Ramalanjaona, G., Kempczinski, R. F., Rosenman, J. E., and Silberstein, E. B.,** The effect of fibronectin coating on endothelial cell kinetics in polytetrafluoroethylene grafts, *J. Vasc. Surg.,* 3, 264, 1986.

26. **Seeger, J. M.,** Improved endothelial cell seeding density after flow exposure in fibronectin-coated grafts, *Surg. Forum,* 35, 450, 1985.

27. **Arrgeri, M., Dilley, R., Herring, M., Miller, B., Baughman, S., Smith, J., and Glover, J.,** The effect of bacteremia on arterial prostheses seeded with endothelium, *Asaio J.,* 8, 118, 1985.

28. **Rosenman, J. E.,Kempczinski, R. F., Berlatzky, Y., Pearce, W. H., Ramalanjaona, G. R., and Bjornson, H. S.,** Bacterial adherence to endothelial-seeded polytetrafluoroethylene grafts, *Surgery,* 98, 816, 1985.

29. **Ryan, U. S.,** Metabolic activity of pulmonary endothelium: modulations of structure and function, *Annu. Rev. Physiol.,* 48, 263, 1986.

30. **Sentissi, J. M., Ramberg, K., O'Donnell, T. F., Jr., Connolly, R. J., and Callow, A. D.,** The effect of flow on vascular endothelial cells grown in tissue culture on polytetrafluoroethylene, *Surgery,* 90, 337, 1986.

31. **Clarke, J. M., Pittilo, R. M., Nicholson, L. J., Woolfe, N., and Marston, A.,** Seeding Dacron arterial prostheses with peritoneal mesothelial cells: a preliminary morphological study, *Br. J. Surg.,* 71, 492, 1984.

32. **Pearce, W. H., Rutherford, R. B., Whitehall, T., Rosales, C., Ramalajaona, G., and Patt, A.,** Successful endothelial seeding with omentally derived microvascular endothelial cells, *J. Vasc. Surg.,* 5, 203, 1987.

33. **Bar, R. S., Dolash, S., Dake, B. L., and Boes, M.,** Cultured capillary endothelial cells from bovine adipose tissue: a model for insulin binding and action in microvascular endothelium, *Metabolism,* 35, 317, 1986.

34. **Jarrell, B. E., Williams, S. K., Stokes, G., Hubbard, F. A., Carabasi, R. A., Koolpe, E., Greener, D., Pratt, K., Moritz, M. J., Radomski, J., and Speicher, L.,** Use of freshly isolated capillary endothelial cells for the immediate establishment of a monolayer on a vascular graft at surgery, *Surgery,* 100, 392, 1986.

35. **Greisler, H., Klosak, J., Dennis, J., Kim, D., Burgess, W., and Maciag, T.,** Endothelial cell growth factor attachment to biomaterials, *Trans. Am. Soc. Artif. Intern. Organs,* 32, 346, 1986.

36. **Emerick, S., Herring, M. N., Arnold, M., Baughman, S. L., and Glover, J. L.,** Leukocyte depletion enhances cultured endothelial retention on vascular prosthesis, *J. Vasc. Surg.,* 5, 342, 1987.

37. **Hollier, L. H., Fowl, R. J., Heck, C. F., Winter, K. A., Fass, D. N., and Kaye, M. P.,** Are seeded endothelial cells the origin of neointima on prosthetic vascular grafts?, *J. Vasc. Surg.,* 3, 65, 1986.

38. **Pennell, R. C., Hollier, L. H., Solis, E., and Kaye, M. P.,** Xenograft seeding of Dacron grafts in dogs, *J. Surg. Res.,* 40, 332, 1986.

39. **Zamora, J. L., Navarro, L. T., Ives, C. L., Weilbaecher, D. G., Gao, Z. R., and Noon, G. P.,** Seeding of arteriovenous prostheses with homologous endothelium, *J. Vasc. Surg.,* 3, 860, 1986.

40. **Fox, P. L. and DiCorleto, P. E.,** Regulation of production of a platelet-derived growth factor-like protein by cultured bovine aortic endothelial cells, *J. Cell. Physiol.,* 121, 298, 1984.

41. **Herring, M. B., Gardner, A., and Glover, J.,** Seeding human arterial prostheses with mechanically derived endothelium. The detrimental effect of smoking, *J. Vasc. Surg.,* 1, 279, 1984.

42. **Herring, M., Baughman, S., and Glover, J.,** Endothelium develops on seeded human arterial prosthesis: a brief clinical note, *J. Vasc. Surg.,* 2, 727, 1985.

Chapter 39

CUTANEOUS MICROVASCULAR ENDOTHELIAL CELLS

Thomas J. Lawley and Yasuo Kubota

TABLE OF CONTENTS

I. INTRODUCTION

Cutaneous microvascular endothelial cells form the barrier between the intravascular circulation and the stroma of the largest organ in the body. These cells must present a nonthrombogenic luminal surface and participate in controlling the vascular tone of an organ subjected to great variations in ambient temperature. Endothelial cells also play central roles in a variety of different types of cutaneous inflammation. These range from physiologic events such as wound healing to pathologic states such as vasculitis, graft rejection, and benign and malignant tumor formation.

Until recently, it has been difficult to study human cutaneous (dermal) microvascular endothelial cells in detail because they could not be isolated in pure form, and because of their rigorous growth requirements. However, breakthroughs have been made in both of these areas, allowing a variety of detailed studies to be performed, and obviating the need to substitute large-vessel endothelial cells derived from human umbilical veins or microvascular cells derived from omental fat for true dermal microvascular cells.[1,2] In attempting to understand the cell biology and immunology of the vascular cells of the skin, it is important not only to use human microvascular endothelial cells, but also microvascular endothelial cells derived from the skin itself. Differences between large-vessel endothelial cells and capillary endothelial cells have been documented by a number of investigators and include differences in ultrastructure, capacity to form tubules, and their ability to secrete mediators of inflammation.[3-5] Moreover, it seems likely that capillary endothelial cells from different organs have different physiologic and pathophysiologic potentials.[6]

Finally, it should be noted that although we will discuss human microvascular endothelial cells as if they were monomorphous, it is clear, at least from ultrastructural mapping studies of the cutaneous vasculature, that they are not.[7] For example, precapillary arterioles, capillaries, and postcapillary venules all exist in the most superficial portion of the dermis with somewhat larger vessels making up the subpapillary plexus. It is not known whether the endothelial cells that make up these multiple types of blood vessels in the cutaneous microvasculature are all the same or whether undiscovered differences exist.

II. ISOLATION AND CULTURE OF CUTANEOUS MICROVASCULAR ENDOTHELIAL CELLS

There are now several methods available for the isolation and purification of microvascular endothelial cells from skin.[1,2,8-10] Moreover, not only can these cells be isolated, but they can be induced to proliferate by a variety of substances and growth conditions. However, one difficulty encountered by a number of investigators attempting to work with these cells has been contamination of cultures by fibroblasts, the predominant cell of the dermis. Since fibroblasts are hardy, rapidly growing cells that can rapidly overgrow microvascular endothelial cells, a number of stategies have been worked out to either eliminate these cells from the primary isolates from skin or to selectively remove these cells from culture.

Davison et al. was one of the first groups to report successful isolation and culture of microvascular endothelial cells from skin.[1] They used neonatal foreskins as their tissue source (Figure 1). Briefly, their technique involved removal of the epidermis by keratome, cutting the foreskin into small squares followed by mild trypsinization and then expression of microvascular segments from the dermis by blunt pressure with the side of a scalpel blade. These cells were then cultured on culture dishes coated with fibronectin in media consisting of Eagle's Minimal Essential Medium supplemented with 10 to 50% pooled human serum. In these authors' hands, fibroblast contamination has not been a problem.

Several groups have published alternative methods of cutaneous microvascular endothelial cell isolation.[2,8-10] In general, these methods involve dissociation of endothelial cells and

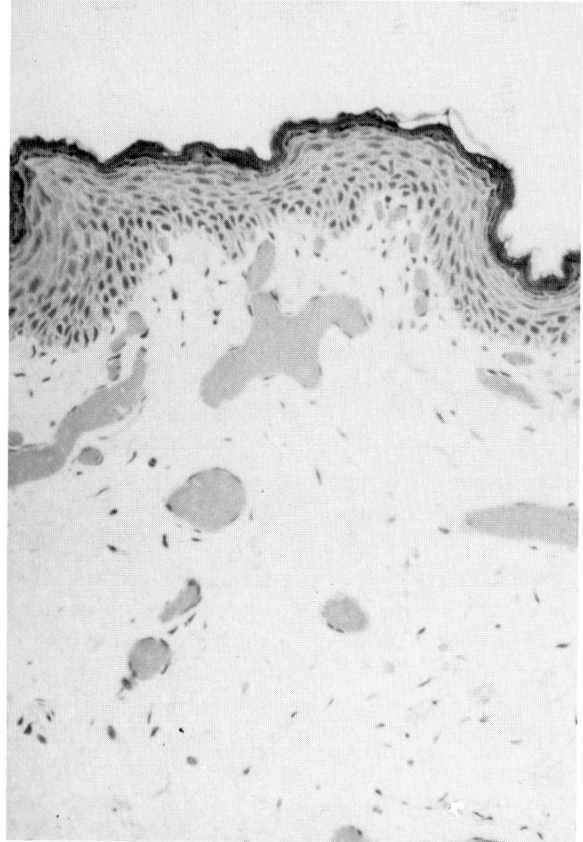

A

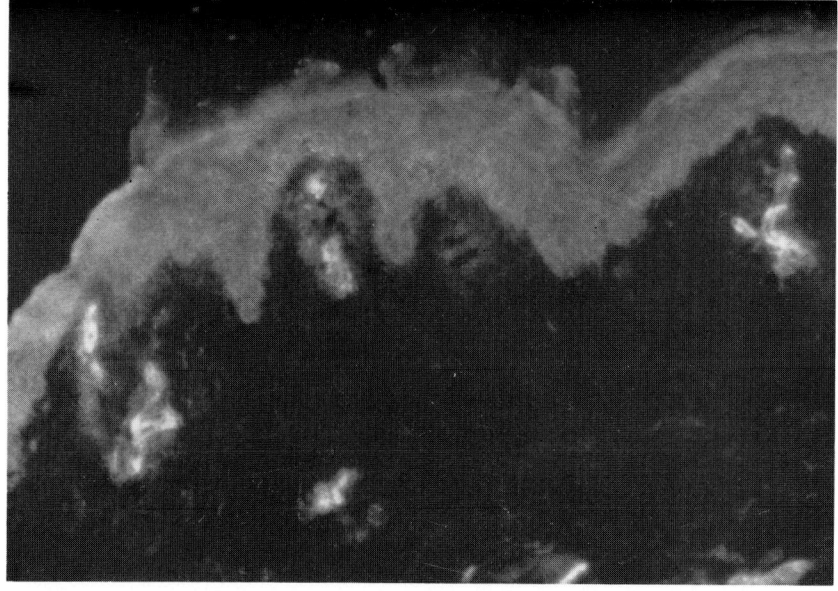

B

FIGURE 1.   (A) A biopsy of human foreskin showing prominent dilated blood vessels in the dermis. This tissue is often used as a source of cutaneous microvascular endothelial cells. (B) Direct immunofluorescence of a biopsy of human foreskin stained with antibody against human Factor VIII-related antigen. The endothelial cells in the small dermal blood vessels stain brightly.

microvascular segments from nonvascular tissue components by enzymes such as trypsin-EDTA, collagenase, or a collagenase-dispase mixture. This is often followed by a step designed to separate microvascular segments from contaminating nonvascular tissue or cells. Sieving the enzymatically dissociated tissue through nylon mesh of varying pore size is commonly used. Since most microvascular segments have a density of 1.048 g/mℓ, density gradient centrifugation through 5% BSA or 35% Percoll in HBSS has been used either in combination with nylon sieving or alone for further purification of endothelial cells from cells such as keratinocytes, fibroblasts, and pericytes.

Microvascular endothelial cells have been grown in a variety of media including Eagle's Minimal Essential Medium (MEM), Dulbeco's Modified Eagle's Essential Medium (DMEM), and Medium 199. However, it seems that the nutritional supplements and growth factors added to the medium of choice are critical. Tumor-conditioned media were used in early studies and proved to be successful. Folkman et al.[9] used media conditioned by sarcoma 180 while Davison et al.[1] used an EHS sarcoma-conditioned medium. High concentrations of serum in the growth media are also important in stimulating microvascular endothelial cells to proliferate. The use of 50% human serum seems optimal.[2] Recently, it has been reported that the use of prepartum maternal serum could reduce the serum requirement of endothelial cell culture from 50 to 2%.[11] It is not known what substance(s) in prepartum maternal serum is responsible for this. Other growth factors used for cutaneous microvascular endothelial cell culture have included endothelial cell growth supplement,[2,9,10] fibroblast growth factor,[1,9] epidermal growth factor,[12] human thrombin,[9] hydrocortisone,[12] cholera toxin,[10,13] isobutyl methylxanthine,[10,13] and dibutyryl cyclic AMP.[2,13] A number of these agents were directly tested for their effect on proliferation of cutaneous microvascular endothelial cells by Marks et al.[2] A combination of endothelial cell growth substance and heparin resulted in maximal proliferation, but heparin alone, epidermal growth factor, fibroblast growth factor, nerve growth factor, and platelet derived growth factor had no proliferative effect. Moreover, they noted that even though endothelial cell growth substance and heparin were used, this did not obviate the need for high concentrations of human serum (up to 30%) and the need for extracellular matrix, either gelatin or fibronectin.

Finally the need for mechanical removal of contaminating fibroblasts has been noted by several groups (Reference 2 and personal observation). In one technique, petri dishes containing mixtures of endothelial cells and fibroblasts are placed on the stage of an inverted phase microscope within a biological hood. Nonendothelial cells are detached or crushed by means of a 25-gauge sterile needle attached to a syringe. Microvascular endothelial cells can be easily distinguished by their morphology and their tendency to form colonies. The culture dishes must be viewed daily and contaminating cells removed regularly until the colonies of microvascular endothelial cells have grown to confluence in a selected area. At that time, all of the remaining cells on the dish are removed with a rubber policeman. Then the cutaneous microvascular endothelial cells can be grown to confluence. They can be removed with either a trypsin-EDTA mixture or EDTA alone and subcultured in a 25- or 75-cm^2 flask. It takes 2 to 3 weeks for the microvascular endothelial cells to reach confluence in the 75-cm^2 tissue culture flask (Figure 2).

III. MORPHOLOGY AND IDENTIFICATION

Human cutaneous microvascular endothelial cells when added to fibronectin coated dishes in culture medium supplemented with endothelial cell growth supplement and heparin attach and grow rapidly. Initially, they form small circular colonies which eventually merge to form cobblestone-pattern monolayers. The individual cell is large (30 × 40μm) and polygonal in shape with a centrally placed nucleus. Rapidly growing cutaneous microvascular endothelial cells are clearly morphologically distinguishable from epidermal keratinocytes,

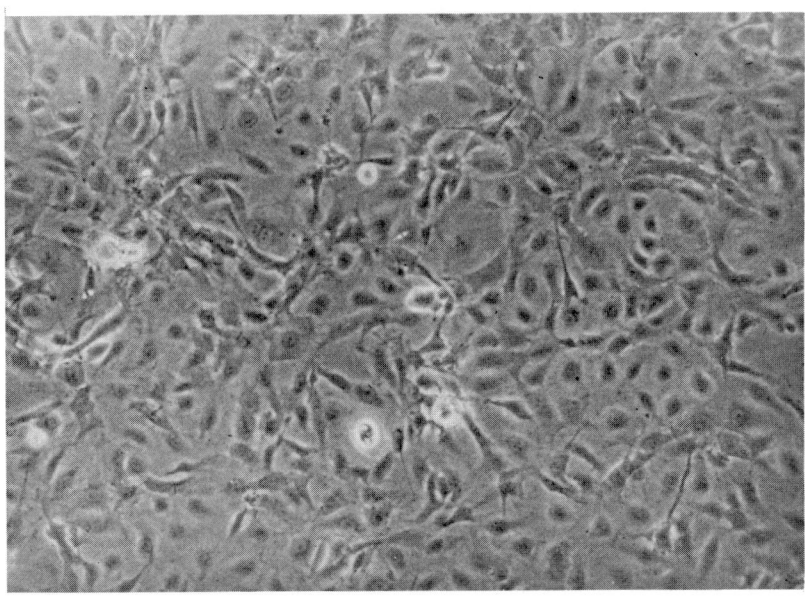

FIGURE 2.   A photomicrograph of a confluent culture of human cutaneous microvascular endoth-elial cells. Most cells are polygonal, but some are beginning to send out cytoplasmic extensions and to elongate.

melanocytes, and fibroblasts by phase contrast microscopy, but do not differ significantly from human umbilical vein endothelial cells in terms of size or shape. If microvascular endothelial cells are plated onto an uncoated culture dish, the cells do not attach rapidly and if they are cultured without endothelial cell growth supplement and heparin, they do not proliferate rapidly. Moreover, the cells in these cultures do not assume a cobblestone con-figuration, but are elongated and irregular in appearance. An additional factor which seems to influence the in vitro growth pattern of microvascular endothelial cells is the presence of high concentrations of human serum or agents that elevate intracellular levels of cyclic AMP such as dibutryl cyclic AMP, cholera toxin, and isobutyl methylxanthine.[1,13,14]

A striking property of cutaneous microvascular endothelial cells is their tendency to form three-dimensional branching networks of tubules in vitro.[4] This tends to occur shortly after cultures become confluent. Folkman and Haudenschild noted that this phenomenon occurred 20 to 40 days after the initiation of primary cultures.[4] More recently, it has also been noted that large-vessel endothelial cells such as human umbilical vein endothelial cells can also form tubules under certain conditions.[15] For instance, when endothelial cell growth factor and heparin are omitted from cultures of human umbilical vein endothelial cells, tubular structures will appear after 4 to 6 weeks.

Cutaneous microvascular endothelial cells in vivo are surrounded by a basement membrane zone and an interstitial subendothelial matrix. Several investigators have recently shown that extracellular matrix proteins can significantly influence the behavior of microvascular en-dothelial cells.[16-18] Several types of matrix components have been used individually or in combination in order to modulate microvascular endothelial cell proliferation, organization, and biosynthetic properties. When microvascular endothelial cells were plated on interstitial collagen (types I, III), they proliferated well and developed a typical cobblestone pattern. If they were cultured for long periods of time (2 to 4 weeks) they formed tube-like structures.[17] Interestingly, when microvascular endothelial cells were grown on basement membrane collagens (types IV, V)[17] or on a basememt membrane component containing gel extracted from EHS sarcoma,[19] they did not proliferate but formed tube-like structures very early in culture (personal observation). When microvascular endothelial cells are grown in collagen gels composed of basement membrane collagens, they also formed tube-like structures shortly after the initiation of culture. Electron-microscopic studies have shown that these tube-like structures contain a narrow lumen.[16,17]

There are a number of markers used for identification of endothelial cells in general which can also be used to identify cutaneous microvascular endothelial cells. Weibel-Palade bodies, first described in 1964,[20] are present in cutaneous microvascular endothelial cells.[1] However, they appear to be less numerous in well differentiated microvascular endothelial cells than in the endothelial cells of large vessels or in rapidly proliferating microvascular endothelial cells.[6] In addition, Weibel-Palade bodies are not as numerous in endothelial cells derived from other species such as rats or cows. One of the most frequently used markers for cutaneous microvascular endothelial cells is Factor VIII-related antigen. While it is not totally specific for endothelial cells, since it is present in platelets and megakaryocytes, Factor VIII-related antigen is very useful in identifying cutaneous microvascular endothelial cells in culture. Factor VIII-related antigen is easily detected by direct immunofluorescence of permeabilized cutaneous microvascular endothelial cells as bright granular cytoplasmic staining. Staining of living cells is less intense, but cell surface staining of cutaneous microvascular endothelial cells by antibody against Factor VIII-related antigen is detectable in most instances. In the development of human cutaneous microvasculature, the expression of Factor VIII-related antigen by microvascular endothelial cells increases with age.[21] Factor VIII-related antigen is focally expressed in the cutaneous blood vessels of neonates, but is nearly confluent in the blood vessels of adult skin.

Recently, it has been noted that certain chemically modified low-density lipoproteins are taken up endothelial cells and macrophages. The uptake is via a scavenger pathway that is different from the receptor for unmodified low-density lipoprotein.[22-24] By attaching a fluorescent label to these types of compounds, followed by incubation with cultured cells, identification of endothelial cells in culture is possible. One such compound now used for identification of endothelial cells in mixed cultures that do not contain macrophages is 1,1′-dioctadecyl-3,3,3′,3′-tetramethyl-indocarbocyanine perchlorate (Dil-Ac-LDL). Cultures of cutaneous microvascular endothelial cells incubated with this material exhibit bright red cytoplasmic fluorescence when viewed in an inverted phase fluorescence microscope, while dermal fibroblasts and epidermal cells do not stain (Figure 3). The viability of labeled cells is not affected by the Dil-Ac-LDL. However, there does appear to be some variability in the ability of endothelial cells to take up acetylated LDL since a recent study has shown that while bovine aortic endothelial cells efficiently take up this label, bovine brain capillary cells do not.[25]

A compound known as *Ulex europaeus*-I (UAE-I), a lectin showing specific affinity for some fucose moieties, also binds to endothelial cells. It has been shown to be a more sensitive marker than Factor VIII-related antigen for small-vessel endothelial cells in a variety of benign and malignant vascular neoplasms of skin.[26]

IV. BIOSYNTHETIC ASPECTS OF CUTANEOUS MICROVASCULAR ENDOTHELIAL CELLS

Endothelial cells in general are metabolically very active. They are known to secrete a variety of extracellular matrix proteins as well as certain types of prostaglandins. Recently, attention has turned to cutaneous microvascular endothelial cells and their ability to synthesize and secrete these types of molecules. When cutaneous microvascular endothelial cells were cultured and then examined by electron microscopy, extracellular matrix was seen in the subendothelial space.[27,28] The ultrastructural characteristics of the subendothelial matrix

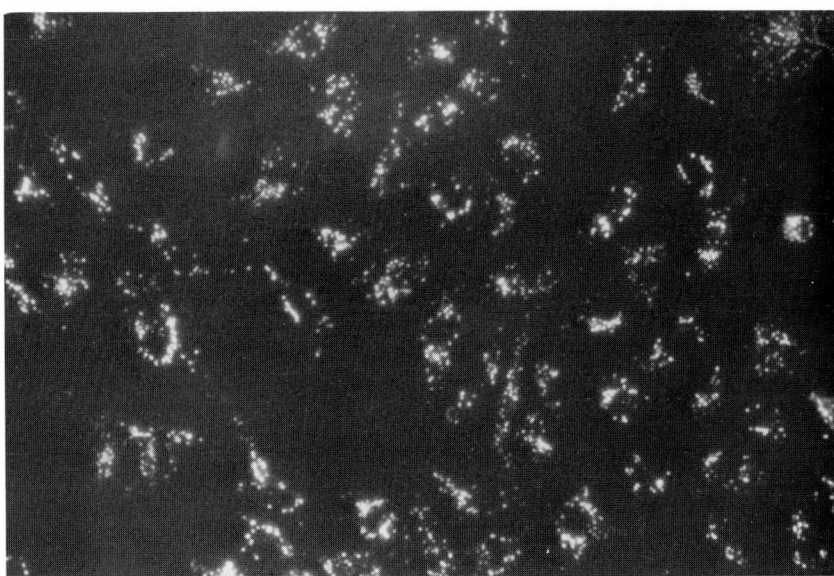

FIGURE 3. A culture of human cutaneous microvascular endothelial cells stained with acetylated low-density lipoprotein conjugated to a fluorescent dye. All cells show bright fluorescence.

varied with the conditions of culture. When cutaneous microvascular endothelial cells were cultured on fibronectin or on cell-free matrices secreted by smooth muscle cells or fibroblasts, the subendothelial matrix secreted by the endothelial cells was polymorphous, multilayered, and discontinuous. However, when these same cells were cultured on a matrix produced by PF HR9 endodermal cells, the subendothelial matrix appeared to be uniform and continuous and also had lamina densa-like and lamina rara-like ultrastructural areas.[27] When the biochemical makeup of the matrix was examined, it was found to consist mainly of type IV collagen and laminin. Although fibronectin is also secreted by these cells, most of it is released into the culture medium and only a small portion is in the subendothelial matrix. Interestingly, when human cutaneous microvascular endothelial cells of newborn and adult skin were compared for thier ability to secrete matrix proteins, it was found that while there was great similarity in the pattern of secretion and localization of type IV collagen, laminin, and fibronectin, thrombospondin was present in large amounts in the neonatal cells but was almost completely absent in the matrix secreted by adult endothelial cells.[28] Moreover, earlier in vitro studies had shown that bovine adrenal capillary endothelial cells secrete mainly collagen types I, III, and V into the subendothelial space.[29] Thus matrix proteins secreted by endothelial cells, vary with the source of the endothelial cells, their age, and the culture conditions.

It is well known that endothelial cells synthesize and secrete prostaglandins.[30] Most studies using large-vessel endothelial cells have shown that prostacyclin (PGI_2) is the major product of arachidonic acid metabolism of endothelial cells. Prostacyclin is a potent inhibitor of platelet aggregation and of platelet binding to endothelial cells as well as being a potent vasodilator.[32,32] Prostacyclin has been shown to be produced by large vessel and microvascular endothelial cells in culture.[5,33] Its production by endothelial cells can be stimulated by a variety of substances including thrombin, trypsin, arachidonic acid, A23187, ionizing radiation, bradykinin, LTD_4, LTC_4, and histamine, and can be inhibited by dexamethasone.[34-39]

However, when cutaneous microvascular endothelial cells and human umbilical vein endothelial cells were both studied for production of arachidonic acid metabolites, major differences were found.[5] Human umbilical vein endothelial cells were found to produce large amounts of PGI_2 and little PGE_2, whereas cutaneous microvascular endothelial cells generated ample amounts of PGE_2 and PGF_2 but almost no PGI_2. The reasons for these findings are not clear, but once again serve to underscore the differences between large- and small-vessel human endothelial cells.

V. IMMUNOLOGY OF CUTANEOUS MICROVASCULAR ENDOTHELIAL CELLS

Although increasing attention has been given to the immunologic capabilities of endothelial cells in general, relatively little is known about the immunology of the microvascular cells of human skin. Furthermore, one must exercise caution when attempting to extrapolate findings from large- to small-vessel endothelial cells, especially when crossing species lines. Most studies have shown that human umbilical vein endothelial cells and human cutaneous microvascular endothelial cells in culture lack class II antigens of the major histocompatibility complex.[40-42] However, pretreatment of both of these types of endothelial cells with human γ-interferon (γ-IFN) results in the appearance of class II antigens that is maximal at 4 to 6 days.[41-43] The class II antigens begin to disappear 3 to 4 days after withdrawal of γ-IFN. Although human cutaneous microvascular endothelial cells lack class II antigens in culture under normal culture conditions, there is evidence to suggest that these cells may possess class II antigens in vivo. A variety of investigators using immunofluorescent techniques have noted positive staining of cells that outline cutaneous vessels with antibody to HLA-DR.[44,45] However, detailed double-staining studies demonstrating that the HLA-DR positive cells are also Factor VIII positive have not been reported. Thus the HLA-DR positive cells, while assumed to be endothelial cells, could be pericytes or another undefined cell type. However, if human cutaneous endothelial cells do display class II antigens in vivo, it indicates that either these cells constituitively express class II antigens when they are fully differentiated or that a γ-IFN-like substnace is present in their environment and promotes class II antigen expression.

A thorough phenotypic analysis of immunologically relevant cell surface antigens and receptors has not been reported and will be an important step forward in our understanding of their immunologic capabilities. However, in addition to class II antigens, these cells also express ABO blood group antigens as well as class I antigens.[40] While human umbilical vein endothelial cells have been reported to be able to activate lymphocytes in allogeneic mixed cell culture reactions and present antigen to antigen specific T cells in an HLA-DR restricted manner, similar studies have not yet been reported for human cutaneous microvascular endothelial cells.[46-48]

VI. THE ROLE OF HUMAN CUTANEOUS MICROVASCULAR ENDOTHELIAL CELLS IN DISEASE

It has been known for many years that human microvascular endothelial cells are damaged and destroyed in areas of necrotizing vasculitis. The damage is mainly mediated by polymorphonuclear neutrophils which infiltrate blood vessel walls following immune complex deposition and complement activation. The details of some of these interactions have been recently examined in vitro by a number of investigators. Early studies using large-vessel endothelial cells suggested that reactive oxygen species, such as H_2O_2, generated by activated neutrophils were responsible for endothelial cell damage as measured by ^{51}Cr release from cultures of large-vessel endothelial cells.[48] Substances shown to be capable of activating polymorphonuclear neutrophils in this way have included C5-derived peptides, the lipid A moiety of lipopolysaccharide, and phorbol myristate acetate.[48,50] In some situations, the effects of these stimulatory substances seem to be additive. More recently, the ability of H_2O_2 to damage human microvascular endothelial cells has also been demonstrated.[51]

Other circulating blood elements can also participate in polymorphonuclear neutrophil-mediated cell damage of endothelial cells since platelet release products have been shown to amplify this type of damage.[52] It has also been recently shown that human cutaneous microvascular endothelial cells are susceptible to damage by cytotoxic T lymphocytes (CTL).[53] In this study, CTL were generated by cocultivation of peripheral blood lymphocytes with irradiated cultures of microvascular endothelial cells and stimulation with IL-2. These CTL efficiently lysed both adult and neonatal cutaneous microvascular endothelial cells, suggesting that this mechanism may be important in such processes as graft rejection and tumor killing.

Another type of endothelial cell damage that may be operative in the skin in certain forms of vasculitis is suggested by recent studies on Kawasaki's syndrome.[54] This is an acute illness of children characterized by fever, mucosal inflammation, cutaneous eruptions, and coronary arteritis. It has been recently shown that patients with Kawasaki's syndrome have an IgM antibody directed against an antigen that can be induced on the surface of human umbilical vein endothelial cells by human γ-IFN. This antigen is not inducible on human fibroblasts or human smooth muscle cells. When human umbilical vein endothelial cells were cultured in the presence of γ-IFN and then exposed to whole sera or purified IgM from patients with acute Kawasaki's syndrome and fresh complement, the cells were killed. The studies have not been carried out with cutaneous microvascular cells, but suggest that antibody- and complement-mediated damage may occur in certain forms of cutaneous vasculitis.

Studies on the pathophysiological mechanisms operating in graft rejection have suggested a role for the microvascular endothelial cell. Dvorak et al. have demonstrated in first set skin graft rejections in humans that widespread endothelial cell injury, characterized by membrane disruption, nuclear pyknosis, and endothelial cells sloughing, is an early event.[55] The vascular damage was preceded by perivascular infiltrates of lymphocytes and monocytes. The endothelial cell damage resulted in narrowing or occlusion of the vascular lumens, loss of perfusion, and ultimately graft rejection. These changes were less severe in larger blood vessels than they were in the cutaneous microvessels. This study suggests that cutaneous microvascular endothelial cells are a major target of the immune response in skin-graft rejection. A similar but more recent study of skin-graft rejection in mice suggests that Ia antigens expressed on murine endothelial cells may be the targets of the immune response in Ia-incompatible skin allograft rejection.[56]

Finally, some mention of the role of cutaneous microvascular endothelial cells in Kaposi's sarcoma should be made. The histopathology of a typical lesion of Kaposi's sarcoma reveals vascular formations consisting of blood vessels lined with prominent endothelial-like cells intermingled with spindle cells with associated vascular slits. The cell of origin of this tumor has been disputed. Studies have suggested either an endothelial cell or pericyte origin, or derivation from cells of the mononuclear phagocyte system.[57,58] Several immunohistochemical studies of skin biopsies of Kaposi's sarcoma obtained from AIDS patients have suggested an endothelial origin for this tumor. One study using monoclonal antibodies directed against endothelial cell-associated antigens and antibody to Factor-VIII-related antigen found that while Factor VIII-related antigen was clearly present in the cells lining the vascular spaces, it was only weakly expressed or absent in the spindle cells. When the monoclonal antibodies were employed, both the cells lining the vascular spaces and the spindle cells were strongly stained with a monoclonal antibody termed E92, which is specific for an endothelial cell-related antigen.[59] Another group has shown that while the spindle cells of nodular lesions of Kaposi's sarcoma show only focal staining for Factor VIII-related antigen, they stain strongly with a monoclonal antibody directed against endothelial cells. However, this group believes that these cells are of lymphatic rather than vascular origin.[59,60]

REFERENCES

1. **Davison, P. M., Bensch, K., and Karasek, M. A.,** Isolation and growth of endothelial cells from the microvessels of the newborn human foreskin in cell culture, *J. Invest. Dermatol.,* 75, 316, 1981.
2. **Marks, R. M., Czerniecki, M., and Penny, R.,** Human dermal microvascular endothelial cells: an improved method for tissue culture and a description of some singular properties in culture, *In Vitro,* 21, 627, 1985.
3. **Kumar, P., Kumar, S., Marsden, H. B., Lynch, P. G., and Earnshaw, E.,** Weibel-Palade bodies in endothelial cells as a marker for angiogenesis in brain tumors, *Cancer Res.,* 40, 2010, 1980.
4. **Folkman, J. and Haudenschild, C.,** Angiogenesis, *in vitro, Nature (London),* 288, 551, 1980.
5. **Charo, I., Shak, S., Karasek, M. A., Davison, P. M., and Goldstein, I. M.,** Prostaglandin I_2 is not a major metabolite of arachidonic acid in cultured endothelial cells from human foreskin microvessels, *J. Clin. Invest.,* 74, 914, 1984.
6. **Zetter, B. R.,** Culture of capillary endothelial cells, in *Biology of Endothelial Cells,* Jaffe, E., Ed., Martinus Nijhoff, The Hague, 1984, 14.
7. **Yen, A. and Braverman, I. M.,** Ultrastructure of the human dermal microcirculation: the horizontal plexus of the papillary dermis, *J. Invest. Dermatol.,* 66, 131, 1976.
8. **Sherer, G. K., Fitzharris, T. P., Faulk, W. P., and LeRoy, E. C.,** Cultivation of microvascular endothelial cells from human preputial skin, *In Vitro,* 16, 675, 1980.
9. **Folkman, J., Haudenschild, C., and Zetter, B. R.,** Long-term culture of capillary endothelial cells, *Proc. Natl. Acad. Sci. U.S.A.,* 76, 5217, 1979.
10. **Davison, P. M., Bensch, K., and Karasek, M. A.,** Isolation and long-term serial cultivation of endothelial cells from the microvessels of the adult human dermis, *In Vitro,* 19, 937, 1983.
11. **Karasek, M. and Eaton, M. J.,** Long-term growth of neonatal and adult human skin microvascular endothelial cells in cell culture, *J. Invest. Dermatol.,* 86, 484, 1986.
12. **Baskin, J. B. and Ham, R. G.,** Improved culture conditions for human dermal microvascular endothelial cells, *In Vitro,* 21, 18A, 1985.
13. **Davison, P. M. and Karasek, M. A.,** Human dermal microvascular endothelial cells *in vitro:* effect of cyclic AMP on cellular morphology and proliferation rate, *J. Cell. Physiol.,* 106, 253, 1981.
14. **Bensch, K. G., Davison, P. M., and Karasek, M. A.,** Factors controlling the *in vitro* growth pattern of human microvascular endothelial cells, *J. Ultrastruct. Res.,* 82, 76, 1983.
15. **Maciag, T., Kadish, J., Wilkins, L., Stemerman, M. B., and Weinstein, R.,** Organizational behavior of human umbilical vein endothelial cells, *J. Cell Biol.,* 94, 511, 1982.
16. **Montesano, R., Orci, L., and Vassalli, P.,** *In Vitro* rapid organization of endothelial cells into capillary-like networks is promoted by collagen matrices, *J. Cell Biol.,* 97, 1648, 1983.
17. **Madri, J. A., Williams, S. K., Wyatt, T. and Mezzio, C.,** Capillary endothelial cell cultures: phenotypic modulation by matrix components, *J. Cell Biol.,* 97, 153, 1983.
18. **Madri, J. A. and Pratt, B. M.,** Endothelial cell-matrix enteractions: *in vitro* models of angiogenesis, *J. Histochem. Cytochem.,* 34, 85, 1986.
19. **Kleinman, H. K., McGarvey, M. L., Liotta, L., Robey, P. G., Tryggvason, K., and Martin, G. R.,** Isolation and characterization of type IV procollagen, laminin and heparan sulfate proteoglycan from the EHS sarcoma, *Biochemistry,* 21, 6188, 1982.
20. **Weibel, E. R. and Palade, G. R.,** New cytoplasmic components in arterial endothelia, *J. Cell Biol.,* 23, 101, 1964.
21. **Tonnesen, M. G., Jenkins, D., Siegal, S. L., Lee, L., Huff, J. C., and Clark, R. A. F.,** Expression of fibronectin, laminin, and factor VIII-related antigen during development of the human cutaneous microvasculature, *J. Invest. Dermatol.,* 85, 564, 1985.
22. **Goldstein, J. L., Ho, Y. L., Basu, S. K., and Brown, M. S.,** Binding site on macrophages that mediates uptake and degradation of acetylated low density lipoprotein, producing massive cholesterol deposition, *Proc. Natl. Acad. Sci. U.S.A.,* 76, 333, 1979.
23. **Stein, O. and Stein, Y.,** Bovine aortic endothelial cells display macrophage-like properties towards acetylated ^{125}I-labelled low density lipoprotein, *Biochim. Biophys., Acta,* 620, 631, 1980.
24. **Voyta, J. C., Via, D. P., Butterfield, C. E., and Zetter, B. R.,** Identification and isolation of endothelial cells based on their increased uptake of acetylated-low density lipoprotein, *J. Cell Biol.,* 99, 2034, 1984.
25. **Goffney, J., West, D., Arnold, F., Sattar, A., and Kumar, S.,** Differences in the uptake of modified low density lipoproteins by tissue cultured endothelial cells, *J. Cell Sci.,* 79, 317, 1985.
26. **Miettinen, M., Holthofer, H., Lehto, V.-P., Miettinen, A., and Virtanen, I.,** *Ulex europaeus* - I lectin as a marker for tumors derived from endothelial cells, in press.
27. **Kramer, R. H., Bensch, K. G., Davison, P. M., and Karasek, M. A.,** Basal lamina formation by cultured microvascular endothelial cells, *J. Cell Biol.,* 99, 692, 1984.

28. **Kramer, R. H., Fuh, G. M., Bensch, K. G., and Karasek, M. A.,** Synthesis of extracellular matrix glycoproteins by cultured microvascular endothelial cells isolated from the dermis of neonatal and adult skin, *J. Cell. Physiol.,* 123, 1, 1985.

29. **Sage, H., Pritzl, P., and Bornstein, P.,** Secretory phenotypes of endothelial cells in culture: comparison of aortic, venous, capillary and corneal endothelium, *Arteriosclerosis,* 1, 427, 1981.

30. **Levin, R. I., Weksler, B. B., Marcus, A. J., and Jaffe, E. A.,** Prostacylin production by endothelial cells, in *Biology of Endothelail Cells,* Jaffe, E. A., Ed., Martkinus Nijhoff, Boston, 1984, 228.

31. **Monocada, S., Gryglewski, K. R., Bunting, S., and Vane, J. R.,** An enzyme isolated from arteries transforms prostaglandin endoperoxides to an unstable substance that inhibits platelet aggregation, *Nature (London),* 263, 663, 1976.

32. **Kulkarni, P. S., Roberts, R., and Needleman, P.,** Paradoxical endogenous synthesis of a coronary dilating substance from arachidonate, *Prostaglandins,* 12, 337, 1976.

33. **Weksler, B. B., Marcus, A. J., and Jaffe, E. A.,** Synthesis of prostaglandin I_2 (prostacyclin) by cultured human and bovine endothelial cells, *Proc. Natl. Acad. Sci. U.S.A.,* 74, 3922, 1977.

34. **Weksler, B. B., Ley, C. W., and Jaffe, E. A.,** Stimulation of endothelial cell prostacyclin production by thrombin, trypsin and the ionophore A 23187, *J. Clin. Invest.,* 62, 923, 1978.

35. **Goldsmith, J. C., Jafvert, C. T., Lollar, P., Owen, W. G., and Hoak, J. C.,** Prostacycin release from cultured and *ex vivo* bovine vascular endothelium, *Lab. Invest.,* 45, 191, 1981.

36. **Baenziger, N. L., Fogerty, F. J., Mertz, L. F., and Chernuta, L. F.,** Regulation of histamine-mediated prostacyclin synthesis in cultured human vascular endothelial cells, *Cell,* 24, 915, 1981.

37. **Eldor, A., Vlodavsky, I., Hyam, E., Atzmon, R., Fuks, Z.,** The effect of radiation on prostacyclin (PGI_2) production by cultured endothelial cells, *Prostaglandins,* 25, 263, 1983.

38. **Clark, M. A., Littlejohn, D., Mong, S., and Crooke, S. T.,** Effect of leukotrienes, bradykinin and calcium ionophore (A 23187) on bovine endothelial cells: release of prostacyclin, *Prostaglandins,* 31, 157, 1986.

39. **Rosenbaum, R. M., Cheli, C. D., and Gerritsen, M. E.,** Dexamethasone inhibits prostaglandin release from rabbit coronary microvessel endothelium, *Am. J. Physiol.,* 250, 970, 1986.

40. **Nunez, G., Ball, E. J., and Stastny, P.,** Accessory cell function of human endothelial cells, *J. Immunol.,* 131, 666, 1983.

41. **Pober, J. S., Gimbrone, M. A., Cotran, R. S., Reiss, C. S., Burakoff, S. J., Fiers, W., and Ault, K. A.,** Ia expression by vascular endothelium is inducible by activated T cells and by human γ-interferon, *J. Exp. Med.,* 157, 1339, 1983.

42. **Fleck, R., Geppert, T. D., and Sontheimer, R. D.,** Gamma-interferon induction of class II cell surface antigens on cultured human dermal microvascular endothelial cells, *J. Invest. Dermatol.,* 86, 475, 1986.

43. **Collins, T., Korman, A. J., Wake, C. T., Boss, J. M., Kappes, D. J., Fiers, W., Ault, K. A., Gimbrone, M. A., Strominger, J. L., and Pober, J. S.,** Immune interferon activates multiple class II major histocompatibility complex genes and the associated invariant chain gene in human endothelial cells and dermal fibroblasts, *Proc. Natl. Acad. Sci. U.S.A.,* 81, 4917, 1984.

44. **Groenewegen, G., Buurman, W. A., and Van der Linden, C. J.,** Lymphokine dependence of *in vivo* expression of MHC class II antigens by endothelium, *Nature (London),* 316, 361, 1985.

45. **Games, M., Schmitt, D., Dezutter-Dambuyant, C., Capra, J. D., and Thivolet, J.,** Expression of class II HLA antigens (HLA-DR, DQW1, DQW3) on normal and pathological skin, *Pathol. Biol.,* 34, 157, 1986.

46. **Vetto, R. M. and Burger, D. R.,** Endothelial cell stimulation of allogeneic lymphocytes, *Transplantation,* 14, 652, 1972.

47. **Wagner, C. R., Vetto, R. M., and Burger, D. R.,** The mechanism of antigen presentation of endothelial cells, *Immunobiology,* 168, 453, 1984.

48. **Sacks, T., Moldow, C. F., Craddock, P. R., et al.,** Oxygen radicals mediate endothelial damage by complement-stimulated granulocytes: an *in vitro* model of immune vascular damage, *J. Clin. Invest.,* 61, 1161, 1978.

49. **Yamada, O., Moldow, C. F., Sacks, T., Craddock, P. R., Boogaerts, M. A., and Jacob, H. S.,** Deleterious effects of endotoxin on cultured endothelial cells, *Inflammation,* 5, 115, 1981.

50. **Weiss, S. J., Young, J., LoBruglio, A. F., Slivka, A., and Nimeh, N. F.,** Role of hydrogen peroxide in neutrophil-mediated destruction of cultured endothelial cells, *J. Clin. Invest.,* 68, 714, 1981.

51. **Capin, L. R., Tonnessen, M. G., Osborn, R. L., Kissinger, M., and Norris, D. A.,** Comparative cytotoxicity of human melanocytes, keratinocytes, endothelial cells and fibroblasts, *Clin. Res.,* 33, 629A, 1985.

52. **Boogaerts, M. A., Yamada, O., Jacob, H. S., and Moldow, C. F.,** Enhancement of granulocyte endothelial cell adherence and granulocyte-induced cytotoxicity by platelet release products, *Proc. Soc. Nat. Acad. U.S.A.,* 9, 7019, 1982.

53. **Karasek, M., Clayberger, C., and Kremsky, A.,** Immune functions of human skin microvascular endothelial cells, *Clin. Res.,* 33, 653A, 1985.

54. **Leung, D. Y. M., Collins, T., Lapierre, L. A., Geha, R. S., and Pober, J. S.,** Immunoglobulin M antibodies present in the acute phase of Kawasaki syndrome lyse cultured vascular endothelial cells stimulated by gamma interferon, *J. Clin. Invest.,* 77, 1428, 1986.

55. **Dvorak, H. F., Mihm, M. C., Dvorak, A. M., Barnes, B. A., Manseau, E. J., and Galli, S. J.,** Rejection of first-set skin allografts in man, *J. Exp. Med.,* 150, 322, 1979.

56. **de Waal, R. M. W., Bogman, M. J. J., Maass, C. N., Cornelissen, L. M. H., Tax, W. J. M., and Koene, R. A. P.,** Variable expression of Ia antigens on the vascular endothelium of mouse skin allografts, *Nature (London),* 303, 426, 1983.

57. **Hashimoto, K. and Lever, W. F.,** Kaposi's sarcoma. Histochemical and electron microscopic studies, *J. Invest. Dermatol.,* 43, 539, 1964.

58. **Dayan, A. D. and Lewis, P. D.,** Origin of Kaposi's sarcoma from the reticulo-endothelial system, *Nature (London),* 213, 889, 1967.

59. **Rutgers, J. L., Wieczorek, R., Bonetti, F., Kaplan, K. L., Posnett, D. N., Friedman-Kien, A. E., and Knowles, D. M.,** The expression of endothelial cell surface antigens by AIDS associated Kaposi's sarcoma, *Am. J. Pathol.,* 122, 493, 1986.

60. **Russell-Jones, R., Spaull, J., Spry, C., and Wilson-Jones, E.,** Histogenesis of Kaposi's sarcoma in patients with and without acquired immune deficiency syndrome (AIDS), *J. Clin. Pathol.,* 39, 742, 1986.

Index

INDEX